SECOND EDITION

Frying of Food

Oxidation, Nutrient and Non-Nutrient Antioxidants, Biologically Active Compounds, and High Temperatures

Edited by

Dimitrios Boskou

Ibrahim Elmadfa

CRC Press is an imprint of the
Taylor & Francis Group, an **informa** business

CRC Press
Taylor & Francis Group
6000 Broken Sound Parkway NW, Suite 300
Boca Raton, FL 33487-2742

First issued in paperback 2019

© 2011 by Taylor and Francis Group, LLC
CRC Press is an imprint of Taylor & Francis Group, an Informa business

No claim to original U.S. Government works

ISBN-13: 978-1-4398-0682-1 (hbk)
ISBN-13: 978-0-367-38317-6 (pbk)

This book contains information obtained from authentic and highly regarded sources. Reasonable efforts have been made to publish reliable data and information, but the author and publisher cannot assume responsibility for the validity of all materials or the consequences of their use. The authors and publishers have attempted to trace the copyright holders of all material reproduced in this publication and apologize to copyright holders if permission to publish in this form has not been obtained. If any copyright material has not been acknowledged please write and let us know so we may rectify in any future reprint.

Except as permitted under U.S. Copyright Law, no part of this book may be reprinted, reproduced, transmitted, or utilized in any form by any electronic, mechanical, or other means, now known or hereafter invented, including photocopying, microfilming, and recording, or in any information storage or retrieval system, without written permission from the publishers.

For permission to photocopy or use material electronically from this work, please access www.copyright.com (http://www.copyright.com/) or contact the Copyright Clearance Center, Inc. (CCC), 222 Rosewood Drive, Danvers, MA 01923, 978-750-8400. CCC is a not-for-profit organization that provides licenses and registration for a variety of users. For organizations that have been granted a photocopy license by the CCC, a separate system of payment has been arranged.

Trademark Notice: Product or corporate names may be trademarks or registered trademarks, and are used only for identification and explanation without intent to infringe.

Library of Congress Cataloging-in-Publication Data

Frying of food : oxidation, nutrient and non-nutrient antioxidants, biologically active compounds and high temperatures, second edition / [edited by] Dimitrios Boskou, Ibrahim Elmadfa. -- 2nd ed.
p. cm.
Includes bibliographical references and index.
ISBN 978-1-4398-0682-1 (hardback)
1. Oils and fats, Edible--Effect of temperature on. 2. Frying. I. Boskou, Dimitrios. II. Elmadfa, I.

TP670. F79 2010
664--dc22 2010043667

Visit the Taylor & Francis Web site at
http://www.taylorandfrancis.com

and the CRC Press Web site at
http://www.crcpress.com

Contents

Preface

In the last three decades there has been an interest in the relationship of dietary antioxidants and other bioactive ingredients to the possible prevention of a number of diseases, in the etiology of which oxidation mechanisms are involved. As a result, naturally occurring nutritive and nonnutritive antioxidants have become a major area of scientific research. In the fields of nutrition and biosciences, phytochemicals such as phenols and certain hydrocarbons are now examined and discussed as food antioxidants. In addition, changes in lifestyles in the modern world have triggered a growing awareness that particular ingredients in foods may favorably modify diet-related problems. This requires that valuable constituents of food are preserved during processing. It is, therefore, interesting to examine the frying of food from the point of view of changes due not only to triacylglycerols but also to non-glyceride components that are also of biological importance.

Among the various biologically active ingredients present in oils and fats, antioxidant vitamins seem to be very important. However, other minor constituents such as phytosterols, phospholipids, and certain hydrocarbons probably affect the performance of heated oil and the nutritional value of fried food. Some of these minor constituents, such as phytosterols, carotene, and squalene may be categorized as "functional." Finally, polar phenolic compounds, which naturally occur in certain oils or may be obtained from herbs, spices, or other natural sources, have been shown to have an impact on the stability of oils at high temperatures and may influence the rate of oxidation of nutrients.

This new edition of *Frying of Food* deals with some important chemical, biochemical, and nutritional aspects of frying. Frying is a rapid and convenient method to produce a palatable product with favorable color and flavor; but during frying, many changes take place and breakdown products are formed. The deterioration of the fat at elevated temperatures is influenced by various factors, such as the nature of the cooking fat, the conditions of frying, the kind of heat transfer, the fryer removal constructions, and the use of antioxidants and other additives. Most of these factors have been discussed in detail, and information can be found in excellent reviews already available in the literature. This book concentrates mainly on the nature of the heated fat, the presence of oxidation retardants, especially those naturally occurring in oils, and the stabilization of frying oils by novel antioxidants obtained from natural sources. From this point of view, frying oil is not seen just as a medium to fry food for many frying operations, but also as a source of antioxidants and bioactive ingredients that may have a beneficial health effect. Other issues addressed, related also to minor constituents, are interactions between frying oils and natural components present in food and substances formed during frying. This feature, the focus on minor constituents and their role in the performance of frying oil and its biological value, sets this book significantly apart from others in the field.

Specific topics examined in the book are fat and nutrition, oxidation products and metabolic processes, formation of radicals and protection mechanisms *in vitro* and *in vivo*, changes in nutrients due to frying, high temperatures and antinutritional factors, enzymes and digestion of thermally oxidized oils and fats, determination of oxidation compounds and polymers, nutrients and nonnutrient antioxidants, healthy stable oils rich in oxidation inhibitors, frying oils as sources of natural antioxidants, phytosterols and their effect on the performance of a frying oil, strategies to reduce *trans* fatty acids in fried food, phytosterol oxidation products, food hazards associated with frying and, finally, critical control points, preventive measures, and safety and reliability during frying.

It is hoped that all those interested in the frying of food will find this an essential reference book. It is also anticipated that some readers will recognize that the book not only presents current facts about the frying of food but also tracks some lines for future research.

Dimitrios Boskou
Ibrahim Elmadfa

About the Editors

Dimitrios Boskou, Ph.D., earned his doctorate in chemistry from the School of Chemistry, Aristotle University of Thessaloniki, Thessaloniki, Greece, his PhD degree from the University of London, and his Doctor of Science degree from the School of Chemistry, Aristotle University. He served as an assistant, lecturer, assistant professor, associate professor, professor, and head of the Laboratory of Food Chemistry, Aristotle University, from 1970 to 2006. From 1986 to 1998 he was a member of the IUPAC Oils, Fats and Derivatives Commission; from 1995 to 2005 he served as a member of the Supreme Chemical Council; and since 1995 he has been a member of the Scientific Committee on Food of the European Commission and member and expert of the Food Additives Panel of the European Food Safety Authority.

Dr. Boskou has written more than 85 published papers and reviews and authored and edited five books (*Olive Oil*, AOCS Press, 1996; *Frying of Food*, Technomic Publishing Co., 1999; *Olive Oil*, Second Edition, AOCS Press, 2006; *Natural Antioxidant Phenols*, Research Signpost, 2006; *Olive Oil, Minor Constituents, and Health*, 2009, Taylor & Francis). He was lead author of more than 10 chapters in books related to heated fats, natural antioxidants, and olive oil chemistry. He has contributed to international scientific encyclopedias and the *Lexicon of Lipid Nutrition*, a joint IUPAC/IUNS work.

Ibrahim Elmadfa, Ph.D., is currently president of the International Union of Nutritional Science (IUNS), vice-president of World Public Health Nutrition Association (WPHNA) and president of the Austrian Nutrition Society. He is trained in food science and nutritional sciences. He earned his PhD degree (1970) and habilitation (1975) in human nutrition at the University of Giessen, Germany. He served as lecturer and assistant professor (1971–1979) and professor (1980–1990) of human nutrition at the University of Giessen. Since 1990 he has been professor for human nutrition and director of the Institute of Nutritional Sciences at the University of Vienna

He served as a scientific expert to the European Commission (1995–2000) and as a member of the Scientific Committee on Food (different functions). He is a member of Codex Alimentarius Austria (Novel Food) and the National Nutrition Committee. He is a member of the WHO Nutrition Guidance Expert Advisory Group (NUGAG) and the International Advisory Council of WHO Global Non-Communicable Disease Network (NCDnet).

His main research interests include nutrient requirements in health and disease, bioavailability of nutrients; nutrition and immune function, food safety and quality and monitoring of nutrition and health status.

Prof. Elmadfa is editor-in-chief of *Annals of Nutrition and Metabolism* and *Forum Nutrition* (since 2000) and editorial board member of several scientific journals. He has published more than 400 papers in international scientific journals and is editor and author/co-author of 24 books and reports and several book chapters. He was coordinator and author of the Austrian Nutrition Report 1998, 2003 and 2008 as well as the European Nutrition and Health Report 2004 and 2009.

Contributors List

Nikolaos K. Andrikopoulos
Laboratory of Food Chemistry–Biochemistry–Physicochemistry
Department of the Science of Dietetics–Nutrition
Harokopio University
Athens, Greece

William E. Artz
Department of Food Science and Human Nutrition
University of Illinois
Urbana, Illinois

Sara Bastida
Departamento de Nutrición y Bromatología I (Nutrición)
Facultad de Farmacia
Universidad de Alcalá
Madrid, Spain

Juana Benedi
Departamento de Farmacología
Universidad Complutense de Madrid
Madrid, Spain

Georgios Blekas
Laboratory of Food Chemistry and Technology
School of Chemistry
Aristotle University of Thessaloniki
Thessaloniki, Greece

Dimitrios Boskou
Laboratory of Food Chemistry and Technology
School of Chemistry
Aristotle University of Thessaloniki
Thessaloniki, Greece

George Boskou
Department of the Science of Dietetics–Nutrition
Harokopio University
Athens, Greece

M. Carmen Dobarganes
Instituto de la Grasa (CSIC)
Department of Food Characterization and Quality
Sevilla, Spain

Jana Dostálová
Department of Food Chemistry and Analysis
Prague Institute of Chemical Technology
Prague, Czech Republic

Ibrahim Elmadfa
Department of Nutritional Sciences
University of Vienna
Vienna, Austria

Maria J. González-Muñoz
Departamento de Nutrición, Bromatología y Toxicología
Universidad de Alcalá
Madrid, Spain

Margit Kornsteiner-Krenn
Department of Nutritional Sciences
University of Vienna
Vienna, Austria

Symon M. Mahungu
Department of Dairy and Food Science and Technology
Egerton University
Egerton, Kenya

Susana Marmesat
Instituto de la Grasa (CSIC)
Department of Food Characterization and Quality
Sevilla, Spain

Gloria Márquez-Ruiz
Instituto de Ciencia Tecnologia de Alimentos y Nutrition (CSIC)
Sevilla, Spain

Malgorzata Nogala-Kalucka
Department of Biochemistry and Food Analysis
Faculty of Food Science and Nutrition
Poznan University of Life Sciences
Poznan, Poland

Raul Olivero-David
Departamento de Nutrición y Bromatología I (Nutrición) and Departamento de Nutrición, Bromatología y Toxicología
Facultad de Farmacia
Universidad de Alcalá
Madrid, Spain

Mary N. Omwamba
Department of Dairy and Food Science and Technology
Egerton University
Egerton, Kenya

Jan Pokorný
Department of Food Chemistry and Analysis
Prague Institute of Chemical Technology
Prague, Czech Republic

Francisco J. Sánchez-Muniz
Departamento de Nutrición y Bromatología I (Nutrición)
Facultad de Farmacia
Universidad de Alcalá
Madrid, Spain

Aleksander Siger
Department of Biochemistry and Food Analysis
Faculty of Food Science and Nutrition
Poznan University of Life Sciences
Poznan, Poland

Joaquín Velasco
Instituto de la Grasa (CSIC)
Department of Food Characterization and Quality
Sevilla, Spain

Karl-Heinz Wagner
Institute of Nutritional Sciences
University of Vienna
Vienna, Austria

1 Fat and Nutrition

Ibrahim Elmadfa and Margit Kornsteiner-Krenn

CONTENTS

1.1 INTRODUCTION

Chemically, fats comprise a nonhomogeneous group of different substances that have some physical and chemical characteristics in common. They were defined as substances that are insoluble in water and soluble in organic solvents. The main components of fats are triacylglycerols (98%–99%), and only 1%–2% are unsaponifiable components such as sterols and fat-soluble vitamins.

The suggested new classification scheme is based on chemistry and determined by the distinct hydrophobic and hydrophilic elements of the individual lipid. This classification makes it possible to categorize lipids according to their structure and their properties. Categories of lipids include fatty acyls (FA), glycerolipids (GL), glycerophospholipids (GP), sphingolipids (SP), sterol lipids (ST), prenol lipids (PR), saccharolipids (SL), and polyketides (Fahy et al., 2005).

In the past, lipids were assumed not to be essential constituents of food. The human organism was supposed to remain healthy even if no lipids were supplied, as long as the requirement of food energy was met. Today, it is well known that the polyunsaturated fatty acids are essential and that a balance between unsaturated and saturated fatty acids is crucial for the normal metabolic function in health and disease. The fact that lipids make up an important constituent of the cell membrane as proteins underlines their essential character.

In nutrition and dietetics, a distinction is made between visible and invisible fats. Visible fats are clearly apparent to the consumer (spreads, cooking oils, or the fat contained in the meat). Most of the fat in many consumed foods, however, is hidden as a natural component of the raw material, through incorporation during the cooking or frying process (cakes, fried potatoes, french fries) or as a result of the formation of emulsions, such as mayonnaise. During frying, the lipid component may undergo qualitative and quantitative changes and exchanges with the fatty acid pattern of fried food. Therefore, it is important to understand the factors affecting the stability of oils and fats at high temperatures as well as the extent to which nutritionally important lipids are deteriorated.

1.2 DIETARY FAT—NOMENCLATURE, STRUCTURE, AND METABOLISM

Fatty acids differ in carbon chain length, which varies from 2 to 30 or more, but common dietary fatty acids occur between C4 (in milk fat) and C22 (in fish oil). The carbon-to-carbon bonds can be fully saturated (no double bonds), monounsaturated (one double bond), or polyunsaturated (more than one double bond) (Elmadfa and Leitzmann, 2004; Nicolaou and Kokotos, 2004).

There are different systems of nomenclature for fatty acids available, but some of these do not provide sufficient information about the structure of fatty acids. It is suggested that a chemical name must describe the chemical structure unmistakably (Ratnayake and Galli, 2009). Therefore, the International Union of Pure and Applied Chemistry (IUPAC-IUB Commission on Nomenclature, 1978) recommends that fatty acids should be named only on the basis of the number of carbon atoms, the number and position of unsaturated fatty acids relative to the carboxyl group (IUPAC, 1978). The arrangement of double bonds, location of branched chains and hetero atoms, and other structural characteristics are also recognized. Fatty acids are made up of a hydrocarbon chain with a methyl group (-CH_3) at one end and a carboxyl group (-COOH) at the other end, which is the number 1. The double bond has to be identified by the lower number of the two connected carbons. In addition, the double bonds are labeled with Z or E, which have been practically replaced by the terms *cis* and *trans*. For instance, the systematic IUPAC nomenclature of linoleic acid is Z-9, Z-12-octadecadienoic acid, or *cis*-9, *cis*-12-octadecadienoic acid (IUPAC, 1978; Ratnayake and Galli, 2009).

Along with the IUPAC nomenclature, additional names from "trivial" or historical names and shorthand notations have become accepted in scientific writings. For instance, unsaturated fatty acids are classified by the location of the first double bond counted from the methyl terminus of the acyl chain and the total number of double bonds. They are described in Table 1.1 by their common names, chain lengths, double bonds, systematic names, melting points, and occurrence. Saturated fatty acids are usually solid at room temperature, and major sources are animal and dairy products. The carbon number, on which the nearest double bond is located, is called

TABLE 1.1
Characterization of Important Fatty Acids in Foods

Common Name	CL	DB	Symbol	Systematic Names	MP	Occurrence
Butyric acid	4	0	C4:0	*n*-Butanoic acid	–8	Milk fat
Caproic acid	6	0	C6:0	*n*-Hexanoic acid	–2	Milk fat
Caprylic acid	8	0	C8:0	*n*-Octanoic acid	16	Milk fat
Capric acid	10	0	C10:0	*n*-Decanoic acid	31	Milk fat
Lauric acid	12	0	C12:0	*n*-Dodecanoic acid	44	Cocos fat
Myristic acid	14	0	C14:0	*n*-Tetradecanoic acid	54	Animal fats
Palmitic acid	16	0	C16:0	*n*-Hexadecanoic acid	63	Animal fats
Palmitoleic acid	18	1	C16:1n7	*cis*-9-Hexadecenoic acid	1	Animal fats, fish oils
Stearic acid	18	0	C18:0	*n*-Octadecanoic acid	70	Animal fats
Oleic acid	18	1	C18:1n9	*cis*-9-Octadecenoic acid	13	Fats and oils
Vaccenic acid	18	1	C18:1n7	*trans*-11-Octadecenoic acid	40	Summer butter
Linoleic acid	18	2	C18:2n6	all *cis*-9,12-Octadecadienoic acid	–6	Phosphatides
γ-Linolenic acid	18	3	C18:3n6	all *cis*-6,9,12-Octadecatrienoic acid		Plant oils
α-Linolenic acid	18	3	C18:3n3	all *cis*-9,12,15-Octadecatrienoic acid	14	Plant oils
Arachidic acid	20	0	C20:0	*n*-Eicosanoic acid	76	Animal fats
Gadoleic acid	20	1	C20:1n9	*n*-11-Eicosenoic acid		
Arachidonic acid	20	4	C20:4n6	all *cis*-5,8,11,14-Eicosatetraenoic acid	–50	Phosphatides
Eicosapentaenoic acid	20	5	C20:5n3	all *cis*-5,8,11,14,17-Eicosapentaenoic acid		Fish oil, phosphatides
Behenic acid	22	0	C22:0	*n*-Docosanoic acid	80	Cerebrosides
Erucic acid	22	1	C22:1n9	*cis*-13-Docosenoic acid	35	

(*Continued*)

TABLE 1.1 (Continued)
Characterization of Important Fatty Acids in Foods

Common Name	CL	DB	Symbol	Systematic Names	MP	Occurrence
Docosapentaenoic acid	22	5	C22:5n3	all *cis*-7,10,13,16,19-Docosapentaenoic acid	—	Fish oils, phosphatides
Docosahexaenoic acid	22	6	C22:6n3	all *cis*-4,7,10,13,16,19-Docosahexaenoic acid	—	Fish oils, phosphatides
Lignoceric acid	24	0	C24:0	*n*-Tetracosanoic acid	84	Phosphatides
Nervonic acid	24	1	C24:1n9	*cis*-15-Tetracosenoic acid	40	Cerebrosides, phosphatide
Cerebronic acid	24	0	C24:0	2-Hydroxytetracosanoic acid	100	Cerebrosides
Hydroxynervoic acid	24	1	C24:1n9	2-Hydroxy-15-Tetracosenoic acid	6	Cerebrosides

Source: Modified from Elmadfa, I. and Leitzmann. 2004. *Ernährung des Menschen*. 4th ed. Eugen Ulmer Stuttgart.
Note: CL = chain length, DB = double bonds, MP = melting point (°C).

TABLE 1.2
Different Subclasses of Saturated and Unsaturated Fatty Acids

Subclasses of Saturated Fatty Acids	Length of Carbon Atoms
Short-chain fatty acids	3–7
Medium-chain fatty acids	8–13
Long-chain fatty acids	14–20
Very long-chain fatty acids	>21
Subclasses of Unsaturated Fatty Acids	
Short-chain unsaturated fatty acids	<19
Long-chain unsaturated fatty acids	20–24
Very long-chain unsaturated fatty acids	>25

Source: Adapted from Ratnayake, W. M. and Galli, C., *Ann. Nutr. Metab.* 55:8–43.

n-x, or ω-x. In this case, the essential α-linolenic acid is described as 18:3n-3 or C18:3ω-3 (not recommended by IUPAC-IUB Commission on Nomenclature, 1978); this nomenclature deviates from the systematic name *cis*-9,12,15-octadecatrienoic acid (Ratnayake and Galli, 2009).

The delta (Δ) system is also broadly used. The categorization is based on the number of carbon atoms introduced between the carboxyl carbon and the nearest double bonds to the carboxylic group. The advantage of this system is that it is able to specify the position of all the double bonds and their *cis*/*trans* configuration. For example, the shorthand notation for rumenic acid is expressed as 18:2Δ9c,11t, which is a conjugated fatty acid isomer of linoleic acid existing in dairy fats (Ratnayake and Galli, 2009).

Saturated and unsaturated fatty acids are also classified into different subclasses according to chain lengths (Table 1.2). Due to heterogeneous definitions in the literature, the FAO/WHO expert consultation (2008) recommends to use the definitions of Table 1.2 (Ratnayake and Galli, 2009).

Both saturated and monounsaturated fatty acids are nonessential for humans and can be biosynthesized in the body by the addition of 2-carbon units to the acyl chain. Monounsaturated fatty acids (MUFAs) can be synthesized by desaturation of saturated fatty acids (with Δ-9 desaturase). The best-known kind of desaturation is the transformation of stearic acid (C18:0) to oleic acid (C18:1n-9), by the insertion of a *cis* double bond between carbons 9 and 10 (Nicolaou and Kokotos, 2004). Unsaturated fatty acids may be *cis* or *trans*, whereas the majority of naturally occurring unsaturated fatty acids have *cis* rather than *trans* configuration. *Trans* double bonds can occur naturally as intermediates in the biosynthesis of fatty acids in ruminant fats or are industrially produced during hydrogenation of polyunsaturated oils (Calder, 2008).

Only plant organisms have the required enzymes (Δ-12 and Δ-15 desaturases) to introduce double bonds to carbon atoms beyond carbon 9 in the acyl chain counting

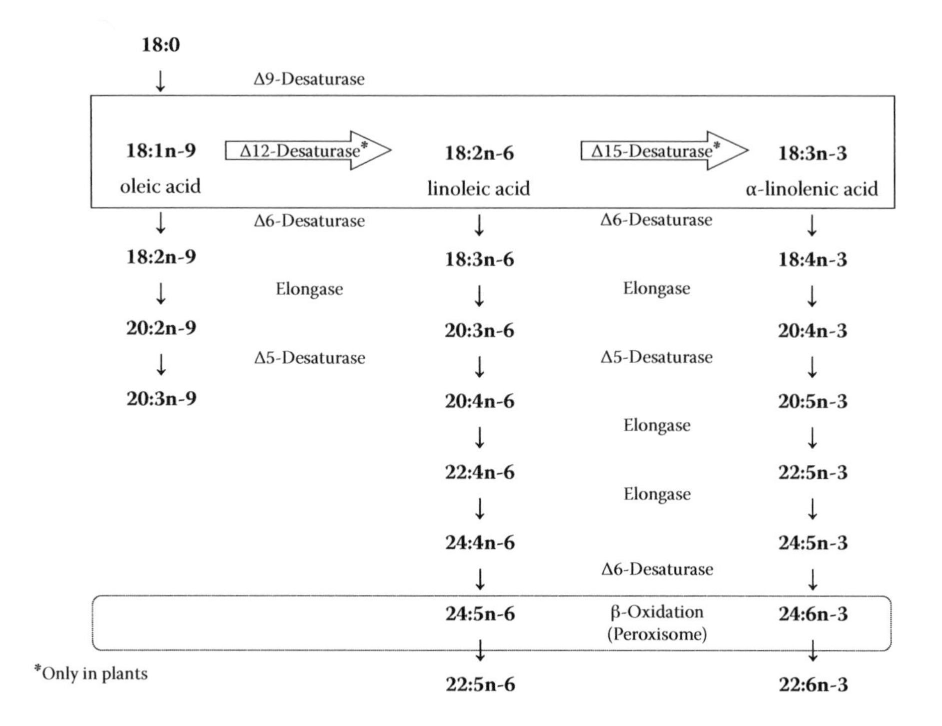

FIGURE 1.1 Biosynthesis of essential and nonessential long-chain polyunsaturated fatty acids.

from the carboxyl carbon (Nicolaou and Kokotos, 2004). Therefore, linoleic acid (C18:2n-6) and α-linolenic acid (C18:3n-3) are essential for humans. Oleic, linoleic, and α-linolenic acid can be converted to their longer chain derivates with more than one double bond, which are also named polyunsaturated fatty acids (PUFAs) (Figure 1.1).

For instance, α-linolenic acid (ALA), the most abundant n-3 fatty acid in the human diet, is a precursor for the synthesis of long-chain n-3 polyunsaturated fatty acids (LCPn-3, >20 carbon atoms). By the action of Δ6-desaturase, ALA is converted to C18:4n-3. This is a rate-limiting step. Further elongations and desaturations (Δ5-desaturase) lead to the synthesis of eicosapentaenoic acid (EPA, C20:5n-3). Docosapentaenoic acid (C22:5n-3, DPAn-3) is elongated to C24:5n-3. Further desaturation is suggested by the action of Δ6-desaturase activity to form C24:6n-3. The intermediate C24:6n-3 is translocated from the endoplasmic reticulum to the peroxisome. There, the fatty acid (C24:6n-3) is shortened by one cycle of β-oxidation to form docosahexaenoic acid (DHA, C22:6n-3). However, the precise regulation of these delicate steps (translocation, β-oxidation) in the pathway regulation has still to be elucidated (Burdge and Calder, 2005).

Similar steps of bioconversion can also be observed for the parent n-6 linoleic acid to docosapentaenoic acid (C18:2n-6 → C18:3n-6 → C20:3n-6 → C20:4n-6 → C22:4n-6 → C24:4n-6 → C24:5n-6 → C22:5n-6) (Nicolaou and Kokotos, 2004; Calder, 2005; Elmadfa and Leitzmann, 2004).

The extent of the conversion from ALA into EPA (eicosapentaenoic acid; C20:5n-3), DPAn-3 (docosapentaenoic acid; C22:5n-3), and DHA (docosahexaenoic acid; C22:6n-3) is a matter of debate. Different investigations have estimated that the bioconversion rate for ALA to EPA ranges from 0.2% to 21%, and for ALA to DHA from 0% to 9% (DeFilippis and Sperling, 2006).

The reason for this is that this bioconversion is influenced by multiple factors including timing of the sample collection, sex, negative feedback inhibition of desaturase by EPA and DHA, and competitive inhibition of desaturase (DeFilippis and Sperling, 2006). In addition, in the liver, ALA, LA, and oleic acid compete for the same series of enzymes as demonstrated in Figure 1.1. The favored fatty acid for the Δ6-desaturation is ALA, followed by LA and then oleic acid. However, the simplest n-6 fatty acid, linoleic acid (LA) is much more widespread in most human diets than ALA. Therefore, the metabolism of n-6 fatty acids dominates the synthesis of n-3 long-chain polyunsaturated fatty acids (LCPn-3) (D-A-CH, 2000; Burdge and Calder, 2005). In addition, the bioavailability of dietary ALA for the conversion to LCPn-3 is limited by the efficiency of absorption across the gastrointestinal tract, uptake and partitioning toward β-oxidation, and incorporation into structural lipids (e.g., phospholipids) and storage pools (e.g., adipose tissue) (Burdge and Calder, 2005).

1.3 FATTY ACID COMPOSITION AND UNSAPONIFIABLE COMPONENTS IN FOODS

Apart from the content of short- and medium-chain fatty acids and the relation of saturated to unsaturated fatty acids, the content of unsaponifiable fat-attendant substances is important for the evaluation of the quality and physiological effects of dietary fat. Fat-soluble vitamins, antioxidants (e.g., carotenoids), taste, and flavor substances as well as sterols belong to this group. Animal fats, except fish oils, consist predominantly of saturated and monounsaturated fatty acids and contain only small quantities of polyunsaturated fatty acids. A low content of unsaponifiable components is common to animal fats such as lard, tallow, and butter. Plant oils and fats—except coconut oil, palm seed oil, and olive oil—have a high content of unsaponifiables. The highest concentrations of unsaponifiables are found in wheat germ oil, rice oil, and corn oil (Table 1.3). The presence of natural antioxidants in the unsaponifiable fraction of vegetable oils is an advantage (longer shelf life).

The ratio of SFAs:MUFAs:PUFAs but also the ratio of SFAs:UFAs can describe the impact on health. For primary health care and for therapeutic use as well, a ratio SFAs:MUFAs:PUFAs of about 1:1:1, which is equivalent to the ratio of SFAs:UFAs 1:2, is accepted as beneficial (Elmadfa and Kornsteiner, 2009).

1.4 FAT DIGESTION, ABSORPTION, AND METABOLISM

Triacylglycerols (TAGs) make up the major lipid component (98%–99%) in foods used in human nutrition (e.g., vegetable oils); only a minor proportion is contributed by mono- and diacylglycerols, free fatty acids, phospholipids, and unsaponifiable compounds. Normally, these components do not exceed 2% of the total lipid composition (Elmadfa and Leitzmann, 2004). The intake of total fat (mainly from TAGs)

TABLE 1.3
Fatty Acid Pattern and Unsaponifiable Matter of Some Oils and Fats

	SFA					MUFA		PUFA				Total Grams SFA:MUFA: PUFA	USP (g/100g)
	4:0	14:0	16:0	18:0	20:0	16:1	20:1	18:2	18:3	20:4	22:5		
Chain Length	12:0				22:0	18:1	22:1			20:5	22:6		
Animal Fats (g%)													
Butter	13	12	26	11	0	31	0	2	T	0	0	62:31:2	0.4
Lard	T	2	27	11	1	58	T	11	T	T	T	41:58:11	0.3
Tallow	T	1–6	20–37	6–40	0.5	27–59	T	0.5–5	2.5	T	T	27.5:27–59:3–7.5	0.4
Salmon	0	3	11	4	0	30	1	5	5	8	15	18:31:33	
Mackerel	T	8	16	2	T	22	25	1	1	8	8	26:47:18	0.7–1
Cod liver oil	T	4	14	3	T	34	23.5	1	T	8	7	21:57.5:16	1
Plant Fats and Oils (g%)													
Coconut oil	63	16	9	2	T	7	0	2	0	0	0	90:7:2	0.2
Palm oil	T	1	42	4	0	43	0	8	T	0	1.0	47:43:8	1.0
Olive oil	0	T	12	2	0	72	0	11	1	0	0	14:72:12	0.8
Peanut oil	T	1	11	3	0	49	0	29	1	0	0	15:49:30	0.7
Rapeseed oil	T	T	3	1	1	16	55	14	10	0	0	5:71:24	0.7–1.1
Lupine oil	0	1	13	5	T	44	0	34	3	0	0	19:44:37	1.04
Sesame seed oil	T	T	9	6	T	38	T	45	1	0	0	15:38:46	1.0

Soybean oil	T	T	10	4	0	25	0	52	7	0	0	14:25:59	1.2
Safflower oil													
Rich in linoleate	0	T	6.5	3	T	14	T	75	1	0	0	9.5:14:76	0.8
Rich in oleate	0	T	6	2	1	74[a]	T	15.8	1	0	0	9:74:15.8	0.8
Corn oil	T	T	13	2.5	T	30.5	T	52	1	0	0	15.5:30.5:53	1.6
Rice germ oil	0	T	16	2	0	42	T	37	1	0	0	18:42:38	5.0
Wheat germ oil	0	T	12	2	0	20	T	61	5	0	0	14:20:66	6.0

Source: Modified from Elmadfa, I., and C. Leitzmann. 2004. *Ernährung des Menschen*. 4th ed. Eugen Ulmer Stuttgart.

[a] Only C18:1n9, T traces, SFA:MUFA:PUFA saturated:monounsaturated:polyunsaturated fatty acids, USP unsaponifiable matter.

TABLE 1.4
Digestion of Lipids

Organ	Enzyme	Effect
Mouth	Lingual lipase	Burst of cell walls and mechanical dispersion, active function of the lipase in the stomach (acidic pH value)—neonatal fat digestion is aided
Stomach	Gastric lipase (tributyrinase)	Breakdown of some triacylglycerols (TAG) and di (DG)- and mono (MG)-glycerides; decomposition of medium-chain fatty acids
Small intestine	(After emulsifying of fats by bile acids)	
	Pancreatic lipase	TAG → DG + MG + free fatty acids
	Cholesterol esterase	Free cholesterol + free fatty acids (FFA) + glycerol
	Phospholipase	Lecithin → glycerol + FFA + phosphoric acid + choline

can vary largely from about 11% (China) up to 50% (rural dwellers in Nigeria) of total energy (Elmadfa and Kornsteiner, 2009).

The first step in the digestion of TAGs, which goes toward absorption, is an incomplete hydrolysis into diacylglycerols (DAGs) and free fatty acids (FFAs), which takes place in the stomach (10%–30 %) and is carried out by gastric lipase (adult) or lingual lipase (infancy) (Table 1.4).

The gastric lipase still plays a role in the TAG hydrolysis into adulthood, which preferentially hydrolyzes the sn3-ester bond, resulting in formation of sn1,2-DAGs (Mu and Hoy, 2004; Nicolaou and Kokotos, 2004; Ramirez et al., 2001). However, the main enzyme is the pancreatic lipase, which only acts with the colipase secreted from the pancreas. In the small intestine, lipids are emulsified by bile salts; this results in the formation of large molecule aggregates called *mixed cells*. The pancreatic lipase hydrolyses the fatty acids from the sn-1 and sn-3 positions, which leads to 2-monoacylglycerols. Unsaturated fatty acids are specifically esterified on this important sn-2 position, which conserves essential fatty acids during the whole digestion (Nicolaou and Kokotos, 2004; Dubois et al., 2007).

Cholesterol esters and phospholipids have to be hydrolyzed by cholesterol esterases and phospholipase A_2. Human lipid absorption takes place mainly in the small intestine, where hydrolyzed lipid products of digestion are absorbed into the enterocytes. Fat absorption is influenced by fatty acid chain length and unsaturation. Medium-chain fatty acids (<12 carbon atoms) are bound to albumin and can be transported directly to the liver via the portal blood. Therefore, they are used for diet therapy with severe malabsorptions (e.g., pancreas insufficiency). Long-chain saturated fatty acids (C16:0, C18:0) are only moderately absorbed from the lumen due to their higher melting point (above body temperature). In addition, long-chain saturated fatty acids tend to form insoluble calcium soaps with divalent cations in the alkaline environment of the small intestine. The fatty acid position in the glycerol structure influences the absorption and metabolism. Unsaturated fatty acids (e.g., AA, EPA, and DHA) are located at the sn-2 position, which remains as 2-monoacylglycerols after pancreatic hydrolysis. Moreover, longer-chain fatty acids have to be re-esterified into TAGs and phospholipids before they can be absorbed into the blood-stream. The generation of transportable lipids called *chylomicrons* is necessary for the aqueous environment of the blood. Chylomicrons are made up of phospholipids, lysophospholipids, apolipoproteins, TAGs, and fat-soluble vitamins. They are produced in the enterocytes and enter the bloodstream via the lymphatic circulation. (Nicolaou and Kokotos, 2004; Ramirez et al., 2001). Therefore, dietary lipids are carried by chylomicrons (main component TAGs), which bind to the lipoprotein lipase (LPL) on the endothelial surfaces of blood capillaries, principally in adipose tissue, but also in muscle and other organs. The LPL hydrolyzes the main component TAGs and releases fatty acids, which are re-esterified inside the target tissue. Chylomicron remnants are rich in cholesterol esters, but have not enough TAGs to compete effectively for LPL. They are taken up by a receptor-mediated mechanism in the liver. The turnover of chylomicrons is about 70–150 g/24 h (Nicolaou and Kokotos, 2004; Elmadfa and Leitzmann, 2004).

Fatty acids can be carried in the blood in lipoproteins, or as nonesterified fatty acids (NEFAs) transported by serum albumin. Main NEFAs arise from the hydrolysis of TAGs in adipose tissue, or to a lesser extent during hydrolysis of chylomicron and very low density lipoprotein triacylglycerols. They are taken up by the liver and primarily used as an energy source (Nicolaou and Kokotos, 2004; Ramirez et al., 2001).

Fat oxidation (up to 18 carbon atoms) occurs primarily in the mitochondria. Under prolonged fasting conditions and high intakes of medium-chain fatty acids, the production of large amounts of acetyl CoA exceeds the capacity for the citric cycle and ketones (e.g., acetoacetate and β-hydroxybutyrate) are built. These ketones can become an important energy source for the brain and muscles during starvation and low carbohydrate intake. Fatty acids of more than 18 carbon atoms need to be shortened in peroxisomes before they can enter the mitochondrial β-oxidation (IOM, 2002).

Lipoproteins have various density classes due to the different composition in their ratio of lipid to protein, their proportions of TAGs, esterified and nonesterified cholesterol, phospholipids, as well as their metabolic functions. These compositional differences influence the density of the lipoprotein, which is the base for the classification into chylomicrons, very-low-density lipoproteins (VLDLs), low-density lipoproteins (LDLs), and high-density lipoproteins (HDLs).

VLDLs are enriched with TAGs and generated in the liver from circulating NEFAs, chylomicron remnants, directly absorbed fatty acids, and from de novo synthesis from glucose. The functions and catabolism of VLDLs are similar to that of chylomicrons and result in VLDL remnants or intermediate-density lipoproteins (IDLs); the turnover of VLDLs is about 25–70 g/24 h. IDLs are transformed to LDLs, which are the major carriers of plasma cholesterol esters in humans. The surface of LDLs has apolipoprotein B, which plays an important role in the recognition of LDLs by cells and following uptake and metabolism by the cells. The reverse cholesterol transport from the peripheral cells to the liver is performed by HDLs. HDLs, which contain cholesterol from cell membranes, are taken up by the liver and degraded. The degradation products are used to synthesize bile acids (Nicolaou and Kokotos, 2004; Elmadfa and Leitzmann, 2004).

Under normal conditions, less than 10% (about 7–8 g/d) of the dietary fat are excreted with the stool (Jeppesen et al., 1997). The remaining fatty acids are normally catabolized entirely by oxidative processes. Important excretion products are carbon dioxide and water, and only small amounts of ketone bodies are formed by fatty acid oxidation and excreted in urine (IOM, 2002).

A sufficient essential fatty acid status depends on adequate intake and absorption. Individuals with a fat malabsorption (e.g., cystic fibrosis) have a higher incidence of an essential fatty acid deficiency (Peretti et al., 2005).

1.5 BIOLOGICAL ROLES OF NONESSENTIAL AND ESSENTIAL FATTY ACIDS (INCLUDING N-6 AND N-3 FATTY ACIDS)

1.5.1 Phospholipids and Membrane Structure

Phospholipids are fat-like substances found in all living cells of animal and vegetable origin. The highest concentrations generally occur in animal products. *Phospholipids* is a loose term used mainly for the class of complex lipids, the phosphoglycerides. There are other classes of phosphorous-containing lipids (e.g., sphingomyelin) that are not phosphoglycerides.

Phosphoglycerides are characterized by a 3-carbon glycerol backbone, where the sn-1 position and the sn-2 position are esterified to hydrophobic fatty acids (Figure 1.2). Unsaturated fatty acids are mainly found in the 2-position of glycerol. Phospholipids have a polar head group, an alcohol, the hydroxyl of which is esterified to the phosphoric acid. The most abundant phosphoglycerides are phosphatidylcholine (or lecithin), phosphatidylethanolamine (ethanolaminocephalin), phosphatidylinositol, phosphatidylserine, and phosphatic acid.

Phospholipids are nature's principal surface-active agents. They have remarkable emulsifying properties and help dietary fat to mix with water. The long-chain fatty acid moieties contribute hydrophobic properties that are counterbalanced by

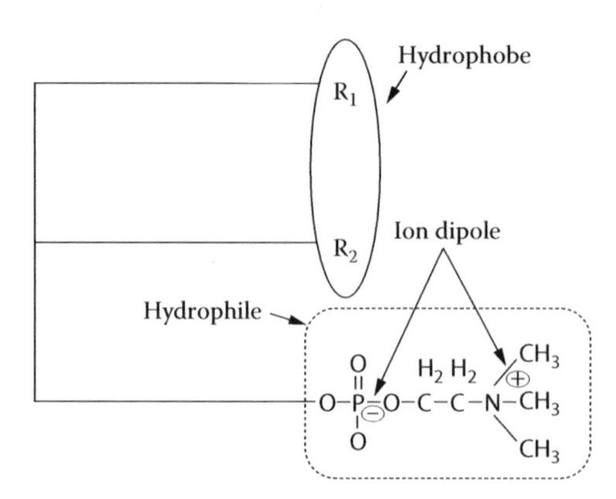

FIGURE 1.2 Lecithin molecule.

the strong hydrophilic character of the phosphate moiety. In an oil–water system, the phospholipid components concentrate at the oil–water interface. The polar parts of the molecule are directed toward the aqueous phase, while the lipophilic hydrocarbon tends toward the oily phase. This concentration of phospholipids at the oil–water interface decreases the surface tension and, thus, emulsions are formed. Especially, these amphipathic properties of phosphoglycerides are important for the formation of the lipid bilayer. For this reason, the lipid bilayer model from Singer and Nicolson still represents a very good overview of the basics of biological membranes. In most cases, the lipids are more or less fluid in the lipid bilayer. In particular, microdomains, such as the so-called *lipid rafts*, contain lipids in a solid-like state. Lipid rafts are very useful in organizing proteins in the membrane that have particular functions. Thus, it is now clear that the lipids have a key role in the function and properties of cell membranes (Lee, 2001). Essential fatty acids are important components of structural lipids, where they ensure optimal environmental conditions for membrane protein function, maintaining membrane fluidity and normal epithelial cell function. Alterations of the phospholipid composition influence cell function in different ways, to include changes in the regulation of gene expression either through effects on receptor activity, on intracellular signaling processes, or on transcription factor activation. As a consequence, the transcription factor activation and gene expression are changed (Calder, 2007, 2008).

Essential fatty acid status can be measured by the fatty acid pattern of serum, erythrocyte, and tissue phospholipids. An essential fatty acid deficiency is characterized by a decrease in the n-6 fatty acids (e.g., linoleic acid, arachidonic acid) and n-3 fatty acids (e.g., α-linolenic acid, eicosapentaenoic acid, docosahexaenoic acid), whereas an increase in nonessential monounsaturated fatty acids and their downstream products (e.g., mead acid) appears (Pettei et al., 1991). Due to the fact that, in the liver, the simplest n-3 (α-linolenic acid), n-6 (linoleic acid), and n-9 (oleic acid) fatty acids compete for the same series of enzymes, the conversion of oleic acid to mead acid (C 20:3n-9) only occurs when both n-3 and n-6 fatty acids are low or absent in the diet (Holman, 1998).

1.5.2 Interactions with Genes

Investigations have shown that free fatty acids from phospholipids released by cellular phospholipases or from the diet are important cell signaling mediators. In addition, emerging evidence shows that fatty acids have the possibility to alter rapidly and directly the transcription of specific genes (Simopoulos, 2008).

PUFAs are able to regulate several expressions of genes such as adipocyte, glucose transporter-4, lymphocyte stearoyl-CoA desaturase 2 in the brain, peripheral monocytes (interleukin [IL]-1b, and VCAM-1 [vascular cell adhesion molecule-1]), and platelets (platelet-derived growth factor [PDGF]) (Simopoulos, 2008). Especially, LCPn-3 fatty acids hold back inflammation, influence cholesterol metabolism, and increase the generation of adiponectin levels that improve insulin resistance. All these effects may be beneficial for protection against coronary heart diseases, prevent the progression of atherosclerosis, and reduce the incidence of type 2 diabetes (Das, 2006).

Thus, various mechanisms of essential fatty acids are exerted through changed gene expression. Nuclear receptors belong to ligand-activated transcription factors that can directly or indirectly regulate various genes of lipid metabolism and inflammatory signaling. Nuclear factor kB (NFkB) is an important transcription factor in a range of inflammatory signaling pathways. Cytokines (e.g., IL-1, IL-2, IL-6, IL-12, tumor necrosis factor-α), chemokines (e.g., IL-8, MIP-1α, MCP1), adhesion molecules (e.g., ICAM, VCAM, and E-selectin), and enzymes (e.g., cyclooxygenase-2) are regulated by NFkB, which can be inhibited by LCPn-3 fatty acids (Schmitz and Ecker, 2008). Peroxysome proliferator-activated receptors (PPARs) are also ligand-activated nuclear transcription factors. They have key roles in cellular differentiation, insulin resistance, atherosclerosis, and various metabolic diseases. PPAR ligands regulate various genes of the lipid metabolism and inhibit NFkB, which induces inflammation. EPA, DHA, and their eicosanoids are more powerful activators of PPAR-α than n-6 fatty acids (Schmitz and Ecker, 2008).

Retinoid X receptors (RXR) and PPAR are involved in cellular mechanisms such as transduction of the retinoid signaling pathway and lipid anabolism and catabolism. DHA is a ligand for RXR, but AA can also bind to RXRα.

The transcription factor sterol regulatory element binding protein 1c (SREBP-1c) has a basic helix-loop-helix leucin zipper. SREBP-1c is expressed in high levels in macrophages, liver, adipose tissue, adrenal gland, and brain. SREBP-1c influences genes of fatty acid metabolism such as fatty acid synthase, stearoyl-CoA desaturase, and ABCG1, which regulates cholesterol efflux. Essential fatty acids, especially LA, EPA, and DHA, have inhibitory effects on the transcription of lipogenic genes by suppressing SREBP-1c gene expression and inhibiting the proteolytic release process of nuclear SREBP-1c. This decrease in SREBP inhibits genes of fatty acid metabolism like fatty acid synthase. It is hypothesized that the liver X receptor, an activator of SREBP-1c, is inhibited. However, the exact mechanism is under investigation. In summary, LCPn-3 fatty acid-mediated activation of PPAR and inhibition of SREBP-1c favor increased lipid degradation and lower lipid biosynthesis (Schmitz and Ecker, 2008).

1.5.3 Essential Fatty Acids and the Immune System

The human immune system defends the host from infectious agents (e.g., pathogenic bacteria, viruses, fungi, and parasites) and from other noxious agents (Calder, 2007). These agents can initiate inflammatory reactions by activating a range of humoral and cellular mediators. Lipid mediators (e.g., prostaglandins, leukotrienes) and interleukins (IL) (e.g., IL-1, IL-6, tumor necrosis factor-α) are released to defend against the invaders (Simopoulos, 2008).

Prostanoids (prostaglandins, prostacyclins, thromboxanes), leukotrienes, lipoxins and resolvins, and neuroprotectin D1 derived from dihomo-γ-linolenic acid (DGLA), arachidonic acid (AA), eicosapentaenoic (EPA), and docosahexaenoic acid (DHA) have a key role in modulating inflammation, cytokine formation, immune response, platelet aggregation, vascular reactivity, and thrombosis (Figure 1.3) (Teraoka et al., 2009; Calder, 2008).

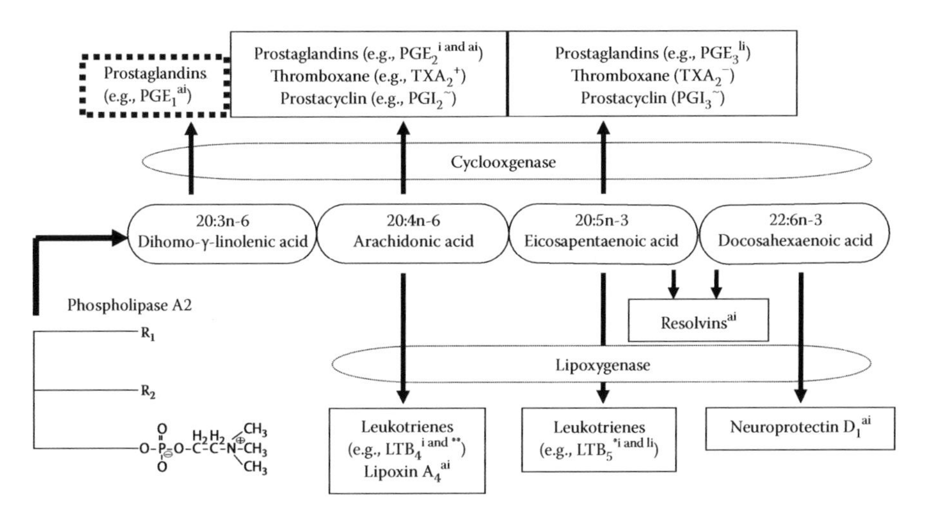

FIGURE 1.3 Pathway of eicosanoid synthesis. +, - With and without proaggregating and vasoconstricting effect. ~- Antiaggregating and vasodilating effect, **-Strong, *-weak chemotaxis, i-Inflammatory, ai-anti-inflammatory and li-less inflammatory manner.

The balance between long-chain polyunsaturated (LCP)n-6 and LCPn-3 fatty acids in phospholipids of neutrophils and monocytes, which is influenced by dietary intake, is suggested to be involved in various pathological processes such as atherosclerosis, coronary heart disease, cancer, diabetes mellitus, bronchial asthma, inflammatory bowel disease, and several other inflammatory conditions (Das, 2006).

Series-1 prostaglandins (PG) from DGLA have mostly inhibitory effects on inflammatory cells. One reason is that DGLA cannot be converted to leukotrienes (LTs), but it can form a 15-hydroxyl derivative that inhibits the transformation of arachidonic acid to LTs (Belch and Hill, 2000). Due to no practical food sources of DGLA, most investigations focus on AA (Teraoka et al., 2009). It produces prostaglandin E_2 (PGE_2), which raises the cardinal signs of inflammation including fever, vascular permeability, and vasodilatation, and enhances pain and edema caused by other agents such as bradykinin and histamine. AA induces formation of thromboxane A_2 (powerful platelet aggregator and vasoconstrictor) as well as prostacyclin I_2 (vasodilator and inhibitor of platelet aggregation). Leukotriene B_4 (LTB_4) is produced via lipoxygenase from AA, which is a potent inducer of inflammation, leukocyte chemotaxis, and adherence. Conversely, investigations have demonstrated that PGE_2 inhibits lipoxygenase and so reduces the formation of LTB_4 and encourages the formation of lipoxins A_4, both of which have anti-inflammatory effects. A decreased production of AA-derived mediators that can be achieved by fish oil consumption has led to the belief that fatty fish has anti-inflammatory effects and that it may be useful in the prevention and therapy of inflammatory conditions (Das, 2006; Simopoulos, 2002; Calder, 2005, 2008).

Eicosapentaenoic acid produces prostaglandin E_3, which has only little inflammatory activity due to a low synthesis rate. It also forms thromboxane A_3 (weak platelet aggregator and vasoconstrictor) and prostacyclin I_3, which leads to an overall rise in total prostacyclin by increasing PGI_3, without a decrease in PGI_2. Both PGI_2 and PGI_3 are active vasodilators and inhibitors of platelet aggregation. EPA modulation induces leukotriene B_5 (LTB_5), which is a weaker inducer of inflammation and chemotactic agents (Simopoulos, 2002).

In addition, EPA and DHA are precursors of different series of resolvins by involving both enzymes cyclooxygenase-2 (COX-2) and 5-lipoxygenase (LOX-5). Resolvins exert potent anti-inflammatory actions and via DHA an additional anti-inflammatory metabolite termed *neuroprotectin D1* can be generated involving 5-lipoxygenase (Das, 2006; Calder, 2008).

In summary, LCPn-3 fatty acids are recognized to have powerful immunomodulatory properties, including anti-inflammatory actions via decreasing leukocyte chemotaxis, adhesion molecule expression, and inflammatory cytokine production. Some of these mechanisms are carried out through reduced activation of the proinflammatory transcription factor Fib and perhaps through raised activation of the anti-inflammatory transcription factor PPAR-γ (Calder, 2008).

1.5.4 Carriers of Fat-Soluble Vitamins

Fats are significant for the transport and absorption of a variety of nonglyceride components that are found in dietary fats and are important for maintaining health. The main fat-soluble, nutritionally important components are retinol, calciferol, tocopherols, and phylloquinone. Many plant oils and products contain considerable amounts of tocopherols. Calciferol occurs in fish oils and fish liver oils. Both vitamins are added to margarine. Processing of oils reduces tocopherol content. Tocopherols are the main fat-soluble antioxidant in the body, present in lipoproteins, especially in LDLs, and in part in HDLs. They are found inside and outside the membranes, protecting the cells against free radical attacks. There, the most important function of vitamin E is to maintain the integrity of long-chain polyunsaturated fatty acids and as a result their bioactivity. Bioactive lipids are among the main signaling molecules and alterations due to oxidation can have an impact on key cellular events (Traber and Atkinson, 2007). Due to their favorable characteristics, it has been suggested that especially vitamin E supplements can reduce cardiovascular diseases and cancer. However, a meta-analysis demonstrated the contrary. Especially, interventions with high dosages of vitamin E (>400 IU/d) increased all-cause mortality (Miller et al., 2005).

Long-chain polyunsaturated fatty acids (n-3 and n-6) influence the synthesis of chylomicrons and VLDLs within the mucosal cells and affect the lymphatic transport of lipid-soluble vitamins within the triglyceride-rich lipoproteins (VLDLs, LDLs). They increase the physiological requirements for vitamin E. Thus, the recommended vitamin E intake is influenced by the PUFAs intake. Diets high in PUFAs but low in vitamin E have adverse effects on the tocopherol status (Horwitt, 1974). If large amounts of PUFAs are ingested, substantial quantities of extra vitamin E are needed to reestablish the plasma tocopherol level (normal plasma vitamin E concentrations in humans range from 12 to 46 μM/L to 0.5–2 mg/dL). Therefore, the intake of 1 g PUFAs (expressed as diene-equivalent using transformation factors as proposed by Horwitt (1974)) requires an additional intake of 0.4 mg α-tocopherol equivalents (d-α-TEQ = d-α-tocopherol equivalents = mg d-α-tocopherol + 0.5 × mg d-β-tocopherol + 0.25 × mg d-γ-tocopherol + 0.01 × mg d-δ-tocopherol).

Fats are also carriers of biologically active ubiquinone. In the human body, the most important type is CoQ10 that varies between 8 μg/g in the lungs and 114 μg/g in the heart. Only small amounts of CoQ9 are found in human tissues. The endogenously synthesized ubiquinone Q participates as an electron carrier in the mitochondrial respiratory chain and with other antioxidants (e.g., vitamin E and C) prevents lipid oxidation (Turunen et al., 2004).

Tocotrienols, transported in fats, exhibit different physiological effects than those observed with tocopherols. As part of the diet in humans and animals, tocotrienols have been manifested to have a hypocholesteremic effect, mediated by their ability to decrease the hepatic HMG-CoA reductase activity (Qureshi et al., 1986; Qureshi et al., 1991).

1.5.5 Cholesterol and Phytosterols

Cholesterol, the main sterol of animal tissues, is the preliminary stage of bile acids. It plays an important role in stabilizing the hydrophobic interactions of the animal cell membrane by inserting itself between the fatty acids in the bilayer. In animal cell membranes, only cholesterol is able to maintain this function. Plant cell membranes, in contrast, contain only a little cholesterol but do contain 4-desmethylsterols, mainly β-sitosterol. The daily intake of cholesterol is up to 800 mg/d, 20%–80% of which is absorbed by passive diffusion. A constant supply of cholesterol for hormone production and cell membranes is provided by endogenous synthesis in the liver (1–1.5 g/d). The cholesterol circulates in the organism through the enterohepatic circle (Figure 1.4).

The liver produces cholesterol from fragments of saturated fatty acids; the liver's ability to regulate the blood cholesterol content by cutting back is limited. Thus, an intake of saturated fatty acids higher than the amount the liver can compensate for may result in an increase of blood cholesterol levels.

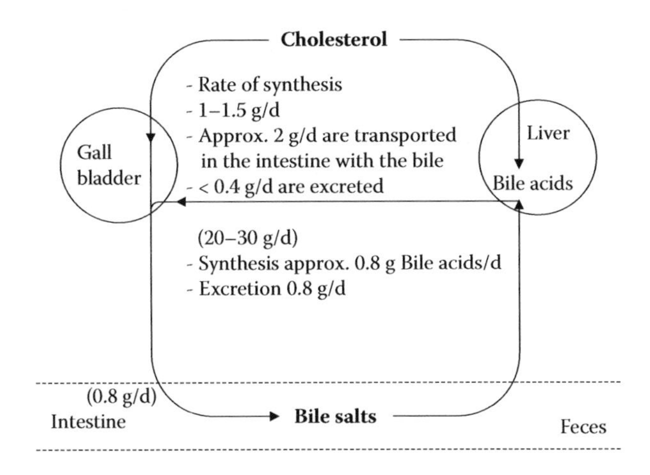

FIGURE 1.4 Enterohepatic circle of cholesterol and bile acids. (Modified from Elmadfa, I., and C. Leitzmann. 2004. *Ernährung des Menschen*. 4th ed. Eugen Ulmer, Stuttgart.)

High cholesterol levels are the main risk factors for cardiovascular diseases. According to epidemiological studies, the increase of LDL-cholesterol in steps of 1 mg/dL enhances the risk of a coronary heart disease by 1%–2% (Elmadfa and Leitzmann, 2004). Saturated fatty acids (especially C12:0–C16:0), which are common in meats, dairy fat, and tropical oils (palm kern oil, coconut oil), increase LDL-cholesterol concentrations (Sacks and Katan, 2002). Replacing these saturated fatty acids with monounsaturated and polyunsaturated fatty acids reduces total cholesterol and LDL-cholesterol (Sacks and Katan, 2002). Therefore, the intake of polyunsaturated fatty acids is inversely associated with the risk of coronary heart disease (Oh et al., 2005).

Plant sterols, also called *phytosterols*, have similar functions as cholesterol in animals, such as the regulation of the membrane fluidity in plants. In human nutrition, plant sterols are nonnutritive compounds that reduce the total cholesterol absorption from diet (between 200 and 500 mg/d) and enterohepatic circulation (around 1g/d). Different dosages ranging from 0.8 to 4.0 g/d of phytosterols are known to lower LDL-cholesterol levels by 10% to 15%. The absorption rate of phytosterols from the intestine is much lower (between 0.04% and 16%) than from dietary cholesterol (55% to 60%) (Brufau et al., 2008), whereas the typical daily intake of sitosterol ranges from 150 to 350 mg/d (Katan et al., 2003).

1.6 DIETARY FAT AND PREVENTION OF NONCOMMUNICABLE CHRONIC DISEASES

1.6.1 Cancer and Dietary Fat

From a global perspective, overweight and obesity are growing health problems where industrialization and westernization of lifestyle takes place. Especially, obesity is related to a number of noncommunicable diseases such as hypertension, cardiovascular diseases, type 2 diabetes, dyslipidemia, metabolic syndrome, and certain types of cancers. Obesity-related breast, prostate, and colon cancers are the most common cancers in industrialized countries. Increased body fat probably influences the development and subsequent progression of different kinds of cancer by releasing hormone-like factors or adipokines, which are suggested to provide a link among cancer, insulin resistance, inflammation, and oxidative stress. In addition, more and more data indicate that insulin resistance and related metabolic syndrome are involved in the pathogenesis of cancer (Murthy et al., 2009). On the other hand, an increase in the intake of plant food is suggested to reduce the risk of major kinds of cancer. This beneficial observation is attributed to an increase in vitamins, minerals, antioxidants, and phytochemicals, while a reduction in total fat intake occurs (Valdes-Ramos and Benitez-Arciniega, 2007).

Especially long-chain polyunsaturated fatty acids are involved in the modulation of cytokine production, lymphocyte proliferation, and expression of surface molecules, phagocytosis, apoptosis, and natural killer cell activity. The rise in n-3 long-chain polyunsaturated fatty acids manages the balance of the generation of pro- and anti-inflammatory eicosanoids as well as cytokines (Valdes-Ramos and Benitez-Arciniega, 2007). Several data indicate that LCPn-3 fatty acids intake from fish or probably from fish oil supplements can reduce the risk of cancer (e.g., colon cancer)

(Gerber, 2009). Nevertheless, the precise impact of total fat and different fatty acids on the development of different kinds of cancer remains an open question.

1.6.2 Atherosclerosis and Cardiovascular Diseases

Coronary heart disease (CHD) has clinical manifestations ranging from angina pectoris to myocardial infarction and sudden death. The primary cause of CHD is coronary atherosclerosis, due to lipid-rich lesions in the intima of coronary arteries.

Different effects of dietary fatty acids on plasma are known. Saturated fatty acids, especially lauric acid C12:0, myristic acid C14:0, and palmitic acid C16:0, raise LDL and HDL cholesterol levels, whereas the effect of stearic acid is much less cholesterol raising compared to lauric, myristic, and palmitic acid. The National Cholesterol Education Program has demonstrated that, for each 1% of energy reduction in SFAs, total plasma cholesterol was reduced by 0.056 mmol/L and LDL-cholesterol by 0.05 mmol/L. In addition, prospective cohort studies confirmed that saturated fatty acid intake was directly related with CHD (Esrey et al., 1996; Kabagambe et al., 2003; Kromhout et al., 1985; McGee et al., 1984). Replacing cholesterol-raising saturated fatty acids with monounsaturated fatty acids lowers total plasma cholesterol and LDL levels (Katan et al., 1994). Polyunsaturated fatty acids are suggested to have cardioprotective effects due to their cholesterol-lowering effects on total cholesterol, LDL cholesterol, and slightly on HDL cholesterol (Kris-Etherton and Yu, 1997). In addition, a meta-analysis of 60 controlled human trials from Mensink et al. demonstrated that inserting PUFAs (as opposed to MUFAs or SFAs) for carbohydrates resulted in a significantly greater reduction in LDL levels and the ratio of total cholesterol to HDL cholesterol, which is a good predictor of CHD risk (Mensink et al., 2003).

The Nurses' Health Study observed that the risk of CHD can be reduced by replacing saturated and especially *trans* unsaturated fatty acids with *cis* monounsaturated and polyunsaturated fatty acids (Hu et al., 1999; Oh et al., 1997, 2005).

In addition, data from primary and secondary prevention studies support the theory that the intake of LCPn-3 (especially eicosapentaenoic and docosahexaenoic acid) reduces all-cause mortality, cardiac and sudden death, and stroke (Wang et al., 2006). The evidence appears especially strong for CHD risk on secondary prevention. LCPn-3 fatty acids appear to confer cardiovascular health benefits mainly through EPA and DHA enrichment of membrane phospholipids. This enrichment can lower abnormal ventricular arrhythmias and blood pressure, improve arterial and endothelial function, lower platelet aggregations, and positively influence autonomic tone (Lee et al., 2008).

1.7 RECOMMENDATIONS FOR DIETARY FAT AND PUFA'S INTAKE

Adequate intake of dietary fat is essential for health. In adults, the acceptable macronutrient distribution range for total fat lies between 20% and 35% of total energy (E), at least 15% E is necessary to meet requirements for essential fatty acids and fat-soluble vitamins. For prevention of atherosclerosis and different metabolic effects, the fatty acid pattern of SFAs:MUFAs:PUFAs should be about 1:1:1. The WHO sets the highest level of saturated fatty acids intake up to 10% of total energy. The acceptable macronutrient distribution range (AMDR) for total PUFA is between 6 and 11% E, whereas 2.5% E from linoleic acid plus 0.5% α-linolenic acid are essential to prevent deficiency symptoms. The AMDR for n-3 FAs ranges from 0.5 to 2% E. The intake of preformed n-3 long-chain polyunsaturated fatty acids (LCPUFAs) from eicosapentaenoic acid and docosahexaenoic acid are recommended between 0.25 and 2 g/d. *trans* FA intake from all sources should be restricted to 1% E (Elmadfa and Kornsteiner, 2009). In case of a higher intake of PUFAs, the intake of antioxidants, especially vitamin E, has to be increased. As carriers of the fat-soluble tocopherols, β-carotene or ubiquinones fats are necessary for transportation and absorption. These compounds are also important antioxidants *in vivo* and may protect cells and cell membranes against oxidative stress.

REFERENCES

Belch, J. J., and A. Hill. 2000. Evening primrose oil and borage oil in rheumatologic conditions. *Am J Clin Nutr* 71:352S–6S.

Brufau, G., M. A. Canela, and M. Rafecas. 2008. Phytosterols: Physiologic and metabolic aspects related to cholesterol-lowering properties. *Nutr Res* 28:217–25.

Burdge, G. C., and P. C. Calder. 2005. Alpha-Linolenic acid metabolism in adult humans: The effects of gender and age on conversion to longer-chain polyunsaturated fatty acids. *Eur J Lipid Sci Technol* 107:426–439.

Calder, P. C. 2005. Polyunsaturated fatty acids and inflammation. *Biochem Soc Trans* 33:423–7.

Calder, P. C. 2007. Immunomodulation by omega-3 fatty acids. *Prostaglandins Leukot Essent Fatty Acids* 77:327–35.

Calder, P. C. 2008. Polyunsaturated fatty acids, inflammatory processes and inflammatory bowel diseases. *Mol Nutr Food Res* 52:885–97.

D-A-CH. 2000. German Nutrition Society (DGE), Austrian Nutrition Society (ÖGE), Swiss Society for Nutrition Research (SGE), Swiss Nutrition Association (SVE): Reference values for nutrient intake. Frankfurt/Main, *Umschau Braus, 2002*].

Das, U. N. 2006. Essential fatty acids: Biochemistry, physiology and pathology. *Biotechnol J* 1:420–439.

DeFilippis, A. P., and L. S. Sperling. 2006. Understanding omega-3's. *Am Heart J* 151:564–70.

Dubois, V., S. Breton, M. Linder, J. Fanni, and M. Parmentier. 2007. Fatty acid profiles of 80 vegetable oils with regard to their nutritional potential. *Eur J Lipid Sci Technol* 109:740–756.

Elmadfa, I., and M. Kornsteiner. 2009. Dietary fat intake—a global perspective. *Ann Nutr Metab* 54:8–14.

Elmadfa, I., and M. Kornsteiner. 2009. Joint FAO/WHO Expert Consultation on fats and fatty acids in human nutrition. Requirements for adults. *Ann Nutr Metab* 55:56–75.

Elmadfa, I., and C. Leitzmann. 2004. *Ernährung des Menschen*. 4th ed. Eugen Ulmer, Stuttgart.

Esrey, K. L., L. Joseph, and S. A. Grover. 1996. Relationship between dietary intake and coronary heart disease mortality: Lipid research clinics prevalence follow-up study. *J Clin Epidemiol* 49:211–6.

Fahy, E., S. Subramaniam, H. A. Brown, C. K. Glass, A. H. Merrill, Jr., R. C. Murphy, C. R. Raetz, D. W. Russell, Y. Seyama, W. Shaw, T. Shimizu, F. Spener, G. van Meer, M. S. VanNieuwenhze, S. H. White, J. L. Witztum, and E. A. Dennis. 2005. A comprehensive classification system for lipids. *J Lipid Res* 46:839–61.

Gerber, M. 2009. Background review paper on total fat, fatty acid intake and cancers. *Ann Nutr Metab* 55:140–61.

Holman, R. T. 1998. The slow discovery of the importance of omega 3 essential fatty acids in human health. *J Nutr* 128:427S–433S.

Horwitt, M. K. 1974. Status of human requirements for vitamin E. *Am J Clin Nutr* 27 (10):1182–93.

Hu, F. B., M. J. Stampfer, J. E. Manson, A. Ascherio, G. A. Colditz, F. E. Speizer, C. H. Hennekens, and W. C. Willett. 1999. Dietary saturated fats and their food sources in relation to the risk of coronary heart disease in women. *Am J Clin Nutr* 70:1001–8.

Hu, F. B., M. J. Stampfer, J. E. Manson, E. Rimm, G. A. Colditz, B. A. Rosner, C. H. Hennekens, and W. C. Willett. 1997. Dietary fat intake and the risk of coronary heart disease in women. *N Engl J Med* 337:1491–9.

IOM. 2002. *Dietary Reference Intakes for Energy, Carbohydrate, Fiber, Fat, Fatty Acids, Cholesterol, Protein, and Amino Acids (Macronutrients) (2005)*. Edited by the Institute of Medicine of the National Academics. Washington, DC: The National Academies Press.

IUPAC. 1978. The nomenclature of lipids (Recommendations 1976) IUPAC-IUB Commission on Biochemical Nomenclature. *Biochem J* 171:21–35.

Jeppesen, P. B., M. S. Christensen, C. E. Hoy, and P. B. Mortensen. 1997. Essential fatty acid deficiency in patients with severe fat malabsorption. *Am J Clin Nutr* 65:837–43.

Kabagambe, E. K., A. Baylin, X. Siles, and H. Campos. 2003. Individual saturated fatty acids and nonfatal acute myocardial infarction in Costa Rica. *Eur J Clin Nutr* 57 (11):1447–57.

Katan, M. B., S. M. Grundy, P. Jones, M. Law, T. Miettinen, and R. Paoletti. 2003. Efficacy and safety of plant stanols and sterols in the management of blood cholesterol levels. *Mayo Clin Proc* 78:965–78.

Katan, M. B., P. L. Zock, and R. P. Mensink. 1994. Effects of fats and fatty acids on blood lipids in humans: An overview. *Am J Clin Nutr* 60:1017S–1022S.

Kris-Etherton, P. M., and S. Yu. 1997. Individual fatty acid effects on plasma lipids and lipoproteins: Human studies. *Am J Clin Nutr* 65:1628S–1644S.

Kromhout, D., E. B. Bosschieter, and C. de Lezenne Coulander. 1985. The inverse relation between fish consumption and 20-year mortality from coronary heart disease. *N Engl J Med* 312:1205–9.

Lee, A. 2001. Membrane structure. *Curr Biol* 11:R811–4.

Lee, J. H., J. H. O'Keefe, C. J. Lavie, R. Marchioli, and W. S. Harris. 2008. Omega3 fatty acids for cardioprotection. *Mayo Clin Proc* 83:324–32.

McGee, D. L., D. M. Reed, K. Yano, A. Kagan, and J. Tillotson. 1984. Ten-year incidence of coronary heart disease in the Honolulu Heart Program: Relationship to nutrient intake. *Am J Epidemiol* 119:667–76.

Mensink, R. P., P. L. Zock, A. D. Kester, and M. B. Katan. 2003. Effects of dietary fatty acids and carbohydrates on the ratio of serum total to HDL cholesterol and on serum lipids and apolipoproteins: A meta-analysis of 60 controlled trials. *Am J Clin Nutr* 77:1146–55.

Miller, E. R., 3rd, R. Pastor-Barriuso, D. Dalal, R. A. Riemersma, L. J. Appel, and E. Guallar. 2005. Meta-analysis: High-dosage vitamin E supplementation may increase all-cause mortality. *Ann Intern Med* 142:37–46.

Mu, H., and C. E. Hoy. 2004. The digestion of dietary triacylglycerols. *Prog Lipid Res* 43:105–33.

Murthy, N. S., S. Mukherjee, G. Ray, and A. Ray. 2009. Dietary factors and cancer chemoprevention: An overview of obesity-related malignancies. *J Postgrad Med* 55:45–54.

Nicolaou, A., and G. Kokotos. 2004. *Bioactive Lipids*. Vol. 17: The Oily Press.
Oh, K., F. B. Hu, J. E. Manson, M. J. Stampfer, and W. C. Willett. 2005. Dietary fat intake and risk of coronary heart disease in women: 20 years of follow-up of the nurses' health study. *Am J Epidemiol* 161:672–9.
Peretti, N., V. Marcil, E. Drouin, and E. Levy. 2005. Mechanisms of lipid malabsorption in cystic fibrosis: The impact of essential fatty acids deficiency. *Nutr Metab (Lond)* 2:11.
Pettei, M. J., S. Daftary, and J. J. Levine. 1991. Essential fatty acid deficiency associated with the use of a medium-chain-triglyceride infant formula in pediatric hepatobiliary disease. *Am J Clin Nutr* 53:1217–21.
Qureshi, A. A., W. C. Burger, D. M. Peterson, and C. E. Elson. 1986. The structure of an inhibitor of cholesterol biosynthesis isolated from barley. *J Biol Chem* 261:10544–50.
Qureshi, A. A., N. Qureshi, J. J. Wright, Z. Shen, G. Kramer, A. Gapor, Y. H. Chong, G. DeWitt, A. Ong, D. M. Peterson, and et al. 1991. Lowering of serum cholesterol in hypercholesterolemic humans by tocotrienols (palmvitee). *Am J Clin Nutr* 53:1021S–1026S.
Ramirez, M., L. Amate, and A. Gil. 2001. Absorption and distribution of dietary fatty acids from different sources. *Early Hum Dev* 65:S95–S101.
Ratnayake, W. M., and C. Galli. 2009. Background review paper on fat and fatty acid terminology, methods of analysis and fat digestion and metabolism. *Ann Nutr Metab* 55:8–43.
Sacks, F. M., and M. Katan. 2002. Randomized clinical trials on the effects of dietary fat and carbohydrate on plasma lipoproteins and cardiovascular disease. *Am J Med* 113:13S–24S.
Schmitz, G., and J. Ecker. 2008. The opposing effects of n-3 and n-6 fatty acids. *Prog Lipid Res* 47:147–55.
Simopoulos, A. P. 2002. Omega-3 fatty acids in inflammation and autoimmune diseases. *J Am Coll Nutr* 21:495–505.
Simopoulos, A. P. 2008. The importance of the omega-6/omega-3 fatty acid ratio in cardiovascular disease and other chronic diseases. *Exp Biol Med (Maywood)* 233:674–88.
Teraoka, N., H. Kawashima, A. Shiraishi-Tateishi, T. Tanaka, J. Nakamura, S. Kakutani, and Y. Kiso. 2009. Oral supplementation with dihomo-gamma-linolenic acid-enriched oil altered serum fatty acids in healthy men. *Biosci Biotechnol Biochem* 73:1453–5.
Traber, M. G., and J. Atkinson. 2007. Vitamin E, antioxidant and nothing more. *Free Radic Biol Med* 43:4–15.
Turunen, M., J. Olsson, and G. Dallner. 2004. Metabolism and function of coenzyme Q. *Biochim Biophys Acta* 1660:171–99.
Valdes-Ramos, R., and A. D. Benitez-Arciniega. 2007. Nutrition and immunity in cancer. *Br J Nutr* 98:S127–32.
Wang, C., W. S. Harris, M. Chung, A. H. Lichtenstein, E. M. Balk, B. Kupelnick, H. S. Jordan, and J. Lau. 2006. n-3 Fatty acids from fish or fish-oil supplements, but not alpha-linolenic acid, benefit cardiovascular disease outcomes in primary- and secondary-prevention studies: A systematic review. *Am J Clin Nutr* 84:5–17.

2 Oxidation Products and Metabolic Processes

Mary N. Omwamba, William E. Artz, and Symon M. Mahungu

CONTENTS

2.1 INTRODUCTION

The first section of this chapter will focus on the oxidation mechanisms of fats, the volatile and nonvolatile products that are formed, and their characterization techniques. The term *fats* will be used to denote both fats and oils. The second section will highlight the possible effects on metabolic processes after ingestion of those products, particularly the nonvolatile products.

Although fat has been portrayed as the "problem" food group of the major food groups, no one can survive without essential fatty acids. Food technologists and dieticians agree that a diet without fat would be less appealing to the palate (Haumann, 1998; Reineccius and Heath, 2006). Fats contribute to the appearance, taste, mouthfeel, lubricity, and flavor of most food products (Akoh, 1995; Artz and

Hansen, 1996a,b; Reineccius and Heath, 2006). The amounts and type of fats present in foods determine the characteristics of that food and consumer acceptance. Nutritionally, fat has different functions. It is a source of energy and of building material for cell components. Fat provides the essential fatty acids linoleic and linolenic to humans; it is a carrier and aid in the absorption of fat-soluble vitamins A, D, E, and K and the carotenoids, and is a factor in the control of serum lipids and lipoproteins. Fats act as building blocks in the formation of biological membranes that surround cells and subcellular particles. Fats in the cell membranes are mainly the phospholipids phosphatidylethanolamine (PE), which predominates in the outer cell membranes (Melton, 1996). Fat acts as a "food solvent," maximizing availability of some phytochemicals and carotenoids such as lycopene from tomatoes (Haumann, 1998; Hu et al., 2008).

During frying, the food is submerged in fat that is heated in the presence of air. Therefore, the fat is exposed to the action of moisture from the foodstuff, oxygen from the atmosphere, and high temperature at which the operation takes place (Krishnamurthy and Chang, 1967; Chang et al., 1978; Gutierrez et al., 1988, Reineccius and Heath, 2006). During deep-fat frying, the oil undergoes a series of complex chemical reactions, such as oxidation, polymerization, hydrolysis, *cis/trans* isomerization, conjugation, pyrolysis, and cyclization (Engelsen, 1997; Choe and Min, 2007). Some of these reactions result in the desirable flavor, color, and texture of the fried food, while others are undesirable from the perspectives of quality, nutrition, and safety (Neff et al., 2000; Márquez-Ruiz and Dobarganes, 2006). Formation of carcinogenic compounds in frying fat has been investigated, and the reported compounds include acrylamide found in high concentration in carbohydrate-rich foods (Mestdagh et al., 2008).

The moisture from the foodstuffs causes hydrolytic reactions giving rise to free fatty acids, monoacyl and diacyl glycerols, and glycerol. The atmospheric oxygen causes oxidative reactions, giving rise to oxidized monomers, dimers, and polymers. Nonpolar dimers and polymers as well as volatile compounds are also produced. The thermal reactions caused by high temperature produce cyclic monomers, dimers, and polymers. The oxidative and thermal degradation take place in the saturated and the unsaturated fatty acid constituents of the triacylglycerols. Thus, the principal components altered are triacylglycerols with at least one of their acyl radicals altered. In addition, the three types of reactions are not only superimposed but also interrelated. High temperature plays a significant role, favoring the formation of oxidative and nonoxidative dimmers and polymers. The free fatty acids produced during hydrolysis are more susceptible to oxidative and thermal changes than when esterified to the glycerol.

Physiological and nutritional effects of frying fats have undergone intensive investigations since the 1950s. Research has focused on whether frying fats are detrimental to human health. The five different stages heated fats go through are break-in, fresh, optimum, degrading, and runaway. Each stage has a distinct flavor and food quality associated with it. A plot of food quality versus fat heating results in a bell-shaped curve with the maximum food quality occurring during the optimum phase (Blumenthal, 1991; Warner, 1999). As the fat is heated, the quality decreases as evidenced by the reduction in heat capacity, and surface and interfacial tension. There is also an increase in specific gravity, viscosity, acid value, and aldehyde and polymer content. Surface tension and interfacial tension are reduced by low-polarity and high-polarity oxidative polymers causing excessive oil pick-up by the food (Blumenthal and Stier, 1991; Dana and Saguy, 2006). Elucidation of different mechanisms under various conditions of oxidation has provided an improved understanding of the products that may be responsible for organoleptic deterioration and biological toxicity (Chang et al., 1978; Kanner, 2007).

From a nutritional perspective, the nonvolatile degradation products of frying fats are the most relevant since they remain in the oil, are absorbed in the food, and are subsequently ingested. Such nonvolatile products include polymeric triacylglycerols, monomeric triacylglycerols containing oxidized or cyclic fatty acid acyls, and significant breakdown products (Johnson et al., 1956, 1957; Alexander, 1978; Gabriel et al., 1978; Márquez-Ruiz and Dobarganes, 2006). These are the compounds that may interfere with the functions of metabolic enzymes in the body after they are ingested. Digestibility of oxidized fats indicated that oxidized fatty acid monomers are nutritionally important. Polymeric fatty acids showed generally poor digestibility, which was attributed to low activity of pancreatic lipase on triacylglycerols polymers (Márquez-Ruitz et al., 1992; Márquez-Ruiz and Dobarganes, 1995).

2.2 LIPID OXIDATION

Fat oxidation to form hydroperoxides takes place by loss of a hydrogen radical in the presence of trace metals, light, or heat. This reaction is described in terms of *initiation*, *propagation*, *branching*, and *termination processes* (Kanner et al., 1987; Kubow, 1992; Choe and Min, 2006). The formation of hydroperoxides can also take place by sensitized photooxidation. Sensitized photooxidation generally involves light excitation of a sensitizer to the singlet state followed by intersystem crossing to the triplet state (Foote, 1968; Mahungu, 1994; Choe and Min, 2006). The hydroperoxide decomposition products depend on temperature, pressure, and the concentration of oxygen. Whereas the volatiles are largely removed from the oil during frying and have implications in the flavor of both the frying oil and the fried food, the nonvolatile compounds remain in the frying oil and are absorbed by the food modifying the oil nutritional and physiological properties.

2.3 VOLATILE PRODUCTS

Breakdown of hydroperoxides gives rise to volatiles and short-chain compounds attached to the glyceridic backbone forming part of nonvolatile molecules. Volatile compounds formed in frying oil include aldehydes, ketones, hydrocarbons, alcohols, acids, esters, and aromatic compounds. The hydrocarbons that have been identified in frying oil include hexene, hexane, heptane, octane, nonane, and decane. Among these decomposition products, aldehydes are the most important because they are the most abundant (Frankel, 1985), and their thresholds are lower than those of other secondary products that characterize the flavor of fried foods and oils.

The formation of volatile decomposition products changes the flavor of the frying fat and contributes significantly to the flavor of the fried food. Volatiles are removed at high temperatures of the frying process, and the composition of volatiles depends on both fatty acid composition and the level of alteration of the oil analyzed. The volatile carbonyl compounds formed during frying depend on the unsaturation level of the oil (Farhoosh and Moosavi, 2008; Choe and Min, 2007). As expected, unsaturated fatty acids contribute significantly more to the formation of volatile compounds than those from the more stable saturated fatty acids, such as palmitic and stearic. The content of linolenic acid is a relevant factor in increasing the off-flavor of the oil during frying (Kalua et al., 2007; Warner, 2008). Morales et al. (1997) applied a thermoxidation process in extra-virgin olive oil to study the evolution of the volatile compounds responsible for off-flavor. They found that the initial volatiles (a total of 60), most of which are responsible for the pleasant sensory characteristics of olive oil, disappear in the first hours of thermoxidation, and that off-flavor compounds are then formed.

The volatiles in the frying oil increase initially, but then fall with increasing frying time. This rise and fall is especially noticeable in french fries, where the maximum level of volatiles occurs at 70 h (Hamilton and Perkins, 1997). It has been found that saturated aldehydes C_6-C_9 in chips reach a maximum between 50 and 70, the enals at 120, and dienals such as *t,t*-2,4-heptadienal at 70 h of frying. One of the complicating factors in frying is that these simple aldehydes are found along with cyclic structures. Haywood et al. (1995) have discussed the toxicological significance of the presence of aldehydes and their precursors in used fats and oils.

2.3.1 Formation Processes/Mechanisms

The mechanism of flavor development in heated oils is essentially that of lipid oxidation. Thermally induced oxidation involves hydrogen radical abstraction, the addition of molecular oxygen to form the peroxide radical, formation of the hydroperoxide, and then decomposition to form volatile flavor compounds. The low-molecular-weight compounds that form due to degradation of the frying oil are considered "volatile," contributing to desirable and undesirable flavors. The decomposition of hydroperoxides to form volatile compounds is through the cleavage between the oxygen and oxygen of R-O-O-H to produce the alkoxy radical, R-O• instead of the hydroperoxy radical, R-O-O• (Frankel, 1985).

2.3.2 Characterization Techniques

Methods that measure volatile decomposition compounds directly or indirectly include peroxide value, gas chromatography analysis of volatile compounds, and sensory analysis. Min and Schweizer (1983) studied lipid oxidation of oil used to prepare potato chips by measuring peroxide value. To evaluate the quality of the aroma of the oil, they analyzed volatile compounds. Tyagi and Vasishtha (1996) studied the chemical changes that took place in oil during the frying of potato chips by evaluating various parameters including iodine index, peroxide value, FFA content, color, and viscosity.

Spectroscopic techniques have been used as a quicker way to study degradation of oil during frying (Baixauli et al., 2002; Muik et al., 2005). Absorbance values in the UV range reflect the presence of conjugated double bonds that are due to the formation of peroxides and other by-products of lipid oxidation. The measurement of absorbance in the visible range indicates the changes that take place in oil color intensity, which are related to other parameters that affect oil quality (Vijayan et al., 1996).

The oxygen disappearance and volatile compound formation in the headspace of the samples can be measured by gas chromatography. Aldehydes, alcohols, esters, hydrocarbons, ketones, furans, and other compounds have been identified and quantitated by gas chromatography–mass spectrometry in good-quality oil (Mahungu et al., 1998; Martin and Ames, 2001). Macku and Shibamoto (1991) collected volatile compounds formed in the headspace of corn oil heated in the presence of glycine. These researchers identified 18 aldehydes, 15 heterocyclic compounds, 13 hydrocarbons, 11 ketones, 4 alcohols, 3 esters, and 7 miscellaneous compounds. With the advent of direct headspace analyses such as atmospheric pressure ionization–mass spectrometry (API-MS), it is now possible to measure mixtures of volatiles in headspace directly with a millisecond time delay. Using carefully controlled conditions in the ion source (Taylor et al., 2000), quantitative real-time analysis of volatile mixtures can be obtained at concentrations around 10 ppbv (nL of vapor in a liter of air). Volatile oxidation compounds have been analyzed by solid-phase microextraction–gas chromatography with flame ionization detector and mass spectrometer in a conjugated linoleic acid-rich oil (García-Martínez et al., 2009).

To develop new knowledge on undesirable flavors affecting the quality of foods containing polyunsaturated lipids, the volatiles in soybean oil oxidized at different conditions were investigated by three capillary gas chromatographic methods as shown in Figure 2.1: (a) direct injection (5 min heating at 180°C); (b) dynamic headspace (purging 15 min at 180°C onto a porous polymer trap, desorbing from trap for 5 min), and (c) static headspace (20 min heating at 180°C, pressurizing for 1 min) (Snyder et al., 1988). A fused silica column was used with bonded polymethyl and phenyl siloxane phase. At peroxide values between 2 and 10, the major volatile products found in soybean oil by the three methods were pentane, hexanal, 2-heptenal, 2,4- heptadienal, and 2,4-decadienal. The intensities of each volatile compound varied with the analytical methods used.

Carbonyl value (CV) determination, which is a quick and simple test, can be considered to be a powerful measure for indicating sensory deterioration of oil during the frying process. Farhoosh and Moosavi (2008) found that in a set of frying oil samples, on average, the CV linearly increased as the frying time increased. There was a linear relationship between the CV and total polar compounds (TPC) throughout the frying process ($R^2 = 0.9747$).

2.3.3 Potential Flavor Effects

Deep-fat-fried foods have their popularity partly due to the characteristic fried flavor that is unique and desirable. The flavor compounds come from thermally induced changes in the food (Maillard reaction) and the frying oil. Wagner and Grosch (1998) studied the key contributors to french fry aroma. The study found the key aroma compounds in french fries to be 2-ethyl-3,5-dimethylpyrazine, 3-ethyl-2, 5-dimethylpyrazine,2,3-diethyl-5-methylpyrazine,3-isobutyl-2-methoxypyrazine, (*E*,*Z*),(*E*,*E*)-2,4-decadienal, *trans*-4,5-epoxy-(*E*)-2-decenal, 4-hydroxy-2,5-dimethyl-3(2H)-furanones, methylpropanal, 2-and

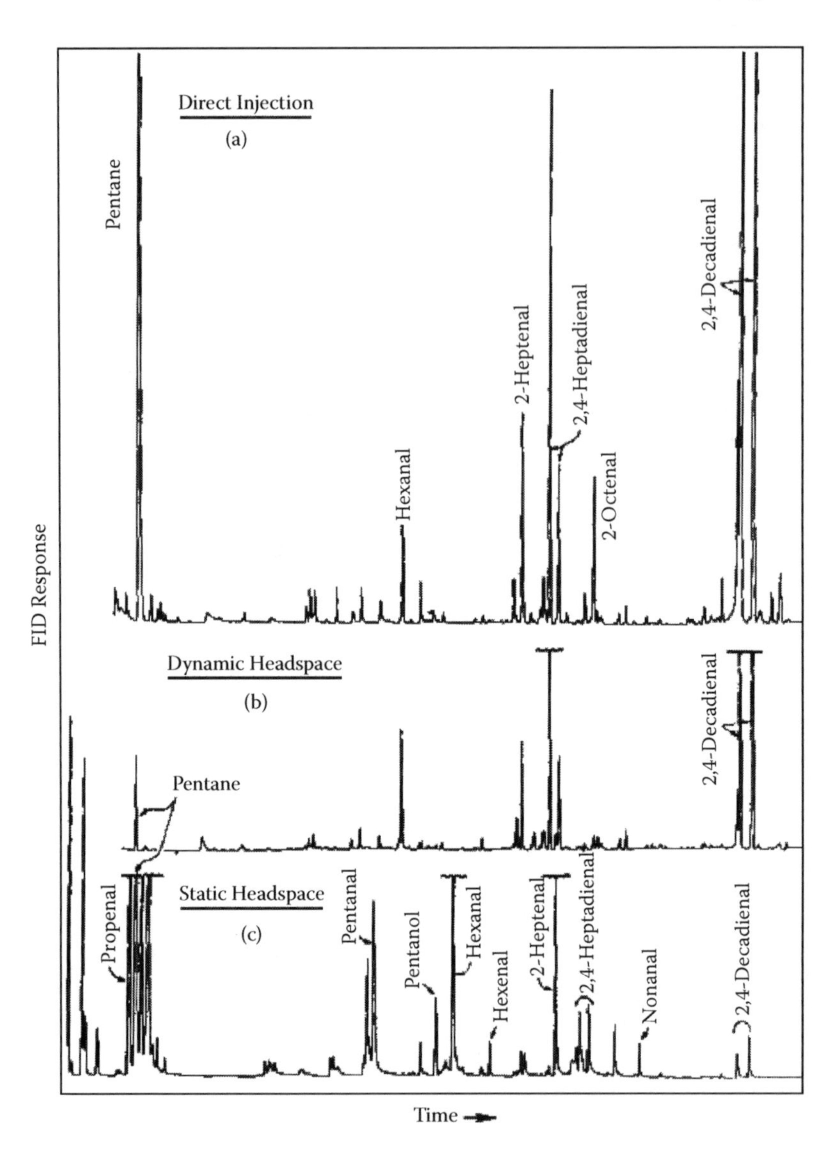

FIGURE 2.1 Comparison of three GC methods for volatile analyses of oxidized soybean oil (PV 9.5) heated to 180°C. (a) Direct injection; (b) dynamic headspace, (c) static headspace.

3-methylbutanal, and methanethiol. Octanal and nonanal derived from oleic acid have a fruity and floral odor, whereas hexanal and 2-hexenal derived from linoleic and linolenic acids have an unpleasant grassy odor (Fujisaki et al., 2002). Neff et al. (2000) working with a triolein and trilinolein model reported the predominant odors, due to heating for 1, 3, and 6 h at 190°C, as fruity and plastic, with other negative odors of acrid and grassy. Some of the volatile compounds that produced negative odors in heated triolein, in order of increasing concentration, were hexanal (grassy), octanal (fruity), (*E*)-2-decenal (plastic), nonanal (fruity), and (*E*)-2-undecenal (plastic). Some of the negative odor compounds in trilinolein, in order of increasing concentration, included (*E*)-2-nonenal (plastic), pentanal (grassy), and hexanal (grassy).

2.4 NONVOLATILE PRODUCTS

The formation of nonvolatile decomposition products is due primarily to thermal oxidation and polymerization of unsaturated fatty acids in fat. The products cause physical changes to fat such as darkening in color, increase in viscosity, and a decrease in smoke point. They also cause chemical changes including an increase in free fatty acid, carbonyl content, cyclic fatty acids, hydroxyl value, and saponification value. The nonvolatile products of major interest in nutrition are the cyclic monomers of fatty acids and the polymeric products, especially the dimers.

2.4.1 Formation Processes/Mechanisms

2.4.1.1 Cyclic Products

Cyclic fatty acid monomers (CFAM) are formed in oils from linoleic and linolenic acids after heat treatment of fats and oils. A typical cyclohexane ring can be formed by cyclization between C_{15} and C_{10}, whereas cyclization between C_{15} and C_{11} will produce a cyclopentane ring. Monounsaturated fatty acids are known to cyclize to monocyclic fatty acids, and when separated by gas chromatography, they elute between methyl palmitate and methyl stearate.

The cyclic monomer formed depends on the type of the fatty acid involved. Linoleic acid in sunflower oil produces cyclic monoenes (Christie et al., 1993; Romero et al., 2003), while linolenic acid in linseed oil produces cyclic dienes (Mossoba et al., 1994; Dobson et al., 1995) on heating. Most CFAM formed from linoleic acid in heated sunflower oil contain a disubstituted cyclopentenyl ring. The spectra of the CFAM picolinyl esters, before and after hydrogenation, indicated the presence of six positional isomers (Figure 2.2a). Four of them had the cyclopentenyl ring at position C_8-C_{12} and

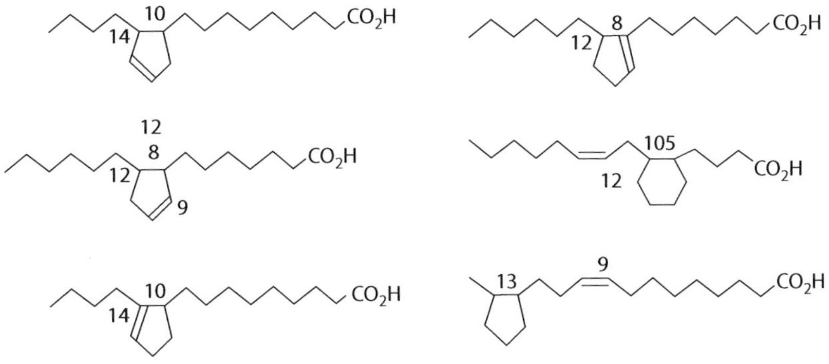

FIGURE 2.2a Main structure of the cyclic fatty acid formed from linoleic acid in heated sunflower oil. (From Le Quere, J.L. and Sébédio, J.L., *Deep Frying: Chemistry, Nutrition and Practical Applications*, ed. E.G. Perkins, and M.D Erickson., pp. 70, 71, 1996. (With permission from the American Oil Chemists' Society, Champaign, IL.)

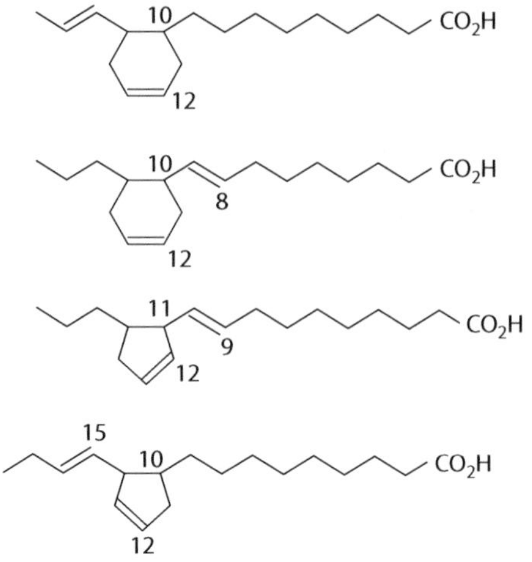

FIGURE 2.2b Main structures of cyclic fatty acids formed from linolenic acid in heated linseed oil.

a double bond at position C_8-C_9; ring at C_{10}-C_{14} and a double bond at position C_{13}-C_{14}; ring at C_8-C_{12} and a double bond at position C_9-C_{10}; and ring at C_{10}-C_{14} with a double bond at position C_{12}-C_{13}. For the other two isomers, one had a cyclopentyl ring at position C_{13}-C_{17} and a double bond at position C_9-C_{10}, while the other had a cyclohexyl ring at position C_5-C_{10} and a double bond at position C_{12}-C_{13} (Christie et al., 1993). The double bonds at C_9 and C_{12} were involved in the cyclization reaction. However, for cyclic dienes isolated from heated linseed oil, cyclization was always directed internally toward other double bonds and never involved the double bond at position C_{12}. Four major structures were identified (Figure 2.2b). Two had a cyclopentenyl ring at C_{11}-C_{15} and C_{10}-C_{12} with double bonds at C_9-C_{10}, $C_{!2}$-C_{13}, and C_{12}-C_{13}, C_{15}-C_{16} , respectively. The other two contained the cyclohexenyl rings at position C_{10}-C_{15}, while the double bonds were at C_8-C_9, C_{12}-C_{13}, and C_{12}-C_{13}, C_{16}- C_{17}, respectively (Dobson et al., 1995). Thus, migration of the ring double bond to the C_{13} position in C_{10}-C_{14} cyclopentenyl fatty acids was observed for cyclic monoenes, but not for cyclic dienes.

2.4.1.2 Polymeric Products

Formation of polymers during frying is associated with the autoxidation process that proceeds via a free radical mechanism. The polymers and dimers represent degradation products that are unique in fried foods, and they are an excellent chemical marker of oil degradation but not food quality. The polymeric products are generally divided into polar dimers, nonpolar dimers, and higher oligomers and polymeric triacylglycerol having molecular weights higher than dimers. The dimeric and higher oligomers are derived from fatty acids and from triglycerides.

The nonpolar dimers from fatty acids are usually formed by radical reactions from allyl radicals. The two fatty acids are linked by the formation of a new bond adjacent to the double bonds forming dehydrodimers. Alternatively, the allyl free radical can add to the double bond in another fatty acid molecule, forming noncyclic dimers. If the free radicals within the newly formed dimeric species combine, a cyclic dimer is formed. Under the frying conditions, the conjugated double bond in one molecule can act as a diene to react in a typical $4n + 2$ reaction with an ethylenic bond in another molecule forming the Diels–Alder dimers. Below 200°C, there is a preferential formation of dehydrodimers. Dehydrodimers from methyl oleate and bicyclic and tricyclic dimers from methyl linoleate, as well as Diels–Alder dimers from methyl linoleate in hydrogenated soybean oil used for frying have been reported (Christopoulou and Perkins, 1989b). The presence of Diels–Alder dimers in the low-polarity fraction isolated from heated soybean oil fatty acid methyl esters had previously been reported by Ottaviani et al. (1979).

Polar dimers are those compounds formed by similar mechanisms to those of nonpolar dimers, but contain oxygenated chains. The structure of these oxy dimers is still largely unknown, but a noncyclic dimer of methyl linoleate containing two hydroxy groups has been isolated as well as some tetra-substituted tetrahydrofuran dimers (Hamilton and Perkins, 1997). Difficulties are primarily due to the heterogeneity of this group of compounds. The oxygenated functions are likely to be present in oxidized monomers before dimer formation or generated by oxidation of nonpolar dimers (Artman and Smith, 1972). The dimer molecule may have more than one functional group, and the oxygen may not be involved in the dimeric linkage. Thus, the large number of possible combinations of polar dimers generated results in a complex mixture that is difficult to separate. The presence of three different polar dimers corresponding to dihydroxy, trihydroxy, and tetrahydroxy dimers with C-C linkages was reported by Perkins and Kummerow (1959) when they subjected corn oil to thermal oxidation (200°C, 48 h) with forced aeration. After concentration of the polymeric compounds by a combination of urea fractionation and molecular distillation, fractions having a molecular mass of 692 to 1600 were isolated.

Higher oligomers have been noted, but their definitive structures have not been elucidated either in methyl esters from frying fats or in model systems. This is due to limitation of techniques for isolation, separation, and identification. Polymeric triglycerides can be formed by similar mechanisms applied to fatty acids.

2.4.2 Characterization Techniques

This section reviews some of the analytical techniques used in the synthesis, isolation, and characterization of the thermally oxidized fat products that have metabolic impact when ingested, primarily the dimers and cyclic fatty acid monomers. The analytical techniques highlighted are meant to give the reader an idea of what is involved in the analysis of oxidation products in heated fats. It is the analytical data obtained that assist in determining the effects of these oxidation products on metabolic processes after ingestion. Polar compound determination constitutes the main basis of the used frying fat regulations with adsorption chromatography.

Direct application of high-performance size-exclusion chromatography (HPSEC) is of special relevance for quality evaluation of used frying fats, where polymers are the most representative group of compounds due to thermoxidative alteration (Márquez-Ruiz and Dobarganes, 1997). The IUPAC Commission on Oils, Fats and Derivatives decided to adopt the method although with limited application to samples containing 3% or more of polymerized triglycerides (Wolff et al., 1991). Refractive index detection permitted direct quantification with no need for extensive calibration or correction factors. This way oxidized monomer, dimer, and polymer fractions were each separated as well as hydrolytic products (see also Chapter 6).

Polar compound determination based on adsorption chromatography was proposed for quality evaluation of used frying fats given that during frying a complex mixture of degradation products of higher polarity than that of the original triglycerides is formed. The methodology based on a combination of silica columns–HPSEC has been revealed as an excellent alternative for used frying oil evaluation (Dobarganes et al., 1988; Arroyo et al., 1992). The standard method uses a silica column with nonpolar and polar fractions being eluted with a mixture of hexane/diethyl ether (87:13) and diethyl ether, respectively. HPSEC is then applied for further analysis of the fractions. The advantages offered by this method are enormous. First, increased possibilities for quantification of polymers are achieved due to the effect of concentration; the limitation of 3% content established for analyses in total samples is overcome. Also, the independent determination of oxidized triglyceride monomers from nonoxidized triglycerides eluting in the nonpolar fraction offers a measurement of oxidative deterioration; finally, the simultaneous quantitation of diglycerides constitutes an indication of hydrolytic alteration, due to lack of any overlap with the most abundant triglyceride peak in the whole sample.

While detection of hydrolytic alteration can only be attained through HPSEC analysis of polar fractions of original oil samples, differences in thermoxidative degradation are better reflected in the values of altered fatty acid methylesters since exclusively the amount of fatty acyl groups affected are measured. Some insight into the complexity of the triglyceride polymer structure can also be revealed by comparing triglyceride and methyl ester dimers and oligomers values.

2.4.2.1 Synthesis and Isolation of Cyclic Products

2.4.2.1.1 Synthesis The cyclic fatty acid monomers (CFAM) have been synthesized as their methyl esters (Vatela et al., 1988; Berdeaux et al., 2009). The synthesis involves a Michael addition of alkylmagnesium bromide to an unsaturated cyclic ester. The resulting *cis* and *trans* ethyl 2-alkylcyclopentane carboxylate are converted to alcohols, then to aldehydes. The aldehydes are condensed with the yield derived from the (co-carboxyalkyl) triphenylphosphonium bromide and methylsulfinylmethide in DMSO (dimethylsulfoxide) to give unsaturated C18 cyclic fatty acid monomers. These fatty acid monomers are further converted to the methyl esters, and the ethylenic bond is reduced by catalytic hydrogenation to give the saturated esters.

The synthesis of methyl 9-(2′-*n*-butylcyclopentyl) nonanoate and methyl 10-(2′-*z*-propylcyclopentyl) decanoate using the Wittig reaction between 2-alkylcyclopentanones and the corresponding co-carbomethoxyalkyl triphenylphosphonium bromides has been reported (Rojo and Perkins, 1989). For example, methyl 9-(2′-*n*-butylcyclopentyl) nonanoate is prepared by a Wittig reaction between the corresponding phosphonium salt and 2-*n*-butylcyclopentanone followed by catalytic hydrogenation of the purified reaction product.

2.4.2.1.2 Isolation The usual procedure for isolation of the cyclic fatty acids involves conversion to methyl esters, then column chromatography on silicic acid. The nonpolar fraction from this separation was fractionated with urea, giving cyclic fatty acid methyl esters of up to 99% purity (Hamilton and Perkins, 1997). Since the number of positional isomers can be overwhelming, it is often necessary to hydrogenate the unsaturated fatty acids. This removes the unsaturation and provides a simpler mixture of compounds for subsequent gas chromatography–mass spectrometry.

Gel permeation chromatography (GPC) and preparative high-performance liquid chromatography (HPLC) (Rojo and Perkins, 1987; Sébédio et al., 1987; Romero et al., 2006) have been used to separate and collect fatty acid monomers for further analysis. The separation of monomeric fractions via GPC is conducted in a column packed with Bio-Beads S-X2 swollen in toluene samples of pure oil (heated and unheated), and methyl esters are then separated in the column. The cyclic fatty acid monomers are determined by preparation of methyl esters, hydrogenation of the methyl esters, and finally, urea fractionation-concentration followed by capillary gas chromatography of the methyl esters.

The preparative HPLC separation can be done on a reverse-phase column for samples of heated or unheated vegetable oils. The oils are saponified with potassium hydroxide, and the unsaponifiables removed by AOCS method Ca-6a-40 (AOCS, 1991). The recovered fatty acids are converted to methyl esters by refluxing with sulfuric acid in methanol and the esters extracted with petroleum ether. The methyl esters are fractionated by column chromatography using silicic acid. The adducts are separated by filtration from the nonurea adduct fraction that contains the cyclic fatty acids. The cyclic fatty acid monomer methyl esters can be hydrogenated using platinum oxide and hydrogen, then the saturated methyl esters are extracted with chloroform after addition of water. The methyl esters are analyzed using gas-liquid chromatography (GLC).

2.4.2.2 Synthesis and Isolation of Polymeric Products

2.4.2.2.1 Synthesis The dimers that are structurally representative of those formed during thermal-oxidative reactions in fats and oils, including the dihydroxy, tetrahydroxy, and the diketo dimers of methyl stearate have been synthesized (Christopoulou and Perkins, 1989a). The tetrahydroxy stearate dimer was prepared by peroxidation of the dehydrodimer of methyl oleate prepared from the reaction of di-*tert*-butyl peroxide and methyl oleate (Perkins and Taubold, 1978). The dimer was purified by column chromatography using chloroform as an eluting solvent. The labeled dimer can be prepared the same way, but labeled methyl oleate is used as a starting material. To prepare the tetrahydroxy stearate, the purified methyl oleate dimer is mixed with formic acid and treated with hydrogen peroxide. Water is added to remove formic acid and, then, the oily phase is extracted with a mixture of ethyl ether and hexane. After solvent evaporation, the residue, which consists of the dimeric hydroformoxy-stearic acids, is heated with aqueous sodium hydroxide, then neutralized with excess hydrochloric acid. The dimer is extracted with a mixture of ethyl ether and hexane.

The dihydroxystearate dimer is prepared by treating a mixture of the methyl oleate dehydrodimer in formic acid with aqueous perchloric acid and refluxing under nitrogen. The excess formic acid is removed by treatment with water. The oily phase is extracted with ethyl ether-hexane, and the solvent is removed using an evaporator. The residue is heated with aqueous sodium hydroxide, and the reaction mixture is neutralized with hydrochloric acid. The product is extracted with ethyl ether-hexane and the solvent is removed by evaporation under vacuum. The diketostearate is prepared by oxidation of the dihydroxystearate with chromic oxide in acetic acid. After heating, the reaction mixture is added to water, extracted with ethyl ether-hexane, dried over sodium sulfate, and the solvent is removed.

The methyl esters of these dimers are prepared according to the AOCS method Ce 1-62 (AOCS, 1991). The trimethylsiloxy (TMS) derivatives are prepared with pyridine and bis(trimethylsily) trifluoroacetamide in glass vial with a Teflon-lined septum. The capped vial is shaken, heated, and the solution of TMS derivatives is used after 4 h.

2.4.2.2.2 Isolation The isolation and structure elucidation of dimers formed as a result of thermal oxidation of fats during deep frying has been of great interest. However, dimers and other polymeric materials present difficulties during isolation and characterization. Dimers were isolated from partially hydrogenated soybean oil that has been used for frying (Christopoulou and Perkins, 1989b). The dimers were separated into their component structures using HPLC and identified by gas chromatography–mass spectrometry. The presence of monohydroxy, dihydroxy, and keto groups in linoleic acid dimers was reported.

Preparative HPSEC has been used to separate polymers (dimers and trimers) in heated olestra/triglyceride blend (Gardner and Sanders, 1990; Romero et al., 2000). Three fractions were collected and the second fraction contained the polymers, dimers, and monomers. This fraction was reinjected into the HPSEC to separate the dimers from the trimers and monomers. The dimeric fraction was methylated and analyzed by gas chromatography.

2.5 METABOLIC PROCESSES

Most of the work reported on the effects of ingesting foods containing lipid oxidation products has been on animals, particularly rats. In animal toxicological tests, the food additives are fed at dietary levels many times in excess of the levels that will appear in foods for human consumption. This is done to provoke potential toxic responses and to establish safety factors. Thus, the responses that may be considered initially to be of toxicological significance may on further investigation be the result of dietary nutrient imbalance or physiological perturbation induced by the test material when fed at excessive exposure levels (Gershoff, 1995). These diets might become unpalatable with a resultant poor growth that might be interpreted as a sign of toxicity.

Assessment of metabolic and toxicological consequences of consuming frying fats offers significant challenges. The presence of a multitude of reaction products makes the analytical evaluation of the fats tested difficult. There are also problems associated with animal trials, for example, formulation of balanced diets, selection of degree of deterioration, determination of appropriate dietary levels of fat, duration of feeding, and extrapolation of the data obtained to humans. Some studies include whole oils, fractions from the oils, or pure altered compounds obtained by synthesis (Kaunitz et al., 1955, 1956; Johnson et al., 1956; Perkins and Kummerow, 1959; Nolan et al., 1967; Shue et al., 1968; Poling et al., 1970; Billek et al., 1978; Miller and Long, 1990; Ruiz-Gutierrez and Muriana, 1992; Márquez-Ruiz and Dobarganes, 2006). Some researchers have used extremely abusive conditions in an attempt to generate sufficient degradative products for their metabolic studies. However, compounds formed in this manner may not be representative of those encountered in frying fats during normal culinary practices.

This section will consider the effects of metabolic processes on ingestion of cyclic fatty acids, polymeric materials (especially dimers), and finally the effects of fats used for frying foodstuffs under restaurant conditions. It is believed that a restaurant frying fat will give a more realistic assessment of what may be the actual effect of metabolic processes on ingestion of oxidative products in frying fats.

During deep frying, several chemical reactions occur within the oil, including the thermal oxidation of unsaturated fatty acids (FA). A heterogeneous mixture of chemically distinct substances is formed during thermal FA oxidation, including hydroperoxy and hydroxy FA such as 13-hydroperoxyoctadecadienoic acid (13-HPODE), 9-hydroxyoctadecadienoic acid (9-HODE), and 13-HODE, as well as cyclic FA monomers (CFAM) (Ringseis and Eder, 2008). During deep-fat frying, a substantial amount of frying oil containing these substances is absorbed by the fried food and thereafter ingested during consumption of the food.

2.5.1 Cyclic Compounds Ingestion

Nutritional studies have shown that adverse physiological conditions can develop when varying amounts of cyclic fatty acid monomers are consumed by experimental animals. The work of Iwaoka and Perkins (1978) using pure ^{14}C-labeled cyclic monomers of methyl linolenate was of great importance in pinpointing the metabolic effects of ingesting cyclic fatty acid monomers. These workers studied the metabolism of radioactive cyclic fatty acids to determine the rate of CO_2 expiration, the rate of appearance of radioactive urinary metabolites, and the distribution of the radioactivity in the lipids of selected tissues. Animals fed radioactive cyclic fatty acid monomers (0.0075%, 0.0225%, and 0.1500% cyclic monomers in the diet) excreted 40% of the total count in the urine and eliminated 13%–15% as $^{14}CO_2$. No significant trend was seen that correlated the previous dietary conditioning of the animal with different levels of cyclic fatty acids and the amount of total radioactivity recovered in the urine or expired as $^{14}CO_2$. The peak expiration took place between 4 and 6 h for both the control and those consuming cyclic fatty acid esters in the diet.

Though the absolute amounts of radioactivity were different for control and test animals, the absorption, transport, and rate of fatty acid catabolism in various tissues were similar since peak expiration values occurred at the same time for both groups. For all sets of animals, over 60% of the total radioactivity in the urine was excreted from 0 to 12 h. The increase of radioactivity of urine in rats suggests that a detoxification mechanism exists and is quite consistent with the preferential incorporation of 5-carbon-membered cyclic monomers in tissue lipids observed in many experiments.

The very low toxicity of 6-carbon-membered cyclic monomers is explained by their rapid excretion (Iwaoka and Perkins, 1978). The radioactivity recovered from the total lipids of selected organs 48 h after administration of labeled cyclic monomers indicated that accumulation of radioactivity in the liver and other selected tissues did not occur. This work showed that very low levels of cyclic fatty acids caused decreased lipogenesis in livers of rats fed 8% and 10% protein and elevated lipogenesis in the adipose tissue of animals fed 10% protein. It has also demonstrated that over 50% of the radioactivity from ingested cyclic acids is either expired as $^{14}CO_2$ or excreted in the urine, and cyclic fatty acids adversely affected animals fed low levels of protein. Thus, dietary protein levels can play a key role in the toxic effects of oxidized frying fats. Previously, the cyclic monomers of methyl linolenate had been found to retard growth. When low levels (0.0075%, 0.0225%, and 0.15%) of cyclic fatty acid methyl esters (>98% pure) were incorporated into diets of weaning rats fed different levels of proteins, decreased weights gains were observed with increasing levels of cyclic esters in the diets (Iwaoka and Perkins, 1976). Liver enlargements due to accumulation of liver lipid were also noted in animals receiving 0.15% cyclic fatty acid esters in their diets.

Purified cyclic fatty acid monomers (CFAM) have been isolated from heated linseed oil and heated corn oil. Generally, CFAM formed from linolenic acid in linseed oil is basically a mixture (1:1) of disubstituted 5- and 6-carbon-membered ring isomers, while those arising from linoleic acid in sunflower oil are mainly 5-carbon-membered ring isomers (Rojo and Perkins, 1987; Sébédio et al., 1989; Christie et al., 1993; Dobson et al., 1995; Mossoba et al., 1994). These molecules have been tested for their physiological effects on rat and heart cells. Cyclic fatty acid monomers are well absorbed and incorporated in rat tissues, both in the neutral lipids and phospholipids, but incorporation is greater in the neutral lipids.

A similar trend is observed for the heart. The 5-carbon-membered ring CFAM is selectively incorporated relative to the 6-carbon-membered ring CFAM (Athias et al., 1992; Ribot et al., 1992; Sébédio and Chardigny, 1996). However, ring size is not the only factor influencing incorporation of CFAM. Such factors as *cis* versus *trans* configuration contribute to the observed differences. Ribot et al. (1992) reported the preferential incorporation of the 5-carbon-membered ring CFAM isomers isolated from heated linseed oil, especially those with a *trans* configuration, in rat heart cell cultures. The 5-carbon-membered ring compounds from heated sunflower oil accumulated more in cellular lipids than the 6-carbon-membered ring compounds from heated linseed oil.

Incorporation of CFAM isolated from sunflower oil in cellular phospholipids affected the spontaneous beating rate of cardiomyocytes, whereas this effect was not observed with CFAM isolated from linseed oil. Activation of the β- and α-adrenergic systems was obtained by addition of isoproterenol and phenylepherine, respectively. Neither the biochemical response (cAMP production) nor the physiological response (chronotropic effect) to the β-adrenergic agonist were altered by either CFAM from sunflower or CFAM from linseed oil, suggesting the β-adrenergic system is not affected by the incorporation of CFAM in cellular lipids. Both biochemical (inositide phosphate production) and physiological (chronotropic effect) responses to the α-adrenergic agonist were significantly reduced in cardiomyocytes containing CFAM from sunflower, but not with CFAM from linseed. This differential sensitivity between the α- and β-adrenergic systems toward CFAM suggests that specific incorporation of CFAM from sunflower oil into phosphatidylinositol may be responsible for the observed effect rather than a more general "membrane effect." The much greater proportion of 5-carbon-membered ring compounds in the fraction arising from *n-6* fatty acids may explain differences in the biological effect of these two types of CFAM (Athias et al., 1992; Ribot et al., 1992).

The effect of cyclic fatty acids on liver enzyme activity has been reported (Lamboni et al., 1998). Cyclic fatty acid monomers fraction was isolated from a batch of partially hydrogenated soybean oil (PHSBO) commercially used for 7 days (group CFA) for frying fish, french fries, and chicken. Results were compared to two other groups of rats, one fed PHSBO after frying (group 7 DH) and the other fed PHSBO before frying (NH). While an increased level of liver protein was observed in both experimental groups (7 DH and CFA) compared to the control (PHSBO), several enzyme activities were altered. For example, the activity of the NADPH-cytochrome P_{450} reductase was significantly increased, suggesting active detoxification by the liver of toxic compounds at high CFAM concentrations. The difference between the two experimental groups, however, also suggested that other components from used frying fat may have an effect in conjunction with CFAM.

The observed significant decrease in content of liver glycogen in experimental groups indicates an increased glycolytic activity leading to the formation of pyruvate, which was subsequently converted to acetyl-CoA via the pyruvate dehydrogenase enzyme complex in the mitochondrial matrix. The acetyl-CoA formed could undergo oxidative degradation through tricarboxylic acid (TCA) cycle or may be utilized for the biosynthesis of long-chain fatty acid such as palmitic acid via fatty acid synthase. The decreased iso-citrate dehydrogenase (ICDH) activity implies an impairment of TCA cycle enzyme activity in the presence of CFAM. Thus, the CFAM generated in fat during the deep-fat frying process is, in part, responsible for the impairment of ICDH in the mitochondrial TCA cycle. Since the first dehydrogenase reaction in the TCA cycle was impaired, it suggested an accumulation of citrate. The excess citrate could have gone through the mitochondrial membrane under passive diffusion and regenerated acetyl-CoA in the cytosol under ATP-citrate lyase activity. The acetyl-CoA in the cytosol could have been converted into fatty acid. This may explain the fatty livers observed in the experimental group. Significantly, depressed activity of glucose 6-phosphate dehydrogenase (G 6-PDH) was also observed for animals of the experimental groups.

2.5.2 Polymeric Compounds Ingestion

The use of polymeric fractions from heated fats in nutritional studies is popular due to their relative ease of isolation and because these compounds represent fat in a state of modification rather than decomposition. The limited absorption of heated fat polymers appears to be largely responsible for the lack of dramatic symptoms in animals after ingestion of such materials. The absorption through collection of lymph from abdominal ducts of rats was reported by Perkins et al. (1970). Corn oil was transesterified with ^{14}C-linoleic acid and subjected to thermal oxidation. The nonurea-adductable fraction of methyl esters was administered to rats via gastric intubation, and lymph was collected after 48 h. Absorption was retarded in comparison to the labeled corn oil used as the control. The peak absorption maxima of nonurea adduct occurred at 24 h, and the global absorption was around 40%. Lymph lipids were analyzed by thin-layer chromatography in which 103 compounds were isolated. Lymph absorption of nonvolatile oxidation products formed in thermoxidized fats and synthetic triglycerides containing a definite aromatic or alicyclic fatty acid showed that 4% of total polymeric acids, 53% of total oxidized monomeric acids, and 96% of total cyclic monomeric acids were recovered in the lymphatic lipids (Combe et al., 1981). Digestibility coefficients have been used as a measure of absorbability and have indicated lower values for thermoxidized oils compared to fresh oils, a fact generally attributed to the presence of the nondistillable nonurea-adductable fraction, which is the polymeric fraction. A significant reduction in the *in vitro* hydrolysis rate of thermoxidized oils due to the difficulties involved in the pancreatic lipase action of complex glyceridic molecules has been reported. Among polymeric fatty acids, the lowest digestabilities have been found for nonpolar dimers.

Perkins and Taubold (1978) reported the nutritional and metabolic effects of noncyclic dimeric fatty acids. The ^{14}C-labeled and unlabeled dehydrodimers of methyl oleate were prepared by treating methyl oleate with di-*tert*-butyl peroxide in a molar ratio of 8.2:1 and heating at 130°C under nitrogen. The product was first distilled at 35°C under reduced pressure to remove the *tert*-butyl alcohol formed and then at 170°C and 0.25 mmHg to remove the unreacted monomer. The ^{14}C-labeled noncyclic dimeric fatty acids were fed in corn oil via gastric intubation to individual male weanling SPF albino rats. These rats had been conditioned by receiving the special diets for 8 weeks. They were immediately placed in all-glass metabolic cages for 48 h after injection of the labeled sample. The expired CO_2, urine, and feces were recovered, and radioactivity was determined by liquid scintillation. The rats were sacrificed after a 48 h period and their livers, stomach, small and large intestines, epididymal fat, and perirenal fat were quickly excised, blotted, weighed, and stored at 0°C. After addition of the appropriate scintillation solvents, the radioactivity was quantitated. The feces, stomach contents, and intestinal contents were extracted with chloroform-methanol, an aliquot of extract evaporated to near dryness, Aquasol was added and radioactivity was quantitated.

The control group was fed 0.0% and experimental groups were fed 0.1%, 1.0%, or 5.0% noncyclic dehydrodimers and the effect on weight gain was determined. There was no significant difference in weight gain for all the groups. Furthermore, no differences in food consumption were observed, indicating that food palatability was not a problem. Higher levels of dimers in diets, such as 12%–20% that are sometimes used, are unrealistically high. Malabsorption and diarrhea may complicate the picture of the actual effects of dimeric fatty acids. There were no significant differences in liver weight/body weight. Thus, while some heated fat isolates do lead to enlarged livers when incorporated into rat diets, the dimeric fatty acids do not lead to enlarged or fatty livers. It is important that absorption of dimeric fatty acids occur if biological effects are to be observed. However, the metabolic data showed that 85% of the recovered radioactivity was found in the gastrointestinal tract and feces, indicating that absorption is minimal. The radioactivity recovered as $^{14}CO_2$ was less than 6% of the total when labeled noncyclic dimers were force fed. The $^{14}CO_2$ recovered, when uniformly labeled methyl oleate was force fed, was 60% to 80% of the total recovered radioactivity. This indicated approximately 10% absorption of the noncyclic dimeric fatty acids that were then metabolized. Thus, noncyclic dimeric fatty acids in the diet do not exhibit any profound effects either on the growth or metabolism of the rat when moderate levels of the compound are present in the diet.

2.5.3 Heated Fats and Oils Ingestion

The greatest concern on nutritional effects of used frying fats has been expressed over intermittent deep-fat frying because these fats are kept at high temperatures for long periods of time, allowed to cool, and then reused. Conversely, there is probably less opportunity for significant alteration both in commercial frying operations, using continuous frying because of rapid oil turnover, and pan-frying, where fats are heated for short periods of time and are rarely reused. The presence of foodstuffs establishes differences between thermally oxidized oils simulating frying and real frying oils. The moisture provided by the food may result in hydrolysis of the fat, and so this reaction is added to those occurring due to the action of high temperature and air. Hydrolysis involves breakage of the triacylglycerol producing hydrolytic products (free fatty acids, monoacylglycerols, and diacylglycerols), which are not of great relevance since they are also generated after pancreatic lipase action prior to intestinal absorption. The most important issue is the interaction of certain nutrients, especially proteins, with oxidized compounds from the frying fat and, hence, the formation of new compounds of potential toxicological or nutritional concern. With a few exceptions, reports on used frying fats are less alarming than those dealing with thermally abused fats. Generally, no evidence of toxicity in terms of biological and histological changes and only slight effects on growth rate and liver enlargement and only mild effects, if any, have been found from studies on actual commercial frying operations (Keane et al., 1959; Poling et al., 1960; Nolan, 1972; Márquez-Ruiz and Dobarganes, 2006; Totani et al., 2008)

Oxidative stress in the liver has been reported by other researchers (Andia and Street, 1975; Ashwin et al., 1991; Quiles et al., 2002; Garibağaoğlu et al., 2007). It was shown by the larger induction of hepatic microsomal activities important to metabolize lipid oxidation products, including cytochrome P_{450}, UDP-glucuronyi transferase, glutathione-*S*-transferase, and superoxide dismutase, observed from the chronic feeding of thermally oxidized oils to rats. Low activities of thiokinase and succinate dehydrogenase have been reported when rats were fed autoxidized safflower oil. This observed metabolic effect was associated with the elevated carbonyl values of the oxidized fat (Yashioka et al., 1974). Feeding experiments using rats, mice, and pigs revealed that ingestion of heated, oxidized fats, compared with fresh fats, provokes a wide array of biological effects in mammals, including oxidative stress, impaired glucose tolerance, and thyroid function and lipid metabolism alterations.

Naruszewick et al. (2009) investigated the possible connection between chronic ingestion of acrylamide-containing potato chips and oxidative stress or inflammation. A significant increase in reactive oxygen radical production by monocytes, lymphocytes, and granulocytes and an increase in CD14 expression in macrophages ($P < 0.001$) were found after the intake of potato chips. Extensive epidemiological studies have found that the risk of hypertension was positively and independently associated with the intake of cooking oil polar compounds (Soriguer et al., 2003). Sanchez-Muniz and Bastida (2003) reported that the triacylglycerol oligomer content in used frying oil gave more precise information about the alteration of the oil and its potential toxicity than polar compounds.

Excretion of unsaponifiables was significantly greater in rats on diets containing thermoxidized oils versus nonheated oils, suggesting that oil alteration may influence endogenous lipid metabolism (Márquez-Ruiz et al., 1992; Edgar, 1999; Ringseis and Eder, 2008). Major sterols and nonpolar fatty acids originating from intestinal bacterial action were quantitated in fecal lipids and were representative of major sources of endogenous lipid products. Fecal endogenous sterol concentrations, particularly for cholesterol, were significantly greater when diets contained thermoxidized oils, and the excretion increased as the percentage of dietary oil alteration increased (Márquez-Ruiz et al., 1992). Increased fecal sterol following ingestion of thermoxidized oils is due to impairment in triacylglycerol hydrolysis and subsequent effect on cholesterol micellar solubilization. Chemical analyses, coefficients of digestibility, and metabolizable energy studies indicate that corn oil, peanut oil, and hydrogenated soybean oil from deep-fat fryers used for frying chicken result in less deterioration than laboratory-heated oils (Alexander et al., 1983). When soybean oil thermally oxidized at 200°C was compared with used frying soybean oil administered to rats at 15% of the diet for 28 days, only the laboratory-heated oil depressed weight gain and feed consumption (Chanin et al., 1988).

Billek et al. (1978) reported that fats heated under usual household or commercial practice contained approximately 10%–20% polar material. Sunflower used for the industrial frying of fish and about to be discarded was fractionated with column chromatography. Fresh oil, frying oil, and both nonpolar and polar fractions of used frying oil were fed to rats for 18 months at 20% in the diet. Only diets containing 20% polar compounds caused a small reduction in growth and increased liver and kidney weights. Changes in clinical or histological parameters were not found, except for increased activities of serum glutamic pyruvic transaminase and glutamic oxaloacetate transaminase. Frying fats were concluded safe, since the animals recovered rapidly with the normal diet. The amount of ingested compounds was excessively higher than normally consumed in foods.

When rats were fed diets containing sufficient vitamin E and 10% unsaturated fats that had been used for frying potatoes, there was no change in nutritional parameters such as growth, plasma lipids, and fatty acid profiles of plasma, liver, and heart lipids (Giani et al., 1985). Similarly, Izaki et al. (1984) showed that liver and serum tocopherol levels decreased and hepatic thiobarbituric acid-reactive substances (TBARS) increased, according to the level of deterioration of the heated frying fats. The liver TBARS were well correlated with petroleum ether-insoluble oxidized fatty acids, polar fraction, separated by column chromatography, glyceride dimer fraction, and carbonyl value. It is well known that normal defense mechanisms against *in vivo* lipid peroxidation include antioxidants (mainly vitamin E) and detoxification enzymes, such as superoxide dismutase, catalase, and glutathione peroxidase.

Lamboni and Perkins (1996) evaluated the effects of dietary heated fats from a commercial deep-fat frying operation on rat liver enzyme activity. The fats were partially hydrogenated soybean oil (PHSBO) used for 4 days (4-DH) and 7 days (7-DH) for frying foodstuffs in a commercial restaurant. These fats were fed to rats in either free access to food or by pair-feeding graded doses. The experiments were conducted with control rats fed nonheated PHSBO (NH) diets. The oil was obtained from a commercial fast-food restaurant deep-fat frying operation producing french fries, fried fish, and fried chicken. The rat enzymes evaluated were carnitine palmitoyltransferase-I (CPT-I), isocitrate dehydrogenase (ICDH), glucose 6-phosphate dehydrogenase (G 6-PDH) and NADPH-cytochrome P_{450} reductase, and cytochrome $P_{450.}$ Feeding rats with commercially heated fats led to growth depression, suggesting that the heated oils do contain substances that prevent animals fed such fats from growing properly. The weight gain declined after 35 days of feeding the experimental diet, indicating that there is a threshold value of accumulated deleterious compounds generated during the commercial deep-fat frying process that must be ingested to exert adverse nutritional effects. There was significant increase in the level of hepatic lipids when rats were fed the 7-DH diet, in both the pair-feeding and free access to food groups due to an accumulation of liver lipid in the experimental group, as compared to the control group. The groups fed the 7-DH diet had an increased content of cytochrome P_{450} due to the presence of xenobiotic materials. There was an impairment of the tricarboxylic acid (TCA) cycle enzyme activity for the 7-DH fed group as indicated by a reduction in the activity of the isocitrate dehydrogenase. Overall, the experimental group fed the 7-DH diet had a larger amount of cytochromes P_{450}, greater activity of NADPH cytochrome P_{450} reductase, lower activities of CPT-I and ICDH, a depressed activity of G 6-PDH, and increased microsomal protein.

Generally, it has been accepted that growth depressing and other nutritionally harmful materials are present in the nonvolatile fraction of heated fats. However, in the literature, only a few reports appear to focus on a direct way to alleviate the deleterious effect of heated edible oils when such fats are fed to laboratory animals. Work on heated fats by Perkins and Lamboni (1998) compared the effects of the heated filtered oil versus heated nonfiltered oil on liver enzyme activity. The partially hydrogenated soybean oil (PHSBO) used for 7 days (7-DH) for frying foodstuffs was obtained from a commercial deep-fat frying operation and treated with magnesium silicate (T-7DH). Isocaloric diets containing 15% of either T-7DH or 7-DH fats were prepared and fed to male weanling rats for 10 weeks in a pair-feeding experiment and compared to control rats fed nonheated PHSBO (NH). The animals fed the T-7DH diet had a greater liver enzyme activity as compared to the NH group, suggesting a positive effect of the treatment on recovering enzyme activity for several metabolic pathways. The carnitine palmitoyltransferase-I, glucose 6-phosphate dehydrogenase, isocitrate dehydrogenase, and NADPH-cytochrome P_{450} reductase systems of the treated groups showed less activity than the untreated group, indicating reduced amounts of harmful compounds in the diet containing the heated filtered oil.

2.6 SUMMARY

The frying conditions, the complexity of the reactions involved, and the great variety of compounds formed make the evaluation of all the changes that take place in fats and oils during frying an enormous task. The oxidation products from heated fats used for frying foodstuffs are probably not detrimental to the metabolic processes in the human body at the concentrations normally encountered in food. The amounts of pure oxidation products used in animals tests, such as fatty acid cyclic monomers (CFAM), are several-fold greater than would be normally encountered and ingested by humans. In addition, the toxic effects of the pure CFAM are substantially reduced by protein concentrations greater than 10%. The polymeric materials are poorly absorbed and hence do not affect metabolic processes significantly. Thus, there is likely no serious detrimental effect on metabolic processes due to ingestion of heated fats. Much work needs to be carried out to evaluate the nutritional implication of oxidized fat products at the levels normally found in used frying fats and oils and the fried food products. It is important to emphasize that frying fats should be discarded at the appropriate time to prevent humans from ingesting high levels of potentially toxic oxidation products, especially the fatty acid cyclic monomers.

REFERENCES

Akoh, C.C. 1995. Lipid based fat-substitutes. *Crit. Rev. Food Sci. Nutr.* 35:405–450.

Alexander, J.C. 1978. Biological effects due to changes in fats during heating. *J. Am. Oil Chem. Soc.* 55:711–717.

Alexander, J.C., Chanin, B.E., and Moran, E.T. 1983. Nutritional effects of fresh, laboratory heated and pressure deep-fry fats. *J. Food Sci.* 48:1289–92.

American Oil Chemists' Society (AOCS). 1991. *Official and Tentative Methods*, Vol. 1, 4th edition (1991). AOCS, Champaign, IL.

Arroyo, R., Cuesta, C., Garrido-Apolonio, C., López-Varela, and Sánchez-Muñiz, F.J. 1992. High-performance size-exclusion chromatography studies on polar components formed in sunflower oils used for frying. *J. Am. Oil Chem. Soc.* 69:557–63.

Andia, A.M.G., and Street, J.C. 1975. Dietary induction of hepatic microsomal enzymes by thermally oxidized fats. *J. Agric. Food Chem.* 23:173–77.

Artman, N.R., and Smith, D.E. 1972. Synthetic isolation and identification of minor components in heated and unheated fat. *J. Am. Oil Chem. Soc.* 49:318–26.

Artz, W.E., and Hansen, S.L. 1996a. Current developments in fat replacers, In: *Food Lipids and Health*, ed. R.E Mc Donald, and D.B Min. Marcel Dekker, New York.

Artz, W.E., and Hansen, S.L. 1996b. The chemistry and nutrition of nonnutritive fats, In: *Deep Frying, Chemistry, Nutrition and Practical Applications*, ed. E.G Perkins, and M.D Eriskson. American Oil Chemists' Society, Champaign, IL.

Ashwin, J.L., Harris, P.J., and Alexander, J.C. 1991. Effects of thermally oxidized canola oil and chronic low ethanol consumption on aspects of hepatic oxidative stress in rats. *Nutr. Res.* 11:79–90.

Athias, P., Ribot, E., and Grynberg, A. 1992. Effects of cyclic fatty acid monomers on the function of cultured rat cardiac myocytes in normoxia and hypoxia. *Nutr. Res.* 737–45.

Baixauli, R., Salvador, A., Fiszman, S.M., and Calvo, C. 2002. Effect of oil degradation during frying on the color of fried, battered squid rings. *J. Am. Oil Chem. Soc.* 79:1127–31.

Berdeaux, O., Dutta, P.C., and Dobarganes, M.C. 2009. Analytical methods for quantification of modified fatty acid and sterols formed as a result of processing. *Food Anal. Methods*. 2:30–40.

Billek, G., Guhr, G., and Waibel, J. 1978. Quality assessment of used frying fats: A comparison of four methods. *J. Am. Oil Chem. Soc.* 55:728–33.

Blumenthal, M.M. 1991. A new look at the chemistry and physics of deep-fat frying. *Food Technol.* 45:68–71.

Blumenthal, M.M., and Stier, R.F. 1991. Optimization of deep-fat frying operations. *Trends Food Sci. Technol.* 2:144–48.

Chang, S.S., Peterson, R.J., and Ho, C. 1978. Chemical reactions involved in the deep fat frying of foods. *J. Am. Oil Chem. Soc.* 55:718–22.

Chanin, B.E., Valli, V.E., and Alexander, J.C. 1988. Clinical and histological observations with rats fed laboratory heated or deep-fry fats. *Nutr. Res.* 8:921–35.

Choe, E., and Min, D.B. 2006. Chemistry and reactions of reactive oxygen species in foods. *Crit. Rev. Food Sci. Nutr.* 41:1–22.

Choe, E., and Min, D.B. 2007. Chemistry of deep fat frying oils. *J. Food Sci.* 72:77–86.

Christie, W,W., Brechany, E.Y., Sebedio, J.L., and Le quere, J.L. 1993. Silver ion chromatography and gas chromatography-mass spectrometry in the structural analysis of cyclic monoenoic acids formed in frying oils. *Chem. Phys. Lipids.* 66:143–53.

Christopoulou, C.N., and Perkins, E.G. 1989a. Dimer acids: Synthesis and mass spectrometry of the tetrahydroxy, dihydroxy, and diketo dimers of methyl stearate. *J. Am. Oil Chem. Soc.* 66:1344–52.

Christopoulou, C.N., and Perkins, E.G. 1989b. Isolation and characterization of dimers formed in used soybean oil. *J. Am. Oil Chem. Soc.* 66:1360–70.

Combe, N., Constantin, M.J., and Entressangles, B. 1981. Lymphatic absorption of nonvolatile oxidation products of heated oils in the rat. *Lipids.* 16:8–14.

Dobarganes, M.C., Pérez-Camino, M.C., and Márquez-Ruiz, G. 1988. High performance size exclusion chromatography of polar compounds in heated and non-heated fats. *Fat Sci. Technol.* 90:308–11.

Dana, D., and Saguy, I.S. 2006. Mechanism of oil uptake during deep-fat frying and the surfactant effect-theory and myth. *Adv. Colloidal Interface Sci.* 128–130:267–272.

Dobson, G., Christie, W.W., Brechany, E.Y., Sebedio, J.L., and Le Quere, J.L. 1995. Silver ion chromatography and gas chromatography-mass spectrometry in the structural analysis of cyclic dienoic acids formed in frying oils. *Chem. Phys. Lipids.* 75:171–82.

Edgar, K. 1999. The effect of dietary oxidized oil on lipid metabolism in rats. *Lipids.* 34:717–25.

Engelsen, S.B. 1997. Explorative spectrometric evaluations of frying oil deterioration. *J. Am. Oil Chem. Soc.* 74:1495–1508.

Farhoosh, R., and Moosavi, S.M.R. 2008. Carbonyl value in monitoring of the quality of used frying oils. *Anal. Chim. Acta.* 617:18–21.

Foote, C.S. 1968. Photosensitized oxygenation and role of singlet oxygen. *Acct. Chem. Res.* 1:104–110.

Fujisaki, M., Endo, Y., and Fujimoto, K. 2002. Retardation of volatile aldehyde formation in the exhaust of frying oil by heating under low oxygen atmospheres. *J. Am. Oil Chem. Soc.* 79:909–14.

Frankel, E.N. 1985. Chemistry of autoxidation: Mechanism, products and flavor significance. In *Flavor Chemistry of Fats and Oils*, ed. D.B. Min, and T.H. Smouse, 1–37. American Oil Chemists' Society, Champaign.

Gabriel, H.G., Alexander, J.C., and Valli, V.E. 1978. Nutritional and metabolic studies of distillable fractions from fresh and thermally oxidized corn oil and olive oil. *Lipids.* 13:49–55.

García-Martínez, M.C., Márquez-Ruiz, G., Fontecha, J., and Gordon, M.H. 2009. Volatile oxidation compounds in a conjugated linoleic acid-rich oil. *Food Chem.* 113:926–31.

Gardner, D.R., and Sanders, R.A. 1990. Isolation and characterization of polymers in heated olestra and olestra/triglyceride blend. *J. Am. Oil Chem. Soc.* 67:788–96.

Garibağaoğlu, M., Zeybek, U., Erdamar, S., Cevik A., Elmacioğlu F. 2007. The hepatotoxic effect of deep fried sunflower oil on rat liver. *Adv. Mol. Med.* 3:35–40.

Gershoff, S.N. 1995. Nutrition evaluation of dietary fat substitutes. *Nutr. Rev*. 53:305–13.

Giani, E., Masi, J., and Galli, C. 1985. Heated fat, vitamin E and vascular eicosanoids. *Lipids*. 20:439–48.

Gutierrez, R., Gonzalez-Quijano, F., and Dobarganes, M.C. 1988. Analytical procedures for the evaluation of used frying fats. In: *Frying of Food. Principles, Changes, New Approaches*, ed. G. Varela, A.E. Bender, and I.D. Morton, 141–54. Ellis Horwood Ltd., Chichester, England.

Hamilton, R.J., and Perkins, E.G. 1997. Chemistry of deep fat frying. In: *New Developments in Industrial Frying*, ed. S.P. Kochhar. PJ Barnes & Associates, 9–33. Bridgwater, U.K.

Haumann, B.F. 1998. The benefits of dietary fats: The other side of the story. *Inform*. 9:366–82.

Haywood R.M., Claxson, A.W.D., Hawkes, G.E., Richardson, D.P., Naughton, D.P., and Combarides, G. 1995. Detection of aldehydes and their conjugated hydroperoxydiene precursors in thermal-stressed culinary oils and fats: Investigations using high resolution proton NMR spectroscopy. *Free Radical Res*. 22:441–82.

Hu, M.Y., Li, Y.L., Jiang, C.H., Liu Z.Q., Qu, S.L., and Huang Y.M. 2008. Comparison of lycopene and fluvastatin effects on atherosclerosis induced by a high-fat diet in rabbits. *J. Nutr.* 24:1030–38.

Iwaoka, W.T, and Perkins, E.G. 1978. Metabolism and lipogenic effects of the cyclic monomers of methyl linolenate in the rat. *J. Am. Oil Chem. Soc.* 55:734–38.

Iwaoka, W.T., and Perkins, E.G. 1976. Nutritional effects of the cyclic monomers of methyl linolenate in the rat. *Lipids*. 11:349–53.

Izaki, Y., Yoshikawa, S., and Uchiyama, M. 1984. Effects of ingestion of thermally oxidized frying oil on peroxidative criteria in rats. *Lipids*. 19:324–31.

Johnson, O.C., Sakuragi, T., and Kummerow, F.A. 1956. A comparative study of the nutritive value of thermally oxidized oils. *J. Am. Oil Chem. Soc.* 33:433–35.

Johnson, O.C., Perkins, E.G., Sugai, T., and Kummerow, F.A. 1957. Studies on the nutritional and physiological effects of thermally oxidized oils. *J. Am. Oil Chem. Soc.* 34:594–97.

Kalua, C.M., Allen M.S., Bedgood, D.R., Bishop, A.G., Prenzler P.D., and Robards, K. 2007. Olive oil volatile compounds, flavour development and quality: A critical review. *Food Chem.* 100:273–86.

Kanner, J. 2007. Dietary advanced lipid oxidation endproducts are risk factors to human health. *Mol. Nutr. Food Res.* 51:1094–101.

Kanner, J., German, J.B., and Kinsella, J.E. 1987. Initiation of lipid peroxidation in biological systems. *Crit. Rev. Food Sci. Nutr.* 25:317–64.

Kaunitz, H., Slanetz, C.A., and Johnson, R.E. 1955. Antagonism of fresh fat to the toxicity of heated and aerated cotton oil. *J. Nutr*. 55:577–87.

Kaunitz, H., Slanetz, C.A., Johnson, R.E., Guilmain, J., Knight, H.B., and Saunders, D.H. 1956. Nutritional properties of the molecularly distilled fraction of autoxidized fats. *J. Nutr.* 60:237–44.

Keane, K.W., Jacobson, G.A., and Kreiger, G.H. 1959. Biological and chemical studies on commercial frying oils. *J. Nutr.* 68:57–74.

Krishnamurthy, R.G., and Chang, S.S. 1967. Chemical reactions involved in the deep fat frying of foods. III. Identification of nonacidic volatile decomposition products of corn oil. *J. Am. Oil Chem. Soc.* 44:136–40.

Kubow, S. 1992. Routes of formation and toxic consequences of lipid oxidation products in foods. *Free Radical Biol. Med.* 12:63–81.

Lamboni, C., Sebedio, J.L., and Perkins, E.G. 1998. Cyclic fatty acid monomers from dietary heated fats effect rat liver enzyme activity. *Lipids*. 33:675–81.

Lamboni, C., and Perkins, E.G. 1996. Effects of dietary heated fats on rat liver enzyme activity. *Lipids*. 31:955–62.

Mahungu, S.M. 1994. Analysis of the volatile components of heated esterified propoxylated glycerols. Ph.D. thesis, University of Illinois, Urbana, IL.

Mahungu, S.M., Hanse, S.L., and Arts, W.E. 1998. Volatile compounds in heated oleic acid—esterified propoxylated glycerol. *J. Am. Oil Chem. Soc.* 75:683–90.

Macku, C., and Shibamoto, T. 1991. Headspace volatile compounds formed from heated corn oil and corn oil with glycine. *J. Agric. Food Chem.* 39:1265–69.

Martin, F.L., and Ames, J.M. 2001. Comparison of flavor compounds of potato chips fried in palmolein and silicone fluid. *J. Am. Oil Chem. Soc.* 78:863–66.

Márquez-Ruiz, G., Peiez-Camino, M.C., Ruiz-Gutiirrez, V., and Dobarganes, M.C. 1992. Absorption of thermoxidized fats. II. Influence of dietary alteration and fat level. *Grasas y Aceites*. 43:198–203.

Márquez-Ruiz, G., and Dobarganes, M.C.1995. Assessments on the digestibility of oxidized compounds from [1-^{14}C]-linoleic acid using a combination of chromatographic techniques. *J. Chromatogr*. 675:1–8.

Márquez-Ruiz, G., and Dobarganes, M.C. 2006. Nutritional and physiological effects of used frying oils and fats. In: *Deep Frying: Chemistry, Nutrition and Practical Applications*, ed. M.D. Erickson, 173–203. American Oil Chemists' Society, Champaign, IL.

Márquez-Ruiz, G., and Dobarganes, M.C. 1997. Analysis of lipid oxidation products by combination of chromatographic techniques. In: *New Techniques and Applications in Lipid Analysis*, ed. R.E. McDonald, and M.M. Mossoba, 216–33. American Oil Chemists' Society, Champaign, IL.

Melton, S.L.1996. Nutritional needs for fat and the role of fat in the diet. In: *Deep Frying: Chemistry, Nutrition and Practical Applications,* ed. E.G Perkins, and M.D. Erickson, 151–59. American Oil Chemists' Society, Champaign, IL.

Mestdagh, F., Castelein, P., Van Peteghem, C., and De Meulenaer, B. 2008. Importance of oil degradation components in the formation of acrylamide in fried foodstuffs. *Food Chem.* 56:6141–44.

Miller, K.W., and Long, P.H. 1990. A 91-day feeding study in rats with heated olestra/vegetable oil blends. *Food Chem. Toxicol.* 28:307–16.

Min, D.B., and Schweizer, D.Q. 1983. Lipid oxidation in potato chips. *Food Chem. Toxicol.* 60:1662–65.

Morales, M.T., Rios, J.J., and Aparicio, R. 1997. Changes in the volatile composition of virgin olive oil during oxidation: Flavors and off-flavors. *J. Agric. Food Chem.*45:2666–73.

Mossoba, M.M., Yurawecz, M.P., Roach, J.A.G., Lin, H.S., McDonald, R.E., Flickinger B.D. et al. 1994. Rapid determination of double bond configuration and position along the hydrocarbon chain in cyclic fatty acid monomers. *Lipids*. 29:893–96.

Muik, B., Lendl, B., Molina-Díaz, A., and Ayora-Cãnada, M.J. 2005. Direct monitoring of lipid oxidation in edible oils by Fourier transform Raman spectroscopy. *Chem. Phys. Lipids.* 134:173–82.

Naruszewick, M., Zapolska-Downar, D., Kosmider, A., Nowicka, G., Kozlowska-Wojciechowska M., Vikström, A.S. et al. 2009. Chronic intake of potato chips in humans increases the production of reactive oxygen radicals by leukocytes and increases plasma C-reactive protein: A pilot study. *Am. J. Clin. Nutr.* 89:773–77.

Neff, W.E., Warner, K., and Byrdwell, W.C. 2000. Odor significance of undesirable degradation compounds in heated triolein and trilinolein. *J. Am. Oil Chem. Soc.* 77:1303–13.

Nolan, G.A. 1972. Effects of fresh and used hydrogenated soybean oil on reproduction and teratology in rats. *J. Am. Oil Chem. Soc.* 49:688–93.

Nolan, G.A., Alexander, J.C., and Artman, N.R.1967. Long-term rat feeding study with used frying oils. *J. Nutr.* 93:337–48.

Ottaviani, P., Graille, J., Perfetti, P., and Naudet, M. 1979. Produits d'alteration thermooxydative des huiles chauffees. II. Composes apolaires ou faiblement polaires. *Chem. Phys. Lipids*. 24:57–77.

Perkins, E.G., and Kummerow, F.A. 1959. The isolation and characterization of the polymers formed during the thermal oxidation of corn oil. *J. Am. Oil Chem. Soc.* 36:371–75.

Perkins, E.G., and Lamboni, C. 1998. Magnesium silicate treatment of dietary heated fats: Effects on rat liver enzyme activity. *Lipids*. 33:683–87.

Perkins, E.G., Vachha, S.M., and Kummerow, F.A. 1970. Absorption by the rat of non-volatile oxidation products of labeled randomized corn oil. *J. Nutr.* 100:725–31.

Perkins, E.G., and Taubold, R. 1978. Nutritional and metabolic studies of noncyclic dimeric fatty acid methyl esters in the rat. *J. Am. Oil Chem. Soc.* 55:632–34.

Poling, C.E., Eagle, E., Rice, E.E., Durand, A.M.A., and Fisher, M. 1970. Long-term responses of rats to heat-treated dietary fats. IV. Weight gains, food and energy efficiencies, longevity and histopathy. *Lipids*. 5:128–36.

Poling, C.E., Warner, W.D., Mone, P.E., and Rice, E.E. 1960. The nutritional value of fats after use in commercial deep-fat frying. *J. Nutr.* 72:109–20.

Quiles, J.L., Huertas, J.R., Battino, M., Ramírez-Tortosa, M.C., Cassinello, M., Mataix, J. et al. 2002. The intake of fried virgin olive or sunflower oils differentially induces oxidative stress in rat liver microsomes. *Br. J. Nutr.* 88:57–65.

Reineccius, G. and Heath, H.B. 2006. *Flavour Chemistry and Technology*, 2nd edition. Boca Raton, FL, CRC Press.

Ribot, E., Grandgirard, A., Sewdio, J.L., Gryberg, A., and Athias, P. 1992. Incorporation of cyclic fatty acid monomers in lipids of rat heart. *Lipids*. 27:79–81.

Ringseis, R., and Eder, K. 2008. Effect of dietary oxidized fats on gene expression in mammals. Examining the central role of peroxisome proliferators-activated receptors. *Inform.* 19:657–59.

Rojo, J.A., and Perkins, E.G. 1987. Cyclic fatty acid monomer formation in frying fats. I. Determination and structural study. *J. Am. Oil Chem. Soc.* 64:414–21.

Rojo, J.A., and Perkins, E.G. 1989. Chemical synthesis and spectroscopic characteristics of CIS 1,2-disubstituted cyclopentyl fatty acid methyl esters. *Lipids*. 6:467–76.

Romero, A., Bastida, S., and Sánchez-Muniz, F.J. 2006. Cyclic fatty acid monomer formation in domestic frying of frozen foods in sunflower oil and high oleic acid sunflower oil without oil replenishment. *Food Chem. Toxicol.* 44:1674–81.

Romero, A., Cuesta, C., and Sánchez-Muniz, F.J. 2000. Cyclic fatty acid monomers and thermoxidative alteration compounds formed during frying of frozen foods in extra virgin oil. *J. Am. Oil Chem. Soc.* 77:1169–75.

Romero, A., Cuesta, C., and Sánchez-Muniz, F.J. 2003. Cyclic fatty acid monomers in high-oleic acid sunflower oil and extra virgin oil used in repeated frying of fresh potatoes. *J. Am. Oil Chem. Soc.* 80:437–42.

Ruiz-Gutierrez, V., and Muriana, F.J.G. 1992. Effect of ingestion of thermally oxidized frying oil on desaturase activities and fluidity in rat-liver microsomes. *J. Nutr. Biochem.* 3:75–9.

Sánchez-Muniz, F.J., and Bastida, S. 2003. Frying oil discarding: Polar content vs. oligomer content determinations. *Forum Nutr.* 56:345–47.

Sébédio, J.L., Prevost, J., and Grandgirard, A. 1987. Heat treatment of vegetable oils. I. Isolation of the cyclic fatty acid monomers from heated sunflower and linseed oils. *J. Am. Oil Chem. Soc.* 64:1026–32.

Sébédio, J.L., Le Quere, J.L., Morin, O., Vatele, J.M., and Grandgirard, A. 1989. Heat treatment of vegetable oils. III. GC-MS characterization of cyclic fatty acid monomers in heated sunflower and linseed oils after total hydrogenation. *J. Am. Oil Chem. Soc.* 66:704–9.

Sébédio, J.L., and Chardigny, J.M. 1996. Physiological effects of *trans* and cyclic fatty acids. In: *Deep Frying: Chemistry, Nutrition and Practical Applications*, ed. E.G. Perkins, and M.D. Erickson., 183–209. American Oil Chemists' Society, Champaign, IL.

Shue, G.M., Douglas, C.D., Firestone, D., Friedman, L., and Sage, J.S. 1968. Acute physiological effects of feeding rats non-urea-adducting fatty acids (urea filtrate). *J. Nutr.* 94:171–7.

Snyder, J.M., Frankel, E.N., Selke, E., and Warner, K. 1988. Comparison of gas chromatographic methods for lipid oxidation compounds in soybean oil. *J. Am. Oil Chem. Soc.* 65:1617–20.

Soriguer, F., Rojo-Martínez, G., Dobarganes M.C., Almeida J.M.G., Esteva, I., Beltrán, M. et al. 2003. Hypertension is related to the degradation of dietary frying oils. *Am. J. Clin. Nutr.* 78:1092–97.

Taylor, A.J., Linforth, R.S.T., Harvey, B.A., and Blake, A. 2000. Atmospheric pressure chemical ionisation mass spectrometry for in vivo analysis of volatile flavour release. *Food Chem.* 71:327–38.

Totani, N., Burenjargal, M., Yawata, M., and Ojiri, Y. 2008. Chemical properties and cytotoxicity of thermally oxidized oil. *J. Oleo Sci.* 57:153–60.

Tyagi, V.K., and Vasishtha, A.K. 1996. Changes in the characteristics and composition of oils during deep-fat frying. *J. Oleo Sci.* 73:499–506.

Vatela, J.M., Sébédio, J.L., and Quere, J.L. 1988. Cyclic fatty acid monomers: Synthesis and characterization of methyl co-(2-alkylcyclopentyl) alkenoates and alkanoates. *Chem. Phys. Lipids.* 48:119–28.

Vijayan, J., Slaughter, D.C., and Singh, R.P. 1996. Optical properties of corn oil during frying. *Int. J. Food Sci. Technol.* 31:353–58.

Wagner, R.K., and Grosch, W. 1998. Key odorants of french fries. *J. Am. Oil Chem. Soc.* 75:1385–92.

Warner, K. 2008. Chemistry of frying oils. In: *Food Lipids Chemistry, Nutrition and Biotechnology*, ed. C.C. Akoh, and D.B Min, 189–200. Boca Raton, FL, CRC Press.

Warner, K. 1999. Impact of high-temperature processing on fats and oils. In: *Impact of Food Processing on Food Safety*, ed. L.S. Jackson, M.G. Knize, and J.N. Morgan, 67–78. Plenum Publishers, New York.

Wolff, J.P., Mordret, F.X., and Dieffenbacher, A. 1991. Determination of polymerized triglycerides in oils and fats by high performance liquid chromatography. *Pure Appl. Chem.* 63:1163–71.

Yashioka, M., Tachibana, K., and Kaneda, T. 1974. Studies on the toxicity of the autoxidized oils. IV. Impairments of metabolic functions induced by autoxidized methyl linoleate. *Yukagaku.* 23:327–31.

3 Free Radicals in Biological and Food Systems

Malgorzata Nogala-Kalucka and Aleksander Siger

CONTENTS

3.1 INTRODUCTION

For most living organisms, oxygen is the crucial element necessary for maintaining their life functions. It is life-giving and widespread, although the wide research on its biological role proves that it can also have adverse effects on living organisms. Different processes and oxygen transformations result in creating free radicals (FR) and other reactive species (RS). The *in vivo* effects exerted by these RS heavily depend on their concentration. These particles that act as signal transmitters and can modify various cellular pathways, take part in gene expression, and influence the cellular metabolism. The role of RS may vary, depending on the type of cell and its internal conditions. It has been reported that low concentrations of molecules can profoundly affect the process of cellular proliferation, whereas high concentrations of molecules can accelerate aging processes that may lead to death. Living organisms developed defensive mechanisms including oxidation-reductive enzymes that scavenge and/or neutralize RS and FR. It has been indicated that, in biological and food systems, such mechanisms are regulated by antioxidant substances.

3.2 TYPES OF FREE RADICALS AND OTHER REACTIVE SPECIES

Free radicals (FR) are molecules, atoms, or ions with an unpaired electron on the valence shell that makes them paramagnetic in nature. In most cases, radicals are electrically neutral and very reactive.

RS are products of excitation or reduction of molecular oxygen. The basic form of oxygen is a triplet with two unpaired electrons ($^3\Sigma_g O_2$). Singlet oxygen is not a free radical; however, the energy of its molecule is much higher than in the triplet form. Total reduction of the oxygen molecule requires acquisition of four electrons and four protons.

$$O_2^- + 4e^- + 4H^+ \rightarrow 2H_2O$$

The reduction of oxygen is not always complete, and RS may be formed. The oxygen molecule reduction in a one-electron reaction results in the formation of superoxide radical anion ($O_2^{\bullet -}$), while in a two-electron reaction hydrogen peroxide (H_2O_2) is formed. Superoxide radical anion can also be protonated and subsequently form the hydroperoxyl radical ($HO_2^\bullet$). Hydroxyl radical ($^\bullet OH$) is formed in the reaction of a three-electron reduction of molecular oxygen (Turrens, 2003).

$$O_2 \xrightarrow{e^-} O_2^{-\bullet} \xrightarrow{e^-,2H^+} H_2O_2 \xrightarrow{e^-,H^+} H_2O + {}^\bullet OH \xrightarrow{e^-,H^+} 2H_2O$$

Other forms of oxygen should also be mentioned, for example, the allotropic form of oxygen such as ozone (O_3) compounds, which are the products of metabolic reactions: nitroxide ($NO^\bullet$), nitrogen dioxide ($NO_2^\bullet$), peroxynitrous acid ($HONO_2$), and hypochlorous acid (HOCl). RS reactions involving organic molecules result in the formation of FR from organic substances: alkoxy radicals ($RO^\bullet$), peroxides ($ROO^\bullet$), and hydroperoxides (ROOH). They are radicals in which the unpaired electron is on the carbon atom or on an atom of another element such as nitrogen.

3.2.1 SUPEROXIDE RADICAL ANION ($O_2^{\bullet -}$)

This reductant is formed in a one-electron reduction of molecular oxygen:

$$O_2 + e^- \rightarrow O_2^{\bullet -}$$

It is a weak oxidant and strong reductant, and its half-life is relatively short: 1×10^{-6} s, at 37°C. It is a precursor of free radicals that are formed *in vivo* such as $HO_2^{\bullet}$, H_2O_2, and $^{\bullet}OH$ (Yu, 1994). In water solutions, $O_2^{\bullet-}$ is in equilibrium with hydroxyperoxide radical.

$$HO_2^{\bullet} \xrightleftharpoons{pk=4,8} O_2^{\bullet-} + H^+$$

Its main source in the cell is the electron leak in the oxidative phosphorylation system that occurs during the oxidation of hemoglobin to myoglobin. In the human body, the oxidation in tissues yields approximately 1.68×10^{23} (Cadens and Davies, 2000), and the hemoglobin oxidation yields approximately 250×10^{15} (Bynoe et al., 1992) molecules of $O_2^{\bullet-}$ in 24 h.

$$Hb(Fe^{2+})O_2 \rightarrow MetHb(Fe^{3+}) + O_2^{\bullet-}$$

In normal conditions, $O_2^{\bullet-}$ is dismutated by superoxide dismutase (SOD) to H_2O_2, which is decomposed into water and oxygen by catalase or glutathione peroxidase by the expenditure of two more electrons. The higher value of redox potential (O_2 / $HO_2^{\bullet}$), approximately 1V, proves that $HO_2^{\bullet}$ is a stronger oxidant than $O_2^{\bullet-}$ ($O_2/O_2^{\bullet-}$), −0.33V. This is reflected in the research results since $HO_2^{\bullet}$, as opposed to $O_2^{\bullet-}$, can attack fatty acids and initiate lipid peroxidation (Halliwell and Gutteridge, 1984).

3.2.2 Hydrogen Peroxide (H_2O_2)

It is classified as RS; it is not a free radical but rather a stable compound; it can be subject to disproportioning catalyzed by transition metal ions:

$$H_2O_2 + H_2O_2 \rightarrow 2H_2O + O_2$$

At pH values that are close to the physiological pH, H_2O_2 oxidizes thiol groups of aminoacids as well as indole, imidazole, phenol, thioester, and methionyl groups. In the presence of transition metal ions, it is decomposed as follows:

$$H_2O_2 \xrightarrow{M^n/M^{n+1}} OH^- + {}^{\bullet}OH$$

Due to the lack of electric charge, H_2O_2 freely diffuses through cellular membranes. It is decomposed by catalase and peroxidases (Afanas'ev, 1989).

3.2.3 Hydroxyl Radical ($^{\bullet}OH$)

Hydroxyl radical is the most active radical among RS. It is formed in Fenton's reaction. It has no capability of diffusing through cellular membranes. Superoxide radical anion and H_2O_2 can be transformed into a very active $^{\bullet}OH$ in the Haber–Weiss reaction. This reaction, first described by Fenton, is catalyzed by the presence of transition metal ions (Fe, Cu, Ti, Co, Mn, Ni, Cr, Ce). Haber–Weiss reaction includes a reduction of Fe^{3+} ions by $O_2^{\bullet-}$ and Fenton's reaction:

$$Fe^{3+} + O_2^{-\bullet} \rightarrow Fe^{2+} + O_2$$

$$Fe^{2+} + H_2O_2 \rightarrow Fe^{3+} + OH^- + {}^{\bullet}OH$$

$$O_2^{-\bullet} + H_2O_2 \xrightarrow{Fe^{2+}/Fe^{3+}} {}^{\bullet}OH + OH^- + O_2$$

Very high reactivity combined with extremely low specificity make ${}^{\bullet}OH$ capable of reacting with most of the important compounds present in natural, biological systems.

3.2.4 Singlet Oxygen ($^1O_2$)

The energy of singlet oxygen amounts to 22.5 kcal, and it is higher than the energy of triplet oxygen. The energy is in existence for long enough to enter a reaction with other molecules in the singlet (Min, 1998). It is not a FR; it can enter a reaction with non-FR compounds, singlet compounds, and electron-rich compounds that contain double bonds. The lifetime of singlet oxygen is 500–700 μs depending on the used system solvent of foods. The rate constants of singlet oxygen and triplet oxygen with linoleic acid are, respectively, $1.3 \times 10^5\ M^{-1}s^{-1}$ and $8.9 \times 10^1\ M^{-1}s^{-1}$. Singlet oxygen can be formed during chemical, enzymatic, and photochemical reactions.

3.2.5 Ozone (O_3)

Ozone is much more reactive than triplet oxygen. During ozone cracking of unsaturated organic substances, the formed ozonides are rapidly degraded to carbonyl compounds. The rate constants of the reactions with polyenoic fatty acids and amino acids such as methionine, tryptophan, and tyrosine amount to 10^6–$10^7\ Lmol^{-1}s^{-1}$, while with cysteine to $10^9\ Lmol^{-1}s^{-1}$. Ozone is formed from molecular oxygen in an endothermic process, under very high voltage or UV radiation, as well as from the decomposition of O_2 and its further recomposition. One of the most significant ways by which ozone affects cellular metabolism is its influence on NADH and NADPH by oxidization. Those two coenzymes take part in such reactions as glycolysis, gluconeogenesis, fatty acid synthesis, beta-oxidation, citric acid cycle, and oxidative phosphorylation.

3.2.6 Nitric Oxide and Peroxynitrite ($NO^{\bullet}$, $ONOO^-$)

Releasing excessive amounts of $NO^{\bullet}$ may result in oxidative stress. Generating $NO^{\bullet}$ is linked to the composition of its other, more active forms such as $NO_2^{\bullet}$, N_2O_3, NO_x, $ONOO^-$, $ONOO^{\bullet}$, ONOONO, NO_3^-, and NO_2^-. Often, these forms are described as reactive nitrogen species (RNS). Nitric oxide can be easily oxidized to very toxic nitric dioxide, which is also a free radical characterized by strong oxidant properties. Then, by reacting with water, it creates nitrate and nitrite ions (Garrel and Fontecave, 1995).

$$2NO^{\bullet} + O_2 \rightarrow 2NO_2^{\bullet}$$

$$2NO_2^{\bullet} + H_2O \rightarrow NO_3^- + NO_2^- + 2H^+$$

A nitric oxide molecule easily enters a reaction with $O_2^{\bullet-}$, creating a highly toxic peroxynitrite (Anggard, 1994). It has strong oxidative properties; it reacts mainly with thiol groups of protein and polyenoic residues of fatty acids and lipids. Oxidizing the SH groups to disulfates results in prooxidative and antioxidative disbalance in the cell and intensification of prooxidative damage. Nitric oxide can react with different compounds that mainly contain unpaired electrons, such as $O_2^{\bullet-}$. This is why it can be seen as a RS remover. Nitric oxide stops the peroxidative damage by entering a reaction with FR that are formed in the process of lipid peroxidation in the cycle of chain reactions (O'Donnele et al., 1997):

$$4NO^{\bullet} + 2ROO^{\bullet} + H_2O \rightarrow 2HNO_2 + RONO_2 + RONO$$

3.2.7 Organic Radicals—Oxidation of Lipids, Proteins, and Carbohydrates

By striving for stability, free radicals, due to paired electrons, in certain conditions inside the cell may take electrons away from biomolecules such as lipids, proteins, carbohydrates, vitamins, and nucleic acids. Especially lipids, which are cellular membrane components, are prone to the attack of free radicals. The loss of one electron initiates a chain reaction of a free radical character. In the course of the reaction, fatty acids are oxidized and peroxides are formed. The products of these reactions are alkyl radicals, alkoxyl radicals, peroxide radicals, hydroperoxide radicals, phenoxyl radicals, thiolic radicals, and other radicals. RS are the key factor responsible for initiating oxidation in food. Classical studies have established the mechanism of autoxidation of lipids as a free radical chain reaction that involves the three stages of *initiation, propagation, and termination.* In order to initiate lipid oxidation, the homolytic bond C-H in the hydrocarbon chain of the polyenoic fatty acid has to be broken. Polyenoic fatty acids (RH) contain reactive methyl residues neighboring two double bonds that easily lose the atom of hydrogenium—the dissociation of the C-H bond (*initiation phase*) in the reaction with oxidative substances with the formation of alkyl radical (R^- contains an unpaired electron in the atom of carbon) (Pryor, 1976; Porter, 1986; Kanner and Rosenthal, 1992; Porter et al., 1995; Kolakowska, 2003).

$$R - H + HO^{\bullet} \rightarrow R^{\bullet} + H_2O$$

Lipid peroxidation can be initiated by O_3, H_2O_2, $^1O_2$, and radicals such as hydroxyl ($^{\bullet}OH$), peroxide ($ROO^{\bullet}$), alkoxy ($RO^{\bullet}$), alkyl ($R^{\bullet}$), hydroperoxyl ($HOO^{\bullet}$), or other species (nitric oxide, nitric dioxide, and chloride anion). In the case of singlet oxygen in the addition reaction of oxygen molecule to the double bond, fatty acid peroxides are formed (the reaction does not involve free radicals), which may subsequently form free radicals. Alkyl radicals are transformed into peroxide radicals (*propagation*), the primary products of fatty acid oxidation that undergo transformation (Yanishlieva-Maslarova, 2001; Choe and Min, 2006).

$$R^{\bullet} + O_2 \rightarrow ROO^{\bullet}$$

$$ROO^{\bullet} + R'H + ROOH + R'^{\bullet}$$

The peroxide radical can be reduced by a fatty acid molecule, which is transformed into an alkyl radical (R$^{\bullet}$) that initiates the chain reactions. Hydroperoxides (ROOH) can be reduced by metal ions to alkoxyl radicals that undergo β-fission and form epoxides, aldehydes, and volatile compounds (Dotan et al., 2004). A reaction between FR results in the formation of a stable compound (*termination*).

$$R^{\bullet} + R^{\bullet} \rightarrow R - R$$

$$ROO^{\bullet} + ROO^{\bullet} \rightarrow R = O + ROH + O_2$$

$$ROO^{\bullet} + R^{\bullet} \rightarrow R = O + ROH$$

In the *termination* reaction, dimers of organic acids are formed, oxo- or hydroxy-fatty acids, etc. During the lipid peroxidation, one lipid radical can lead to the damage of 15 other molecules of polyenoic fatty acids (Chaiyasit et al., 2007).

Another process that takes place during the peroxidation of lipids is *reinitiation*. The more double bonds there are in the molecule of a fatty acid, the easier it is for the peroxidation to occur.

In biological systems, the reaction of peroxidation occurs mainly in the residues of polyenoic fatty acids in phospholipids—the main element of cellular membranes. During the oxidation of membrane lipids, the hydrocarbon chain is degraded. Apart from that, the phosphate residue and the substitute can be detached from the glycerol segment in the molecule. Oxidation leads to the disintegration of the membranes. This results in permeability changes, changes in the flexibility of the cell, and changes in signaling pathways (Schnitzer et al., 2007).

In *in vitro* conditions, the interaction between NO$^{\bullet}$ and polyenoic fatty acids in human and animal cells results in the formation of nitro fatty acids. In low oxygen concentrations, NO$^{\bullet}$ causes the homolytic breakdown of the double bond, and nitro-allyl derivatives are formed. The reaction with the second radical, NO, results in the formation of nitro-alkenes or nitro-alcohols. In higher oxygen concentrations, NO$^{\bullet}$ causes the formation of hydroperoxides of fatty acids and their *cis*-to-*trans* isomerization (Rubbo et al., 2009; Rudolph et al., 2009).

Proteins, nucleic acids, and carbohydrates can also be subject to oxidation. RS reactions with proteins result in modifying the residues of amino acids, prosthetic groups modifications, and aggregation or fragmentation of protein molecules (Wolff and Dean, 1986). The most important factor that initiates protein oxidation is $^{\bullet}$OH. The atom of hydrogen is detached from the protein molecule. The recombination of protein FR results in the formation of covalent protein dimers. RS reactions with proteins result in protein oxidation and the formation of reductive groups in the proteins that can reduce the cytochorme C and metal ions. They can lead to the formation of amino acids and protein peroxides, and damage the nucleic bases and nucleic acids (Stadman and Levine, 2000).

Most saccharides, especially sugars, are prone to oxidation in the presence of RS and trace amounts of transition metal ions (Rice-Evans, 1994). Glucose that reacts with hydroxyl radical is oxidized, and it can form further forms of reactive radicals. Patients with diabetes suffer from systemic oxidative stress because the glucose present in their body fluids may react in the following way:

$$C_6H_{12}O_6 + O_2 + {}^{\bullet}C_6H_{11}O_6 + H^+ + O_2^{\bullet-}$$

This reaction may occur in healthy people as well but, in persons suffering from diabetes, the amount of produced $O_2^{\bullet-}$ is significantly higher. There is a positive correlation between the oxidation processes of glucose and other carbohydrates, the lipid peroxidation level, and the rate of protein glycation. This interdependency is described as *glycoxidation* (glycation and oxidation) (Bilska and Wlodek, 2005).

3.3 MECHANISMS OF FREE RADICAL FORMATION

Homolytic breakdown of the covalent bond results in the formation of radicals. The breakdown can occur between the atoms C, O, H, and halogens. The breakdown of O-O in organic peroxides and hydroxyperoxides requires energy of 159 $kJmol^{-1}$. Homolytic breakdown can also occur within bonds between different atoms such as C-H, O-H, and N-O. The breakdown of a covalent bond can be induced by heating, UV and VIS or ion radiation (Asmus and Bonifacic, 2000).

Radicals can also be generated by catalytic methods. One of them is based on generating hydroxyl radicals in the process of H_2O_2 breakdown catalyzed by Fe^{2+} ions in Fenton and Harber–Weiss reactions described earlier. External factors affecting RS formation follow:

Electromagnetic radiation—As a result of exposure to irradiation, the molecules of water (radiolysis) become ionized and excited, resulting in the formation of $H^{\bullet}$, H_2O_2, hydroxyl radicals, and hydrated electrons (Asmus and Bonifacic, 2000):

$$H_2O \xrightarrow{\text{Irradiation}} e_{eq}^-,\ {}^{\bullet}OH,\ H^{\bullet},\ H_2O_2$$

In the presence of oxygen, superoxide radicals are formed from hydrated electrons and their protonated forms generated from hydrogen:

$$e_{eq}^- + O_2 \rightarrow O_2^{\bullet-} \ \text{k} = 1.9 \times 10^{10}\ \text{Lmol}^{-1}\text{s}^{-1}$$

$$H^{\bullet} + O_2 \rightarrow HO_2^{\bullet}\ \text{k} = 2.1 \times 10^{10}\ \text{Lmol}^{-1}\text{s}^{-1}$$

For instance, it causes the breakdown of H_2O_2 to $2HO^{\bullet}$. UV causes damage to macromolecules and the decrease of glutathione in cells. It can initiate the formation of singlet oxygen and the release of heme from the heme proteins (Basu-Modak and Tyrrell, 1993). UV and VIS radiation can initiate the formation of radicals mainly by photoactivation of photosensitizers (^{1}Sen) that absorb energy and undergo light excitation (^{3}Sen*).

$$^1Sen \xrightarrow{hv} {}^3Sen*$$

There are two types of photooxidation reactions (Gordon, 2001). Type I—the H atom or electron is transferred between the photosensitizers in the state of excitation and the substrate (S), for instance, polyenoic fatty acid, resulting in the formation of FR and ions of free radicals. Transferring the electron to the oxygen molecule results in $O_2^{\bullet-}$ and other RS.

$$^3Sen* + SH \rightarrow H^+ + Sen^{\bullet-} + S^{\bullet}$$

$$Sen^{\bullet-} + O_2 \rightarrow {}^1Sen + O_2^{-\bullet}$$

Type II—photosensitizers, being in the triplet state, easily react with oxygen to form two molecules in the singlet state. Singlet oxygen can further react with polyenoic fatty acids, resulting in the formation of hydroxyperoxides. The reaction of singlet oxygen with fatty acids is more than 1500 times faster than with the oxygen in the triplet state.

Ultrasounds—as a result of water solutions sonication, the amount of formed RS depends on the parameters of the sonication. The ultrasound-induced H_2O breakdown results in the creation of H and $^{\bullet}OH$ atoms, which are later transformed into H_2O_2 and superoxide radical anion. In the presence of oxygen, singlet and atomic forms of oxygen are produced. They react with nitrogen $NO^{\bullet}$ and peroxynitrite (Riesz, et al., 1985).

Transition metals ions—Fenton and Haber–Weiss reactions are main examples of RS-forming reactions catalyzed by transition metals ions. Qian and Buettner (1999) claim that with a high rate of $[O_2]/[H_2O_2]$, the formed RS, such as perferryl ion during oxidation of iron (II) with molecular oxygen, can be competitive or even more efficient than Fenton reaction.

$$Fe^{2+} + O_2 \Leftrightarrow [Fe^{2+}O_2] \Leftrightarrow [Fe^{3+}O_2^{\bullet-}]$$

Fe^{2+} ions can be oxidized in the presence of oxygen, and so, the reaction can be carried out *in vivo*. However, there is still no evidence on the "life span" of formed perferryl ions and their reactivity. It is also unknown whether they are sufficiently reactive to enter into reactions with biomolecules and whether they break down to form superoxides or not.

$$[Fe^{2+}O_2] \Leftrightarrow [Fe^{3+}O_2^{\bullet-}] \Leftrightarrow Fe^{3+} + O_2^{\bullet-}$$

Xenobiotics—numerous compounds, that is, pesticides, herbicides, drugs, food ingredients, ozone, environmental pollutants, etc., can initiate RS generation and the peroxidation of biomolecules such as lipids. There are two main mechanisms of lipid oxidation depending on the xenobiotic:

1. Under the influence of reducing factors such as metal ions, ascorbic acid or glutathione, xenobiotics are transformed into radical cations. Their oxidation results in O_2 formation. Fe-initiated, ascorbate-dependant lipid oxidation can be an interesting example of this mechanism. Ascorbate acts as a prooxidant as it reduces Fe^{3+} ions or coordinates Fe^{2+} ions.
2. When xenobiotics are reduced by endogenous reductants (NADH, NADPH, or other elements of mitochondria or microsomal oxidative multienzyme complex). CCl_4 reduced to $^{\bullet}CCl_3$ by microsomes is a very good example. This particle detaches the H atom from the polyenoic fatty acids, and it initiates their oxidation (Denisov and Afanas'ev, 2005).

3.3.1 The Sources of RS in Biological Systems

- *The oxidation of oxygen-transport proteins*—ferryl and perferryl forms of Fe can be responsible for RS formation. They are formed in the reaction of hemoglobin or methemoglobin oxidation. Ferryl forms and their derivatives can initiate protein and lipid damage.
- *Enzymatic reactions*—the activity of many enzymes causes RS formation, for example, xanthine oxidase and NAD(P)H oxidase in the plasmatic membrane of the cells that carry out phagocytosis (oxidative burst) and produce O_2^-. Another type of reaction is the enzymatic peroxidation in reactions such as the synthesis of prostanoids: prostaglandins, thromboxanes, and leukotrienes from arachidonic acid. Cyclooxygenases and lipoxygenases take part in these reactions.
- *Mitochondrial oxidative phosphorylation enzyme multicomplex*—the main source of superoxide radical anions in most aerobic cells. Most O_2^- is produced as a result of the reaction between ubiquinone and oxygen.

3.4 SOURCES OF NATURAL ANTIOXIDANTS AND THEIR PROPERTIES

Many food ingredients undergo numerous transformations during processing and storage. Very often, the transformations are disadvantageous in terms of nutrition. Fat oxidation is the main example of such transformations. This process can cause changes in color, flavor, taste, structure, and loss of bioactive compounds. In food systems, there are compounds that inhibit fat oxidation by entering into reactions with oxidative agents or with intermediates of oxidation (FR). Those substances are called *antioxidants*. They inhibit or slow down the process of oxidation in very low concentration in comparison to the substrate (Halliwell, 1994). Antioxidants can be natural or synthetic (BHT, BHA) (Yanishlieva-Maslarova, 2001). Natural antioxidants are substances that can be found in food systems, mainly in plant products that are diet components (Nogala-Kalucka, 2002; Boskou, 2006). Classified in this group are substances such as tocochromanols, carotenoids, ascorbic acid, polyphenols, and some aminoacids, peptides, and the intermediates of nonenzymatic browning. Antioxidants in food conduce to preserving better sensory qualities of products and to reducing the loss of some components of high nutritional value. Fat-soluble vitamins are found in animal food as well, but in significantly smaller amounts. Fruit and vegetables mainly contain vitamin C, β-carotene, and phenolic compounds (anthocyanins, flavonoids, phenolic acids) (Kähkönen et al., 2001; Sikora et al., 2008); cereals and leguminous plants contain phenolic acids, isoflavones, and tannins (Lampart-Szczapa et al., 2003); coffee contains chlorogenic acid; tea contains catechins (Cao et al., 1996; Gramza-Michalowska et al., 2007); wine contains polyphenols and resveratrol (Dreosti, 2000); herbs and spices contain phenolic acids, flavonoids, and rosemarinic acid; and olive oil contains hydroxytyrosol, flavonoids, and lignans (Boskou, 2006; Yanishlieva et al., 2006). Both hydrophilic and lipophilic antioxidants can be found in natural products. The best sources of hydrophobic antioxidants are oil plants seeds, fruits (rape, sunflower, flax, maize, olives, and others) (Hall, 2001; Kris-Etherton et al., 2002; Nogala-Kalucka et al., 2003; Boskou, 2006), and oils derived from those plants. Refined oils may lose their native antioxidant properties during refining (Nogala-Kalucka et al., 1993). Cold-pressed oils that have become more and more popular are a good source of antioxidants (olive oil, hemp, black caraway, black currant, carrot, cranberry, pumpkin, marionberry), as they preserve the whole natural composition of antioxidants (Parry et al., 2005; Siger et al., 2005; 2008; Boskou, 2006).

3.4.1 Tocopherols and Tocotrienols

Vitamin E is a term describing four homologous tocopherols (-T) and tocotrienols (-T3). Antioxidative activity of homologous tocopherols *in vivo* can be described as follows: α-T > β-T > γ-T > δ-T; their *in vitro* activity is opposite: α-T < β-T ≈ γ-T < δ-T (Eitenmiller and Lee, 2004). Antioxidative properties of tocochromanols depend on their concentration, type of substrate, other chemical compounds that act prooxidatively and synergistically, type of solvent, light, and temperature. This is shown in numerous studies with various fat substrates in different conditions (temperature, light) (Yanishlieva-Maslarova, 2001). The effectiveness of particular homologs depends on the degree of substitution of the chromanol ring and the side chain. In homogenous solutions, the reaction rate depends mainly on the number of methyl groups in the ring (Azzi and Stocker, 2000). Tocopherols quench $O_2^{\bullet-}$, and they are the most important inhibitors of the free radical chain reaction that occurs during the oxidation of lipids (Schneider, 2005). Tocopherols also quench singlet oxygen ($^3O_2$), and they act as stabilizers of biological membranes. (See also Chapter 7).

3.4.2 Carotenoids

Beta-carotene is the most important carotenoid occurring in higher plants. Carotenoids absorb light quanta and they have a protective role, that is, selective absorption of unwanted radiation and protection from damage during photooxidation; they are also precursors of various apo-carotenoids that take part in the formation of plant hormones (abscisic acid) (Wurtzel, 2004). One of the crucial functions of these compounds is the prevention of excessive RS formation. β-Carotene is symmetric with β-ionone rings on both ends. It can be cleaved into two molecules of retinol, and it has the highest activity as provitamin A. Other carotenoids such as α-carotene or β-cryptoxanthin only have one β-ionic formation. In *in vitro* investigation, it was shown that β-carotene stops the oxidation of lipids if the O_2 concentration is high (Warner and Frankel, 1987). The importance of carotenoid structure in radical scavenging has been studied by Mortensen and Skibsted (1997).

3.4.3 Phenolic Compounds

Presently, over 8,000 different phenolic structures are known. They are very common in plants. The majority of phenolic compounds have high antioxidant and radical scavenging activity under *in vitro* conditions. However, sometimes, similar to vitamin C, plant phenols may stimulate oxidative processes by reducing metals (Sugihara et al., 1999). Some flavonoids in the presence of NO show prooxidative activity (Ohshima et al., 1998). The factors affecting the antioxidative activity of plant phenols are the location and the number of OH groups, polarity, solubility, and their stability during technological processing (Decker, 1998). The energy of the bond between the atoms of hydrogen and oxygen in the –OH group in the aromatic molecule is much lower than in the aliphatic compounds. It results in high H mobility and transformation of the phenolic compounds into phenoxyl radicals and then into phenol derivatives (Rice-Evans et al., 1995). Introduction of additional –OH groups into the aromatic ring of a compound results in the increase of its antioxidative activity. This is a result of the stabilization of phenoxyl radicals by forming a hydrogen bond. The antioxidative properties of flavonoids stem from their ability to trap free radicals as well as the ability to chelate transition metals ions (Arora et al., 1998). The radicals and reactive species that are most easily captured by flavonoids are $O_2^{\bullet -}$, $^{\bullet}OH$, lipid radicals, and $^3O_2$ (Harborne and Williams, 2000). The majority of flavonoid structure chelators form complexes with Fe^{2+} ions in the molar ratio of 3:1. The research on the formation of complexes with transition metal ions by flavonoids showed that both rutin–glycoside and quercetin–aglycone effectively react to create "charge-transfer" complexes (Ryan and Petry, 1993). Polar molecules of antioxidants are very reactive in the single-phase environment (e.g., in oil), and nonpolar compounds in heterogenous environments (e.g., in emulsions). This type of behavior was named the *polarity paradox*. It is an important physicochemical parameter that heavily affects the antioxidative activity of polyphenolic compounds both *in vivo* and ex vivo (Porter, 1993; Frankel et al., 1994).

3.4.4 Ascorbic Acid

Ascorbic acid has strong reductive properties that make it antioxidative. This can be observed in the reactivity of ascorbic acid with $O_2^{\bullet -}$, H_2O_2, $^{\bullet}OH$, HOCl, peroxide radicals, and singlet oxygen. In biological fluids, it is nearly fully dissociated as ascorbic anion. Its reactions with oxidants are either a one-electron reduction that results in nonreactive ascorbyl free radical or a two-electron reduction to dihydroxy ascorbate. Dihydroxy ascorbate is unstable, and it can be easily hydrolyzed to 2,3-diketo-l-gulonic acid, which can be further reduced to form ascorbate via glutathione. Alternatively, the ascorbyl radical may be enzymatically reduced back to ascorbate (NADH-dependent semidehydroascorbate reductase, glutaredoxin, thioredoxin reductase) (Higdon and Frei, 2002). Similarly to α-tocopherol, ascorbic acid reacts with superoxide very slowly ($k = 0.32\ Lmol^{-1}s^{-1}$); however, it reacts relatively fast with hydroperoxyl radical ($k = 1.6 \times 10^4\ Lmol^{-1}s^{-1}$) (Cabelli and Bielski, 1983). Vitamin C also acts as a co-antioxidant by regenerating α-tocopherol from the α-tocopheroxyl radical in lipoproteins and membranes (Zhu and Frei, 2002). However, in *in vitro* experiments, it was proved that vitamin C in low concentrations can show prooxidative properties. Ascorbate stimulates free radical reactions especially in the presence of transition metal ions—apart from $O_2^{\bullet-}$. Ascorbate can reduce Fe^{3+}, and it can stimulate Fenton's reaction (May, 1999).

3.5 ANTIOXIDANTS' FUNCTION AND THEIR MECHANISM OF ACTION IN FOOD SYSTEMS

Frying is an important and perhaps the most widely used process in the preparation and manufacturing of food. Large amounts of oils and fats are used in frying, and they are subject to high temperatures. During heating at frying temperatures, fat oxidation, hydrolysis, polymerization, isomerization, and cyclization occur. The intensity of these processes depends on the type and quality of oil used (Kochhar, 2001). Frying fat can be considered one of the integral food components as the level of absorbed fat ranges from a low percentage to 40% of the mass of cooked product. The fatty acid composition of the oil and the presence of natural antioxidants that significantly affect the stability during frying are of great importance. The following are substances in oils that can be classified as antioxidants: tocopherols (-T) and tocotrienols (-T3), phospholipids (less than 100 mg/kg), carotenoids, sterols and some of their esters, squalene, phenolic acids, and other polyphenols (Kochhar, 2001). It was shown that some phospholipids such as phosphatidylcholine can stop the degradation of tocopherols and their dimers (Judde et al., 2003). Other substances, such as sesamoline (antioxidant precursor), sesamolin and its isomers, sesamol and its dimers (in the sesame oil), and oryzanol (a group of esters of ferulic acid and sterols in the rice bran oil), affect their high stability during frying (Kochhar, 2001). Alpha-T can enter a reaction both with peroxide radicals and with alkoxyl radicals, and the reaction rate equals $2 \times 10^5\ Lmol^{-1}s^{-1}$.

$$LOO^{\bullet} + \alpha - TOH \rightarrow LOOH + \alpha - TO^{\bullet}$$

Ross et al. (2003) reported that, among homologous tocopherols, the α-form passes the H on free radicals first. In the antioxidant reaction, α-T is being transformed into a fairly stable radical α-tocopheroxyl-semichinone (Schneider, 2005). Each molecule of α-T traps two free radicals, so it accomplishes two reactions of oxidation at the same time as apart from tocopherol tocopheroxyl radicals that take part in the reaction of inactivation of the peroxide radicals and oxidation to α-tocopheryl-chinone (α-T = O) (Burton, 1994).

$$\alpha - TO^{\bullet} + LOO^{\bullet} \rightarrow \alpha - T = O + LOOH$$

The tocopheryl-semichinone may still enter various reactions at higher temperature, for example, during frying. In the reaction between radicals γ-T, two of its dimers with antioxidative properties are formed: diphenyl-dimer and γ-T ether-dimer. Alpha-T is decomposed much faster during frying, and it releases four products of oxidation among which only one possesses antioxidative properties—ethane-dimer α-T. Gamma-T is better for oil stabilization during frying as, in its oxidation, more stable products of antioxidative properties are released (Kochhar, 2001). At high temperatures, γ-T is also a more efficient inhibitor of polymerization than α-T (Warner et al., 2003). In membranes, there is one molecule of α-T for every 2000 molecules of phospholipids. This fact explains the rate of peroxide radical reaction with α-T, that is, 10,000 times higher than the reaction between the radical and a fatty acid (Bramley et al., 2000). Tocopherols also have some prooxidative properties. During the reduction of Fe^{3+} to Fe^{2+} or Cu^{2+} to Cu^{1+}, they can stimulate the formation of hydroxyl radicals (Ross et al., 2003).

The mechanism of the reaction between carotenoids and FR depends on their oxidation–reduction potential, half-life, and concentrations. Trapping radicals can be based on donating the electron (carotene radical cation), addition reaction (radical carotene adducts), and addition of electron to the molecule that results in the formation of radical anion (Beutner et al., 2001). Apart from antioxidant properties, their prooxidative properties were also noted. The type of exposed properties depends on the oxygen concentration, carotenes themselves, color of carotenes, chemical structure (functional groups, number of linked bonds, acyclical structure of chains), synergy with other antioxidants, lipid matrix model, and exposure to light (Burton and Ingold, 1984). Similar to tocopherols, carotenoids are effective in quenching singlet oxygen. One molecule of β-carotene is able to quench more than 1000 molecules of singlet oxygen. Reaction rates of singlet oxygen quenching that are the kinetic determinants of antioxidant capacity are very high for carotenoids, and they amount to 10^{10} $M^{-1}s^{-1}$, whereas this value for α-T amounts to 2.7×10^{7} $M^{-1}s^{-1}$, and 10^{6}–10^{8} $M^{-1}s^{-1}$ for flavonoids and phenols (Beutner et al., 2001).

All phenol compounds have the right structure to be FR inactivators. The key feature for antioxidant properties is their hydrophobic and lipophilic character, which is very diverse within the group of phenol compounds due to their structure. Flavonoids and phenol compounds act as good quenching agents of $^{3}O_2$ (Ross et al., 2003). As antioxidants, they can act in various ways, among others, by direct reaction with FR, trapping free radicals, intensifying dismutation to compounds of lower reactivity, and inhibiting enzymes. They can also boost the activity of other oxidants. Derivatives of cinnamic acid have higher antioxidant activity than the derivatives of benzoic acid (Yanishlieva-Maslarova, 2001). The group -CH=CH–COOH is responsible for this higher activity. It ensures the capability of donating H, and the presence of –COOH group ensures the stability of the formed radical (Balasundram et al., 2006). The antioxidant activity of flavonoids is connected to the OH groups in the molecule. The processes based on trapping FR may rely on transferring the H-atom on the radical molecule (monohydroxyflavone), whereas for polyphenols it is transfer of the most weakly bonded H- atom on the radical molecule.

$$Fl-(OH)_n \rightarrow Fl-(OH)_{n-1}O^{\bullet} + H^{\bullet} \qquad (acidic\ pH)$$

$$Fl-(OH)_n \rightarrow Fl-(OH)_{n-2}O_2^{\bullet-} + H^{\bullet} + H^{+} \qquad (basic\ and\ neutral\ pH)$$

Phenol compounds with a bigger number of OH groups have more profound antioxidative properties (Van Acker et al., 1996).

3.6 ANTIOXIDANTS IN BIOLOGICAL SYSTEMS AND THEIR INFLUENCE ON THE PREVENTION OF HEALTH DISEASES

The system of defense from oxidation-caused damage is composed of antioxidants. They can be divided into three groups (Shi et al., 2001): *prevention antioxidants*—inhibiting or stopping the reaction of RS with biologically active compounds through the decomposition of hydroxygens and H_2O_2 without free radicals (e.g., catalase, glutathione peroxidase, phospholipid hydroperoxide, peroxidase), transition metal ion sequestration through chelatation (e.g., transferrin, lactoferrin, hemopexin, ceruloplasmin, albumin) and through quenching active oxygen (superoxide dismutase –(SOD), carotenoids, vitamin E). The second group is formed by *radical-scavenging antioxidants* that stop and inhibit free radical chain reactions (vitamin E, ubiquinol, carotenoids, flavonoids, vitamin C). The last group is composed of *eliminate and repair antioxidants* that remove the outcomes of reactions between RS and biomolecules.

Biological activity of -T and -T3 is absolutely independent from their antioxidative properties. Antioxidative properties of α-T result in lower concentrations of MDA in plasma, slower aging processes, improved mental processes, decreasing the amount of DNA damage and the frequency of tumor transformations, and also inhibition of the development of Alzheimer and Parkinson diseases (Pham and Plakogiannis, 2005). It also prevents the formation of potentially cancer-causing asymmetrical β-carotene breakdown products (β-apo-carotenoids) (Russell, 2002). Tocopherols are used in prophylactic treatments of cardiovascular disease as they prevent the formation of nitric oxide that can lead to endothelial dysfunctions. Both α- and γ-T are capable of peroxynitrite inactivation (Bloodsworth et al., 2000). The formed nitric dioxide nitrates proteins causing nitrosative stress that is equally damaging to oxidative stress. Only γ-T occurs in food naturally. It binds irreversibly formed, active nitric dioxide. The product of this reaction is 5-NO_2-γ-tocopherol or tocored (2,7,8-trimethyl-2(4,8,12-trimethyltridecyl)-5,6-chromaquinone) oxidation products of tocopherols (Christen et al., 1997). Gamma-T prevents nitrating DNA bases and free and protein-bound tyrosine. It is this homolog of vitamin E that slows down the development of atherosclerosis. Used in high doses, α-T replaces γ-T from the plasma and from other tissues excluding its antiatherosclerosis properties. Epidemiologic studies confirm that the chemopreventive properties of α-T are stronger than those of γ-T (Campbell et al., 2003).

Tocopherols apart from antioxidant functions possess a whole array of the so-called *nonantioxidative functions* used in stabilizing the structure of biological membranes. Alpha-T, similar to cholesterol, affects the membrane liquidity, and it can affect its permeability for small ions and molecules. Vitamin E can also act as an antiproliferation agent by affecting the signal transduction with protein kinase C (PKC). Particularly, α-T slows down the aforementioned kinase, growth of some specific cells, and the transcription of CD36 and the genes coding for procollagen. The activation was also observed in relation to protein phosphatase 2A (PP_2A) and the expression of other genes (connective tissue growth factor) (Azzi et al., 2001). The most recent investigations showed the role of tocotrienols, and especially alpha-T3, in the protection of neurons through the mechanism linked to their antioxidative properties. Diet supplementation based on these compounds ensures the immune system protection (Khanna et al., 2003). The role of tocotrienols in the suppression of breast gland epithelial cells CL-S1 as an agent inhibiting breast cancer has been investigated as well (Shah and Sylvester, 2005). All forms of compounds consisting of vitamin E can prevent the formation of mutagenic aldehydes from lipid peroxides (Campbell et al., 2003). It was shown that vitamin E can suppress the transcription of approximately 30 genes. Some of them are very important in the development of atherosclerosis, for example, VCAM-1gene in the human endothelium vascular cells and on the surface of macrophages. Vitamin E can increase the expression of the gene coding for the protein responsible for tocopherol transport (α-Tocopherol Transport Protein) in rats, so it contributes to maintaining the right level of vitamin E in the plasma (Azzi and Zingg, 2006).

Among carotenoids, β-carotene has the highest activity as provitamin A. Lycopene, lutein, zeaxantine, and other compounds also play an important role in biological systems (Fraser and Bramley, 2004). In human blood, carotenoids migrate bound to lipoproteins. They can also be found in the adipose tissue and cellular membranes. They can also bind to some proteins by hydrogen (Zhang and Omaye, 2001). In the animal world, carotenoids occur in high concentrations as dyes fulfilling certain functions; for example, they take part in the process of seeing. It was proved that β-carotene inhibits the oxidation of lipids and protects DNA by protecting it from FR attack.

Vitamin C takes part in many biochemical transformations in the organism: the synthesis of hormones and neurotransmitters in biochemical reactions in the brain and in the immune system. It plays a very important role in the process of aging, as it takes part both in collagen synthesis and in its protection from degradation by increasing the activity of hydrolases that gradually lose their activity as the organism ages. Vitamin C controls the activity of lysosomal phosphatases taking part in the metabolism of proteoglycans containing sulfur that affect the processes of senile osteoporosis (Mates et al., 1999). Vitamin C optimizes the expression of the genes coding for superoxide dismutase and catalase. It also reduces the expression of adhesive molecules in monocytes and expresses cytoprotective properties. It improves the function of endothelium in coronary arteries in patients with high blood pressure and diabetes.

Polyphenol compounds of a different nature, such as quercetin, catechins, phenolic acids, resveratrol, and hydroxytyrosol, are very significant in the prevention of cardiovascular, and other heart diseases. There are some other mechanisms of significant importance that should be mentioned: the ability of limiting inflammatory processes, adhesion, and aggregation of thrombocytes, and vascular smooth muscle proliferation (Lo et al., 2007). In patients suffering from atherosclerosis with higher expression of adhesive molecules ICAM-1, VCAM-1, P- and E-selectin, a significant decrease in the expression of these proteins was observed after treating them with grape seed extract (Kalin et al., 2002). Polyphenols in red wine affect thrombocyte aggregation profoundly. In many investigations both *in vivo* and *ex vivo*, it was proved that polyphenols in red wine protect lipoproteins from oxidation (Rasmussen et al., 2005). The protective effect of phenolic compounds present in olive oil on lipid oxidative damage seems to be established. In recent years, these phenols have been the focus of intensive research in relation to individuals prone to oxidative stress. The efforts aim at identifying the population groups in which ingestion of phenolic compounds from olive oil could provide the greatest benefits (Covas et al., 2009).

Vitamin E, ascorbic acid, carotenoids and enzymes (e.g., glutathione peroxidase), polyphenols, and proteins compose a biological net of antioxidants acting against reactive species (Packer, 1992).

3.7 CONCLUDING REMARKS

Radicals and other reactive species play an important role in biological and food systems. Antioxidants may limit or retard the development of various diseases listed as "civilization diseases" that are inextricably linked to the presence of the reactive species in biological and food systems. This explains why so many researchers, from biologists and geneticists to dietitians, food nutrition specialists, and medical doctors, are so deeply involved in research related to this matter.

The diverse character of reactive species explains the complexity of their reactivity, their role in oxidative stress-associated processes, and their response to changes caused by physical factors. The effectiveness of antioxidants in the human organism is dependent on their bioavailability, which can vary in individuals. The amount of active factors that are delivered to the tissues depends on the entire food chain—from raw material to the final product being consumed. Plant and animal sources are processed for obtaining particular food products. In those processes, antioxidants—the bioactive substances present in food—are destroyed. Proper procedures are applied to improve the food quality and to maintain the antioxidative potential, since antioxidants are expected to act efficiently only when they are delivered to the target tissues in the correct chemical structure and in the right concentration.

REFERENCES

Afanas'ev, I.B. 1989. Deprotonation and deprotonation-oxidation reactions by superoxide ion. In: *Superoxide Ion: Chemistry and Biological Implications*, ed. I.B. Afanas'ev, 63–102. Boca Raton, FL: CRC Press.

Anggard, E. 1994. Nitric oxide: Mediator, murder and medicine. *The Lancet* 343:1199–206.

Arora, A., Muraleedharan, G.N., and G.M. Strasburg. 1998. Antioxidant activities of isoflavones and their biological metabolites in a liposomal system. *Arch. Biochem. Biophys.* 356:133–41.

Asmus, K-D., and M. Bonifacic. 2000. Free radical chemistry. In: *Handbook of Oxidants and Antioxidants in Exercise*, ed. C.K. Pen, L. Packer, and O. Hänninen, 3–45. Amsterdam: Elsevier Sciences.

Azzi, A., Breyer, I., Feher, M., and et al. 2001. Nonantioxidant function of alpha-tocopherol in smooth muscle cells. *J. Nutr.* 131:378S–81S.

Azzi, A., and A. Stocker. 2000. Vitamin E: Non-antioxidant roles. *Prog. Lipid Res.* 39:231–55.

Azzi, A., and J-M. Zingg. 2006. Vitamin E: Textbooks require updating. *Biochem. Mol. Biol. Edu.* 33:184–187.

Balasundram, N., Sundram, K., and S. Samman. 2006. Phenolic compounds in plant and agri-industrial by-products: antioxidant activity, occurrence and potential uses. *Food Chem.* 99:191–203.

Basu-Modak, S., and R.M. Tyrrell. 1993. Singlet oxygen: A primary effector in the ultraviolet A/near-visible light induction of the human heme oxygenase gene. *Cancer Res.* 53:4505–4510.

Beutner, S., Bloedorn, B., Frixel, S., Blanco, I.H., Hoffmann, T., Martin, H-D., and et al. 2001. Quantitative assessment of antioxidant properties of natural colorants and phytochemicals: carotenoids, flavonoids, phenols and indigoids. The role of β-carotene in antioxidant functions. *J. Sci. Food Agric.* 81:559–68.

Bilska, A., and L. Wlodek. 2005. Lipoic acid—the drug of the future? *Pharm. Rep.* 57:570–77.

Bloodsworth, A., O'Donnell, V., and B.A. Freeman. 2000. Nitric oxide regulation of free radical and enzyme mediated lipid and lipoprotein oxidation. *Arterioscler. Thromb. Vasc. Biol.* 20:1707–15.

Boskou, D. 2006. Sources of natural phenolic antioxidants. *Trends Food Sci. Technol.* 17:505–12.

Bramley, P.M., Elmadfa, I., Kafatos, A., Kelly, F.J., Manios, Y., Roxborough, H.E. et al. 2000. Vitamin E. *J. Sci. Food Agric.* 80:913–38.

Burton, G.W. 1994. Vitamin E: molecular and biological function. *Proc. Nutr. Soc.* 53:251–62.

Burton, G.W., and K.U. Ingold. 1984. β-carotene: An usual type of lipid antioxidant. *Science* 244:569–73.

Bynoe, L.A., Gottsch, J.D., Pou S., and G.M. Rosen. 1992. Light-dependent generation of superoxide from human erythrocytes. *Photochem. Photobiol.* 56:353–56.

Cabelli, D.E., and B.H.J. Bielski. 1983. Kinetics and mechanism for the oxidation of ascorbic acid/ascorbate by HO_2/O_2^- (hydroperoxyl/superoxide) radicals. A pulse radiolysis and stopped-flow photolysis study. *J. Phys. Chem.* 87:1809–12.

Cadens, E., and K. Davies. 2000. Mitochondrial free radical generation, oxidative stress, and aging. *Free Radic. Biol. Med.* 29:222–30.

Campbell, S., Stone, W., Whaley, S. et al. 2003. Development of gamma-tocopherol as a colorectal cancer chemopreventive agent. *Crit. Rev. Oncology/Hematology* 47:249–59.

Cao, G., Sofic, E., and R.L. Prior. 1996. Antioxidant capacity of tea and common vegetables. *J. Agric. Food Chem.* 44:3426–31.

Chaiyasit, W., Elias, R.J., Mcclements, D.J., and E.A. Decker. 2007. Role of physical structures in bulk oils on lipid oxidation. *Crit. Rev. Food Sci. Nutr.* 47:299–317.

Choe, E., and D.B. Min. 2006. Chemistry and reactions of reactive oxygen species in food. *Crit. Rev. Food Sci. Nutr.* 46:1–22.

Christen, S., Woodall, A.A., Shigenaga, M.K., Southwell-Keely, P.T., Duncans, M.W., and B.N. Ames. 1997. γ-Tocopherol traps mutagenic electrophiles such as NO_x and complements α-tocopherol: Physiological implications. *Proc. Natl. Acad. Sci. USA* 94:3217–22.

Covas, M-I., Khymenets, O., Fito, M., and R. Dela Tore. 2009. Bioavailability and antioxidant effect of olive oil phenolic compounds in humans. In: *Olive Oil, Minor Constituents and Health,* ed. D. Boskou, 109–28. Boca Raton, FL: CRC Press.

Decker, E.A. 1998. Antioxidant mechanisms. In: *Food lipids. Chemistry, Nutrition and Biotechnology*, ed. C.C. Akoh, and D.B. Min, 397–421. New York: Marcel Dekker.

Denisov, E.T., and I.B. Afanas'ev. 2005. *Oxidation and Antioxidants in Organic Chemistry and Biology*. Boca Raton, FL: Taylor & Francis.

Dotan, Y., Lichtenberg, D., and I. Pinchuk. 2004. Lipid peroxidation cannot be used as a universal criterion of oxidative stress. *Prog. Lipid Res.* 43:200–27.

Dreosti, I.E. 2000. Antioxidant polyphenols in tea, cocoa, and wine. *Nutrition* 16:692–94.

Eitenmiller, R., and J. Lee. 2004. *Vitamin E—Food Chemistry, Composition and Analysis*. New York: Marcel Dekker.

Frankel, E.N., Huang, S.W., Kanner, J., and J.B. German. 1994. Interfacial phenomena in the evaluation of antioxidants: Bulk oils versus emulsions. *J. Agric. Food Chem.* 42:1054–59.

Fraser, P.D., and P.M. Bramley. 2004. The biosynthesis and nutritional uses of carotenoids. *Prog. Lipid Res.* 43:228–65.

Garrel, C., and M. Fontecave. 1995. Nitric oxide: chemistry and biology. In: *Analysis of Free Radicals in Biological Systems*, ed. A.E. Favier, J. Cadet, B. Kalyanaraman, M. Fontecave, and J.-L. Pierre, 21–37. Basel: Birkhäuser Verlag.

Gordon, M.H. 2001. The development of oxidative rancidity in foods. In: *Antioxidants in Food. Practical Applications*, ed. J. Pokorny, N. Yanishlieva, and M. Gordon, 7–21. Boca Raton: CRC Press.

Gramza-Michalowska, A., Korczak, J., and M. Hes. 2007. Purification process influence on green tea extracts' polyphenol content and antioxidant activity. *Acta Sci. Pol., Technol. Aliment.* 6:41–8.

Hall, C. 2001. Sources of natural antioxidants: oilseeds, nuts, cereals, legumes, animal products and microbial sources. In: *Antioxidants in Food*, ed. J. Pokorny, N. Yanishlieva, and M. Gordon, 156–209. Cambridge: CRC Press.

Halliwell, B. 1994. Free radicals, antioxidants, and human disease: Curiosity, cause, or consequence? *Lancet* 344:721–24.

Halliwell, B., and J.M. Gutteridge. 1984. Oxygen toxicity, oxygen radicals, transition metals and disease. *Biochem. J.* 219:1–14.

Harborne, J.B., and C.A. Williams. 2000. Advances of flavonoid research since 1992. *Phytochemistry* 55:481–504.

Higdon, J.V., and B. Frei. 2002. Vitamin C: An introduction. In: *The Antioxidant Vitamins C and E*, ed. L. Packer, M.G. Traber, K, and Kraemer, B. Frei, 1–16. Santa Barbara: AOCS Press.

Judde, A., Villeneuve, P., Rossignol-Castera, A., et al. 2003. Antioxidant effect of soy lecithins on vegetable oil stability and their synergism with tocopherols. *J. Am. Oil Chem. Soc.* 80:1209–15.

Kähkönen, M.P., Hopia, A.I., and M. Heinonen. 2001. Berry phenolics and their antioxidant activity. *J. Agric. Food Chem.* 49:4076–82.

Kalin, R., Righi, A., Del Rosso, A., Bagchi, D., Generini, S., Guiducci, S. et al. 2002. Activin, a grape seed-derived proanthocyanidin extract reduces plasma levels of oxidative stress and adhesion molecules (ICAM-1, VCAM-1 and E-selectin) in systemic sclerosis. *Free Radic. Res.* 36:819–25.

Kanner, J., and I. Rosenthal. 1992. An assessment of lipid oxidation in foods. *Pure Appl. Chem.* 64:1959–64.

Khanna, S., Roy, S., Ryu, H., Bahadduri, P., Swaan, P.W., Ratan R.R. et al. 2003. Molecular basis of vitamin E action: Tocotrienol modulates 12-lipoxygenase, a key mediator of glutamate-induced neurodegradation. *J. Biol. Chem.* 278:43508–15.

Kochhar, S.P. 2001. The composition of frying oils. In: *Frying—Improving Quality*, ed. J.B. Rossell, 87–114. Boca Raton, FL: CRC Press.

Kolakowska, A. 2003. Lipid oxidation in food systems. In: *Chemical and Functional Properties of Food Lipids*, ed. Z.E. Sikorski, and A. Kolakowska, 133–66. Boca Raton, FL: CRC Press.

Kris-Etherton, P.M., Hecker, K.D., Bonanome, A, Coval, S.M, and A.E. Binkoski. 2002. Bioactive compounds in foods: Their role in the prevention of cardiovascular disease and cancer. *Am. J. Med.*113:71–8.

Lampart-Szczapa, E., Siger, A., Trojanowska, K. et al. 2003. Chemical composition and anti-bacterial activities of lupin seeds extracts. *Nahrung/Food*, 47:286–90.

Lo, H-M., Hung, C-F., Huang, Y-Y., and W-B., Wu. 2007. Tea polyphenols inhibit rat vascular smooth muscle cell adhesion and migration on collagen and laminin via interference with cell-ECM interaction. *J. Biomed. Sci.* 14:637–45.

Mates, J.M., Perez-Gomez, C., and I.N. De Castro. 1999. Antioxidant enzymes and human diseases. *Clin. Biochem.* 32:595–603.

May, J.M. 1999. Is ascorbic acid an antioxidant for the plasma membrane? *FASEB J.* 13:995–1006.

Min, D.B. 1998. Lipid oxidation of edible oil. In: *Food lipids. Chemistry, Nutrition and Biotechnology,* ed. C.C. Akoh, and D.B. Min, 283–95. New York: Marcel Dekker.

Mortensen, A., and L.H. Skibsted. 1997. Importance of carotenoid structure in radical scavenging reactions. *J. Agric. Food Chem.* 45:2970–7.

Nogala-Kalucka, M. 2002. Fat-soluble vitamins. In: *Chemical and Functional Properties of Food Lipids*, ed. Z.E. Sikorski, and A. Kolakowska, 109–33. Boca Raton, FL: CRC Press.

Nogala-Kalucka, M., Gogolewski, M., and E. Swiatkiewicz. 1993. Changes in the composition of tocopherols and fatty acids in postdeodorisation condensates during refining of various oils. *Fat Sci. Technol.* 95:144–7.

Nogala-Kalucka, M., Gogolewski, M., Lampart-Szczapa, E., Jaworek, M., Siger, A., and A. Szulczewska. 2003. Determination of vitamin E active compounds as biological antioxidants occurring in oilseeds of the selected rape varieties. *Oilseed Crops* 24:587–96.

O'Donnele, V.B., Chumley, P.H., Hogg, N., Bloodsworth, A., Darley-Usmar, V.M., and B.A. Freeman. 1997. Nitric oxide inhibition of lipid peroxidation. Kinetics of reaction with lipid peroxyl radicals and comparison with tocopherol. *Biochemistry* 36:15216–23.

Ohshima, H., Yoshie, Y., Auriol, S., and I. Gilibetr. 1998. Antioxidant and prooxidant actions of flavonoids: Effects on DNA damage induced by nitric oxide, peroxynitrite and nitroxyl anion. *Free Radic. Biol. Med.* 25:1057–65.

Packer, L. 1992. Interactions among antioxidants in health and disease. Vitamin E and the redox cycle. *Proc. Soc. Exp. Biol. Med.* 200:271–6.

Parry, J., Su, L., Luther, M. et al. 2005. Fatty acid composition and antioxidant properties of cold-pressed marionberry, boysenberry, red raspberry and blueberry seed oils. *J. Agric. Food Chem.* 53:566–73.

Pham, D.Q., and R. Plakogiannis. 2005. Vitamin E supplementation in Alzheimer's disease, Parkinson's disease, Tardive dyskinesia, and cataract. *Ann. Pharmacother.* 39: 2065–71.

Porter, N.A. 1986. Mechanisms for the autoxidation of polyunsaturated lipids. *Acc. Chem. Res.* 19:262–8.

Porter, N.A., Caldwell, S.E., and A.M. Karen. 1995. Mechanisms of free radical oxidation of unsaturated lipids. *Lipids* 40:227–90.

Porter, W.L. 1993. Paradoxical behavior of antioxidants in food and biological systems. *Toxicol. Ind. Health* 9:93–122.

Pryor, W.A. 1976. The role of free radical reactions in biological systems. In: *Free Radicals in Biology*, ed. W.A. Pryor, 1–50. New York: Academic Press.

Qian, S.Y., and G.R. Buettner 1999. Iron and dioxygen chemistry is an important route to initiation of biological free radical oxidations: An electron paramagnetic resonance spin trapping study. *Free Radic. Biol. Med.* 26:1447–56.

Rasmussen, S.E., Frederiksen, H., Strintze-Krogholm, K., and L. Poulsen. 2005. Dietary proanthocyanidins: Occurrence, dietary intake, bioavailability and protection against cardiovascular disease. *Mol. Nutr. Food Res.* 49:159–74.

Rice-Evans, C.A. 1994. Formation of free radicals and mechanism of action in normal biochemical processes and pathological states. In: *Free Radical Damage and Its Control,* ed. C.A. Rice-Evans, and R.H. Burdon, 131–51. Amsterdam: Elsevier Sciences.

Rice-Evans, C.A., Miller, N.J., and G. Paganga. 1995. Structure-antioxidant activity relationship of flavonoids and phenolic acids. *Free Radic. Biol. Med.* 20:933–56.

Riesz, P., Berdahl, D., and C.L. Christman 1985. Free radical generation by ultrasound in aqueous and nonaqueous solutions. *Environ. Health Persp.* 64:233–52.

Ross, L., Barclay, C., and M.R. Vinqvist. 2003. Phenols as antioxidants. In: *The Chemistry of Phenols,* ed. Z. Rappoport, 839–908. Chichester: John Wiley & Sons.

Rubbo, H., Trostchansky, A., and V.B. O'Donnell. 2009. Peroxynitrite-mediated lipid oxidation and nitration: mechanisms and consequences. *Arch. Biochem. Biophys.* 484:167–72.

Rudolph, V., Schopfer, F.J., Khoo, N.K.H., Rudolph, T.K., Cole, M.P., Woodcock, S.R. et al. 2009. Nitro-fatty acid metabolite: saturation, desaturation, beta-oxidation, and protein adduction. *J. Biol. Chem.* 284:1461–73.

Russell, R.M. 2002. Beta-carotene and lung cancer. *Pure Appl. Chem.* 74:1461–7.

Ryan, T.P., and T.W. Petry. 1993. The effects of 21-aminosteroides on the redox status of iron in solution. *Arch. Biochem. Biophys.* 300:699–704.

Schneider, C. 2005. Review: Chemistry and biology of vitamin E. *Mol. Nutr. Food Res.* 49:7–30.

Schnitzer, E., Pinchuk, I., and D. Lichtenberg. 2007. Peroxidation of liposomal lipids. *Eur. Biophys. J.* 36:499–515.

Shah, S., and P.W. Sylvester. 2005. Gamma-tocotrienol inhibits neoplastic mammary epithelial cell proliferation by decreasing Akt and nuclear factor kB activity. *Exp. Biol. Med.* 230:235–41.

Shi, H., Noguchi, N., and E. Niki. 2001. Introducing natural antioxidants In: *Antioxidants in Food*, ed. J. Pokorny, N. Yanishlieva, and M. Gordon, 147–58. Cambridge: CRC Press.

Sikora, E., Cieślik, E., and K. Topolska. 2008. The sources of natural antioxidants. *Acta Sci. Pol., Technol. Aliment.* 7:5–17.

Siger, A., Nogala-Kałucka, M., Lampart-Szczapa, E., and A. Hoffmann. 2005. Antioxidant activity phenolic compounds of selected cold-pressed and refined plant oils. *Oilseed Crops* 26:549–59.

Siger, A., Nogala-Kałucka, M., and E. Lampart-Szczapa. 2008. The content and antioxidant activity of phenolic compounds in cold-pressed plant oils. *J. Food Lipids* 15:137–49.

Stadman, E.R., and R.L. Levine 2000. Protein oxidation. *Ann. N.Y. Acad. Sci.* 899:191–208.

Sugihara, N., Arakawa, T., Ohnishi, M., and K. Furuno. 1999. Anti- and pro-oxidative effects of flavonoids on metal-induced lipid hydroperoxide-dependent lipid peroxidation in cultured hepatocytes loaded with alpha-linolenic acid. *Free Radic. Biol. Med.* 27:1313–23.

Turrens, J.F. 2003. Mitochondrial formation of reactive oxygen species. *J. Physiol.* 522:335–44.

Warner, K., and E.N. Frankel 1987. Effects of beta-carotene on light stability of soybean oil. *J. Am. Oil Chem. Soc.* 64:213–8.

Warner, K., Neff, W.E., and F.J. Eller. 2003. Enhancing quality and oxidative stability of aged fried food with gamma-tocopherol. *J. Agric. Food Chem.* 51:623–7.

Wolff, S.P., and R.T. Dean. 1986. Fragmentation of proteins by free radicals and its effect on their susceptibility to enzymic hydrolysis. *Biochem. J.* 234:399–403.

Wurtzel, E.T. 2004. Genomics, genetics and biochemistry of maize carotenoid biosynthesis. In: *Secondary Metabolism in Model System: Recent Advances in Phytochemistry, 38*, ed. J.T. Romeo, 85–110. Amsterdam: Elsevier.

Van Acker, S.A.E., Van Den Berg, D-J., Tromp, M.J.L., Griffioen, D.H., Van Bennekom, W.P., Van Der Vijgh, W.J.F. et al. 1996. Structural aspects of antioxidant activity of flavonoids. *Free Radic. Biol. Med.* 3:331–42.

Yanishlieva, V.N., Marinova, E., and J. Pokorny. 2006. Natural antioxidants from herbs and spices. *Eur. J. Lipid Sci. Technol.* 108:776–93.

Yanishlieva-Maslarova, N.V. 2001. Inhibiting oxidation. In: *Antioxidants in Food: Practical Applications*, ed. J. Pokorny, N. Yanishlieva, and M. Gordon, 22–70. Boca Raton, FL: CRC Press.

Yu, B.P. 1994. Cellular defenses against damage from reactive oxygen species. *Pharmacol. Rev.* 74:139–62.

Zhang, P., and S.T. Omaye. 2001. Antioxidant and prooxidant roles for beta-carotene, alpha-tocopherol and ascorbic acid in human lung cells. *Toxicol. In Vitro* 15:13–24.

Zhu, B.Z., and B. Frei. 2002. Biochemical and physiological interactions of vitamin C and iron: Pro- or antioxidant? In: *The Antioxidant Vitamins C and E,* ed. L. Packer, M.G. Traber, K. Kraemer, and B. Frei, 32–49. Santa Barbara: AOCS Press.

4 Changes in Nutrients, Antinutritional Factors, and Contaminants at Frying Temperatures

Jan Pokorný and Jana Dostálová

CONTENTS

4.1 INTRODUCTION

Both deep-fat frying and pan frying systems usually consist of two phases: frying oil and fried material. Changes during deep-fat frying are different from those seen in other cooking techniques (Bognar, 1998). The frying temperature and frying time are of great importance Another factor is the composition of fried food (foods of

TABLE 4.1
Processes and Changes during Frying

Type of Change
Physical Processes
Loss of moisture and volatiles
Migration between frying oil and fried food
Chemical Processes
Dehydration
Oxidation and reactions of oxidation products
Interactions between various components

plant origin behave differently from foods of animal origin), and that of frying oil (García-Arias et al., 2003). Several factors that influence one another affect food frying (Pokorný, 1998).

At high temperatures (130°C–200°C) and for a short frying time (5–10 min), only the surface layer of fried food is affected. The inner layers are rarely heated up to 100°C during the normal frying process. The fried material nearly always contains rather high, but sometimes only small, amounts of water. Water present in the surface layer is converted into steam, and causes hydrolysis of both frying oil and various components of fried food. Hydrolytic products are usually more reactive than the original structures. Atmospheric oxygen dissolved in fried material or penetrated by diffusion through frying oil causes oxidation. The effect of air is more important with pan frying than with deep-fat frying. The nutritive value of fried food is changed both by physical and chemical processes (Table 4.1).

The physical processes involve loss of water, especially from the surface layer of fried food (proteins are denatured during heating); migration of fat into fried food and in the reverse direction from fried food migration into frying fat (frying oil always contains some typical fish lipids after frying fish); and migration of other substances from fried food, such as essential oils or other volatiles.

Changes caused by chemical processes involve dehydration of fried food components, which causes structural rearrangement of proteins and other molecules; oxidation reactions, limited only by moderate access of air by diffusion from the atmosphere; and interaction between food, frying oil, and their reaction products. This point will be considered later in this chapter. The maximum concentration of the chemically changed products is found at the interface between the two phases. The interaction products usually have a negative effect on nutritive value. On the other hand, the sensory value is usually much better than that of the original material.

Frying oil and its oxidation products interact almost exclusively in the surface layer of food components. In rare cases, original food contains some oxygen. The oxygen diffusion from the food surface into inner layers is usually too slow to substantially influence the composition during a relatively short time. Crust formation on the interface may prevent a reactive substance from coming in contact with frying oil oxidation products. Processes are different in a microwave oven combined with a grill, where the whole material is heated to the frying temperature. For this reason, the sensory profile character is different.

In order to analyze changes of the chemical composition accurately, it is desirable to remove the surface layer from the interior of fried food, and to weigh and analyze both fractions separately. Analytical data are then combined, taking into consideration the amounts of inner and surface layer.

4.2 CHANGES IN PROTEINS, PEPTIDES, AND AMINO ACIDS DURING FRYING

Proteins are important nutrients. They are always present in fried food in greater or lesser amounts. Their reactivity was reviewed by Gardner (1979). Amino acids and peptides are present in much smaller quantities in foods than proteins, but they are more reactive. Therefore, they are usually used up during frying by various interactions. The main reaction of proteins is denaturation. Most proteins denature at temperatures lower than 100°C, so the protein fraction denatures even in the inner layers. Denatured proteins are generally better digested than native proteins. The process is thus desirable from the standpoint of nutrition. Some biologically active proteins, such as enzymes, negatively affect the stability of food during storage. They are deactivated under frying conditions. Therefore, fried products are usually more stable on storage than raw foods.

Antinutritional proteins, such as avidin in egg white or trypsin inhibitors in legumes, are deactivated, at least partially. Trypsin inhibitors in tempeh, a fermented product from soybeans, or in other soybean products, were found to be completely free of the inhibitor in the course of deep-fat frying (Fukuda et al., 1989).

On the contrary, the soluble protein fraction decreases during tortilla chip preparation, and the process is accompanied by unfavorable changes in protein digestibility (Vivas-Rodríguez et al., 1990). Similar protein changes are even more pronounced in the case of sorghum products.

Substantial protein changes were observed upon frying falafel (bean dough), by degradation of the high-molecular-weight protein fraction (Hamza et al., 1987). *Papads* are a traditional Indian protein food. If digestibility is tested by pepsin and pancreatin in pan frying, moderately higher losses in protein take place than in the case of roasting or microwave cooking (Shashikala and Prakash, 1995). Changes are similar both in products from black gram (*Phaseolus mungo*) and green gram (*Phaseolus radiatus*).

Changes during the frying are small, but significantly different for fish species (Amin and Hainida, 2004). No significant difference was detected between the amino acid composition of fried and boiled fish. Battered catfish fried at 159°C had about the same protein quality (PER = protein efficiency ratio) as catfish roasted in a rotating hot air oven (Ibrahim and Unclesbay, 1986). Various fish species were cooked, smoked, or fried. Protein digestibility and NPU (= net protein utilization) were very satisfactory. Deep frying has not been found to significantly affect the protein fraction. It was about the same as with other cooking methods (Steiner-Asiedu et al., 1991). The effect of frying on protein digestibility depends on the type of food. It increases when frying mushrooms. It decreases when frying soft cheese. A combination of defrosting and frying showed that frying without previous defrosting resulted in lower quality fish fillet (Krbavčič and Barič, 2004).

The nutritional value is influenced mostly by reactions of individual amino acids (Table 4.2).

Lysine is the most reactive amino acid. Lysinoamino acids are thus formed (Figure 4.1), because lysine contains free primary amino group (in the 6- or ε-position). Average losses of amino acids bound in the protein of potatoes did not exceed about 7%. Losses of both total and available lysine in soy products were of the same order under common frying conditions (Fukuda et al., 1989). However, the losses were greater than during baking if free amino acids have been added.

Another heat-sensitive amino acid is tryptophan, because of the indole group present. Tryptophan losses range between 65.6% in pork and 46.2% in white chicken muscle, respectively (Ribarova et al., 1994). Its losses during deep frying

TABLE 4.2
Reactions of Amino Acids during Frying

Amino Acid	Type of Change
Lysine	Dehydration, condensation
Tryptophan	Thermolysis, condensation
Cysteine, cystine	Oxidation, dehydration, cleavage
Serine	Dehydration
Methionine	Oxidation, cleavage

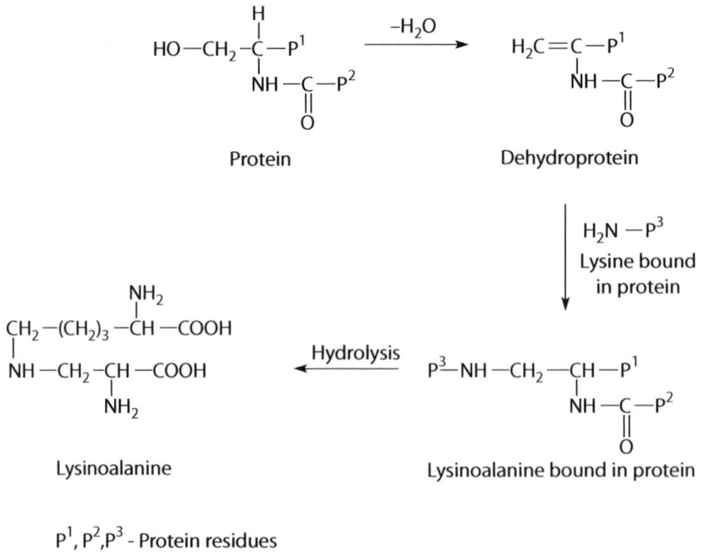

Protein $\xrightarrow{-H_2O}$ Dehydroprotein $\xrightarrow{H_2N-P^3 \text{ (Lysine bound in protein)}}$ Lysinoalanine bound in protein $\xrightarrow{\text{Hydrolysis}}$ Lysinoalanine

P^1, P^2, P^3 - Protein residues

FIGURE 4.1 Formation of lysinoanalanine.

were lower than during grilling or microwave heating. On the contrary, losses in turkey meat were higher during frying than during grilling or microwave heating.

Sulfur-containing amino acids are also unstable against high heating temperatures. Cysteine losses were significantly higher during frying than with other amino acids as the thiol group of cysteine is easily cleaved off. In the surface layer, where the temperature is very high (between 140°C and 180°C), bound or free serine and cysteine are converted into a protein intermediate; then they react with the 6-amino group of lysine to form lysinoalanine (Figure 4.1), an unavailable and potentially moderately toxic compound (Borowski at al., 1986).

The racemization of l-amino acids into their d-amino acid analogues is negligible under deep frying conditions, as the conversion would require higher temperature. From the nutritional standpoint, it is unfortunate that just those amino acids are attacked that are essential and often deficient in human nutrition.

4.3 FORMATION OF CARCINOGENS (MUTAGENS) FROM AMINO ACIDS

The formation of carcinogens from amino acids and proteins was intensively studied in frying oils, where their concentration increases only a little due to their low content (Table 4.3). Carcinogens in fried foods are more dangerous, but they may partially migrate into frying oil. The active carcinogens are heterocyclic conjugates formed from proteins or mixtures of proteins and carbohydrates at temperatures higher than 180°C (Table 4.3). They are mainly found in drippings, charred pieces of bread and batter, or in overheated pan-fried products.

Carcinogenic heterocyclic compounds were detected in fried meat and fish. Their precursor is creatine, which is converted into creatinine at high temperatures. Free amino acids, such as threonine, serine, phenylalanine, leucine, and tyrosine, are sensitive to the formation of carcinogens (Övervik et al., 1989). Different additions of creatine to model emulsions similar to sausages and meat balls increased the mutagenic

TABLE 4.3
Carcinogens and Mutagens Formation during Frying

Type of Compounds	The Most Important Representatives
Heterocyclic conjugates	Imidazoquinolines
	Imidazoxyquinolines
	Imidazopyridines
	Aminoimidazoquinolines
Nitrosamines	*N*-nitrosoproline
	N-nitrososarcosine
	N-nitrosodimethylamine
Amides	Acrylamide
Various other classes of compounds	Polycyclic aromatic hydrocarbons

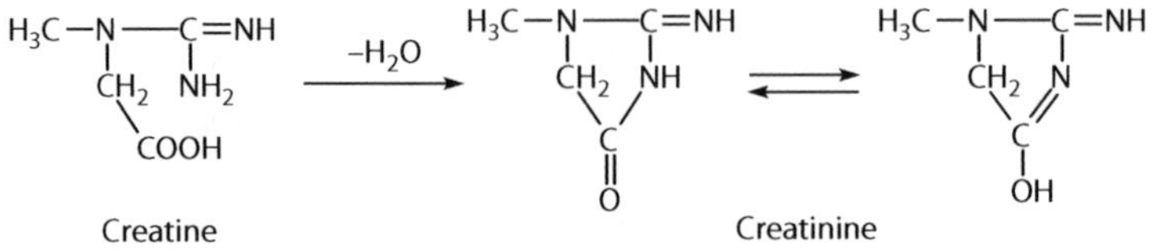

Creatine Creatinine

FIGURE 4.2 Transformation of creatine into creatinine.

Aminoimidazopyridines

R^1 = phenyl, PhIP

Aminoimidazoquinolines

R^1 = H, IQ

R^1 = CH_3, MeIQ

Aminoimidazoquinoxalines

R^1= H, R^2 = H, R^3 = H, IQx
R^1= CH_3, R^2 = H, R^3 = H, MeIQx
R^1= CH_3, R^2 = CH_3, R^3 = H, 4,8-DiMeIQx
R^1= H, R^2= H, R^3 = CH_3, 7-MeIQx
R^1= CH_3, R^2 = H, R^3 = CH_3, 4,7-DiMeIQx

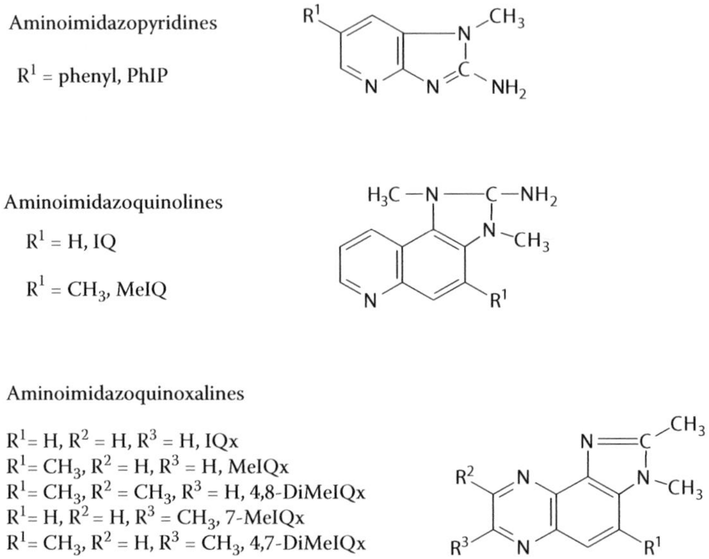

FIGURE 4.3 Mutagens formed in fried food.

activity of fried crust, after transformation of creatine to creatinine (Figure 4.2). It is possible that Maillard products are first formed by reaction with glucose (always present in small amounts in meat) with amino acids. These precursors of creatinine produce imidazoquinolines and imidazoquinoxaline mutagens (Figure 4.3). On the contrary, an excess of glucose and other reducing sugars blocks the transformation of creatine into creatinine, thus inhibiting the mutagenic activity of the fried product. Glycogen was observed by Olsson et al. (2002) to be the main precursor of mutagens in fried meat (obviously after hydrolysis to glucose). Liver and kidney, which are low in creatine, show nearly no mutagenic activity after frying (Chen et al., 1990).

DNA adducts were identified as an important source of mutagens in fried poultry (Brockstedt and Pfau, 1998). The addition of carbohydrates, especially starch, inhibited mutagen formation on frying (Persson et al., 2004).

Levels of heterocyclic amines decrease by deep frying in virgin olive oil (which is rich in antioxidants) compared with refined olive oil, but the addition of rosemary extract, which is a powerful inhibitor of lipid oxidation, had no effect (Persson et al.,

2003). On the contrary, extracts from black or green tea leaves inhibited the formation of mutagens in fried beef patties (Weisburger et al., 2002). Organic sulfur compounds, including cysteine and cystine, reduced the mutagen content in frying beef patties by up to 70% (Han-Seung et al., 2002) by reaction with free radicals.

The most important carcinogenic heterocyclic compounds belong to the classes of imidazopyridines, imidazoquinoxalines, and imidazoquinolines (Figure 4.3). Among these aminoimidazo azarenes, the mutagenic activity increases with number of hetero atoms, with the methyl substitution both on nitrogen and carbon atoms (Hatch et al., 1991). Creatine is responsible for the formation of the imidazole ring. The amounts of all the foregoing types of carcinogens rise with frying time and frying temperature, especially at temperatures exceeding 180°C–200°C, as observed in frying beef patties and in model experiments. During frying of lean ground pork in a pan with no fat added, the mutagenic activity was doubled with each 50°C increase in frying temperature (Nielsen et al., 1988).

The mutagenic activity is, naturally, maximum in the crust, where the exposure of reactants to heat is very intensive. Flavor substances are produced at high temperature, too. Therefore, it is necessary to choose a compromise between the flavor intensity and mutagenic activity. When lean pork was pan-fried in different fats, the mutagenic activity was nearly independent of fat unsaturation. The mutagenicity was very pronounced, however, in samples fried in fat than in samples fried without fat addition. In beef patties with different fat content, the relationship between fat content and mutagenic activity was different in the case of different frying fats. Mutagens formed in food have low polarity, so they become liposoluble. They can migrate into frying oil in the case of deep-fat frying, thus increasing its mutagenicity.

The most intensive research has been performed on pan-fried food. During frying in a pan, the safe temperature can be exceeded more easily than in deep-fried samples. Of course, small amounts of carcinogens are produced even during deep-fat frying if the temperature exceeds 180°C. In a model system of glycine, creatinine, and glucose heated to 180°C, several aminoimidazoquinoxalines were detected.

As the reaction involves some free radical steps, traces of iron, copper, and manganese come into consideration. Traces of transient valency metals or other prooxidants contribute to mutagen formation. Oxidized fat free of iron ions had little effect on mutagenic activity of the fried product, but ferrous sulfate substantially increased the rate of mutagen formation (Johansson and Jägerstad, 1993). The catalytic action of heme pigments cannot be excluded. Several mutagens were detected in fried poultry meat; the main precursors were creatine, amino acids, and carbohydrates (Murkovič et al., 1997). Beef burgers containing various levels of fat were deep-fried in a pan fryer. Several antioxidants were present in all samples at levels up to 2.3 ng/kg (Persson et al., 2003 and 2003a). The content of carcinogens was much higher (up to 9.27 ng/kg) in meat products fried at 150°C–230°C (Abdulkarim and Smith, 1998). Various meat products available in the market were fried at 150°C–230°C. A study of meat products and consumption in New Zealand showed daily consumption of 1.0 μg carcinogens per person and day; reduction of pan frying frequency and grilling is suggested to reduce the mutagen intake (Thomson et al., 1996). The same recommendation was made by another laboratory (Robbana-Bornat et al., 1996).

The highest values of heterocyclic amines were determined in Italian sausage and smoked sausage (Abdulkarim and Smith, 1998).

The effect of antioxidants is controversial. Natural concentrations of tocopherols and tocotrienols in fried food had no evident effect. Similarly, *tert*-butyl hydroxytoluene and polyphosphate had no effect (Chen, 1990). Other antioxidants, such as *tert*-butyl hydroxyanisole, propyl gallate, citrates, or tocopherols, added in addition to those naturally present inhibited the formation of mutagens.

Nitrosamines belong to another group of mutagens (Figure 4.4). They are formed during frying of meat, fish, and seafood. Their precursors are *N*-nitroso derivatives formed by interaction of amine groups present in proteins with nitrites used for curing meat. Sucuk is a typical Turkish food containing nitrite, but no nitrosamines were detected in cooked samples. There are obviously other pathways of nitrite degradation than transformation into nitrosamines. Nitrites are usually mixed with nitrates, which are reduced to nitrites in the body. Typical precursors include secondary amines, such as *N*-nitrosoproline and *N*-nitrososarcosine. Smoking meat and meat products increased the content of nitrosoproline. Frying of boiled sausages and ham further enhance the formation of carcinogens.

N-Nitrosodimethylamine was formed from minced Alaskan pollock and surimi for the preparation of frankfurters. The content of *N*-nitrosodimethylamine increased much more rapidly during boiling and frying than during microwave cooking (Fiddler et al., 1992). Other nitrosoamines were detected in fried bacon (Gloria et al., 1997), such as *N*-nitrosodimethylamine, *N*-nitrosodibutylamine, *N*-nitrosopyridine,

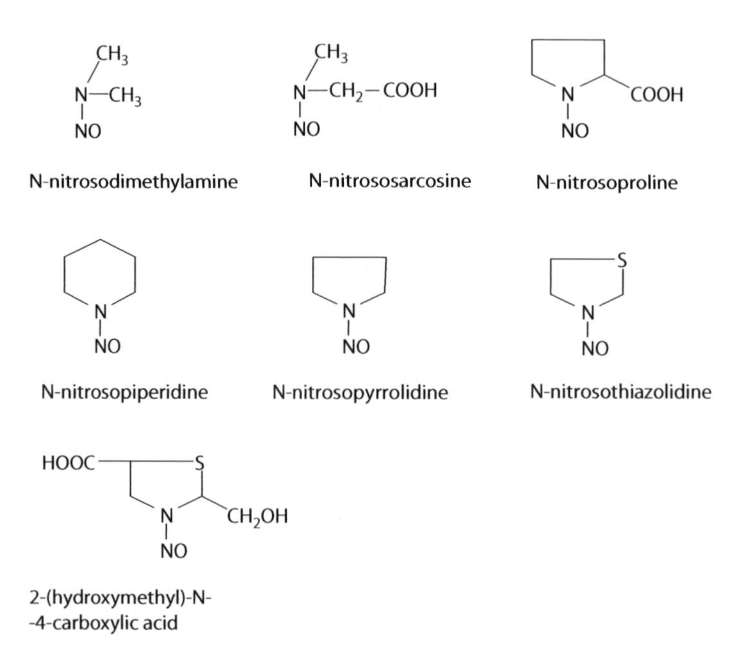

FIGURE 4.4 Nitrosamines formed during frying food.

N-nitrosopiperidine, and *N*-nitrosothiazolidine. Substantially higher levels of *N*-nitrosopyrrolidine were found in bacon fried in a pan than in microwave-cooked samples.

Several nitrosamines are volatile, so they are not present in the final product. Some nitrosamines are nonvolatile, such as 2-hydroxymethyl-*N*-nitrosothiazolidine, and are detected in cured and smoked meat, fish, and seafood, but their formation during frying was only occasional. The mutagenic activity of cured bacon is significantly reduced by adding α-tocopherol prior to frying (Buckley et al., 1989).

4.4 FORMATION OF ACRYLAMIDE DURING FRYING

Acrylamide (Figure 4.5) is another undesired compound formed during frying. The carcinogenicity of acrylamide is low, but its safe level is not yet quite certain. Therefore, it is important to study the various factors influencing its formation in order to minimize the content. Three mechanisms of its formation have been reviewed (Gertz and Klostermann, 2002).

Acrylamide is not present in raw food, but is formed during heating. It was detected in baked or fried foods rich in carbohydrates and fats (Giese, 2002). It can migrate from fried food into frying oil. The presence of acrylamide in fried potato products was of major interest.

Many factors affect the formation of acrylamide. The composition of fried food is of great importance. A high content of reducing sugars and of amino acids resulted in a high content of acrylamide (He et al., 2007). Reducing sugars are necessary precursors. Soaking and blanching potatoes before frying reduced the amount of acrylamide because glucose was leached out. In an acidic medium, the content of acrylamide decreased (Matthäus et al., 2004; Kita et al., 2004).

Another important factor is amino acids, especially asparagine. High color intensity resulting from Maillard reactions increased with an increase in the rate of acrylamide formation (Fiselier et al., 2004). Maillard products then probably participate in the formation of acrylamide.

The formation of acrylamide depends on the parameters of the frying process. Higher temperature (especially >175°C) leads to higher acrylamide content (Matthäus et al., 2004). The reaction is not linear, but rather complicated. Frying time was exponentially related to acrylamide content (Romani et al., 2008), but not under all conditions. It was found useful to heat chips first to a higher temperature for a short time, and then to continue at a lower temperature. Vacuum frying substantially decreases acrylamide formation. Oxidation is then an important step in the process (Granada et al., 2004), but it is not necessary in high-starch foods. However, flavonoids (but not all antioxidants) reduced acrylamide content in potato chips. The

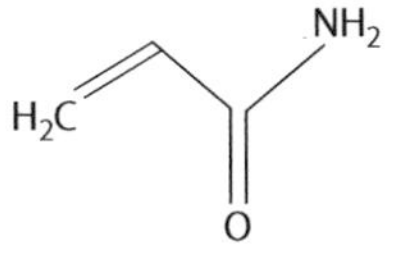

FIGURE 4.5 Chemical structure of acrylamide.

presence of polar substances, such as diacylglycerols, increases acrylamide formation (Gertz et al., 2003), but not in all experiments.

The application of batter helps reduce acrylamide content, but the effect depends very much on batter composition (Shih et al., 2004). The effect is similar to the effect of batter on the penetration of oil into food (Pedreschi et al., 2004).

The real danger of acrylamide consumption is still debatable (Mitka, 2002). The results on the rate of acrylamide formation have sometimes generated controversy, so it is difficult to predict acrylamide formation from the literature data only. Acrylamide content obviously depends on the complex interactions of several factors. In spite of intensive research in the last few years, much more effort will be needed before acrylamide formation becomes fully understood and predictable.

4.5 CHANGES IN SUGARS DURING DEEP FRYING

The main reaction of sugars is the Maillard (or, better, nonenzymic browning) reaction. Reducing sugars react with free amino acids or with free amino groups of proteins or peptides (Figure 4.6). Reducing sugars, such as glucose and fructose, react directly as they have a free carbonyl group. Sucrose reacts only after previous hydrolysis into reducing sugars; however, the hydrolysis is very rapid compared with the subsequent browning reactions. The browning proceeds at the same rate with glucose and with fructose. The effect of sucrose on browning was confirmed in potato chips (Leszkowiat et al., 1990).

The most important intermediary products are Amadori products, called premelanoidins, which are colorless. They are rapidly transformed into brown melanoidins under frying temperature (higher than 150°C). Melanoidins are partially cleaved with formation of volatile products, imparting specific flavor notes to the fried food.

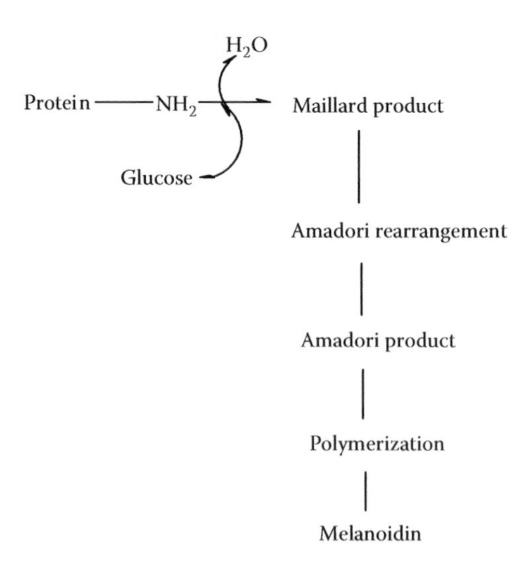

FIGURE 4.6 Mechanism of the Maillard reaction.

From the standpoint of nutrition, Maillard products are undesirable as they reduce the bioavailability, affecting mainly bound lysine. The sensory quality and type of flavor depend on the fried material. Deep brown coloration of the crust is favorable in bakery products as it conveys crispness and crunchiness. It is also preferable in fried meat and meat products. In contrast, it is considered negative in french fries or potato chips, particularly in the United States. Their production requires special potato cultivars with low glucose content. In some other countries, consumers prefer golden or reddish brown fried potato products because of their agreeable appearance and a more intense flavor. Countries preferring light-colored chips predominate. Glucose can be leached out from the surface layer with water. Sugars may be removed by lactic acid fermentation, too. Potato strips were soaked in corn syrup, and the product heated to 160°C–180°C. Crust color changes followed first-order kinetics, so prediction of crust color changes during deep frying was possible (Moyero et al., 2002). Color development was tested in fried battered squids. The color intensity depended on the polarity of frying oil during repeated fryings (190°C, 30 min) for 40 consecutive days (Baixauli et al., 2002).

Color development does not depend on sugars only, but also on free amino acids. Lysine, γ-aminobutyric acid, and glycine produce a more intense brown color than glutamic acid. In fried meat, such as light-colored chicken patties, an orange-to-brown crust is desirable, which is achieved by lower frying temperature and longer frying time (Yi and Chen, 1987). At higher frying temperature, color formation is an exponential function of time, while at lower temperature the browning rate becomes first order. In breaded fish fillet, color development depends on the batter type. With the addition of wheat flour, the color of fried product becomes lighter. Dipping raw material in sugar solution before frying darkens the color of fried battered products.

The brown crust on meat products is not mainly due to Maillard's reaction, as the glucose content in meat is very low. Interaction of oxidized lipids from frying fat with amino acids is more important. Polyenoic fatty acids are more active than monoenoic or saturated fatty acids.

Other reactions eliminate some pigments on frying. During frying of peppers and other vegetables, carotenoid pigments may be decomposed into light-colored or colorless products.

4.6 FORMATION OF FLAVOR COMPOUNDS IN FRIED FOOD

Because of an agreeable flavor, frying has become a widespread cooking process, readily accepted by consumers. The typical fried flavor originates from frying oil degradation, but it is modified by mixtures of volatiles formed by interaction between frying oil oxidation products and components in fried food. Therefore, it is possible to distinguish flavors of fried foods from one another. Some examples of volatiles produced by reactions of various functional groups of food components are given in Table 4.4. Other examples of volatile products in frying oil may be found in Chapters 2 and 11 of this book.

In model experiments, using a method developed by S. S. Chang, cotton balls were impregnated with a solution of glucose and different amino acids. The mixture was then fried at 180°C in vegetable oil of controlled composition (Chun and Ho, 1997). Among 29 nitrogen-containing compounds, alkylated pyrazines and pyridines predominated. They are formed in other thermic processes such as roasting

TABLE 4.4
Formation of Volatile Flavor Compounds in Fried Foods

Precursor	Process
Sugars	Pyrolysis, Maillard reactions
Amino acids and proteins	Deamination, interaction with aldehydes, Maillard reaction, Amadori reaction
Sulfur compounds	Cleavage, oxidation, interaction with aldehydes and amines
Fried food lipids	Oxidation, cleavage, reaction with amines and aldehydes
Phenolics	Oxidation, pyrolysis
Terpenes	Oxidation, condensation, polymerization

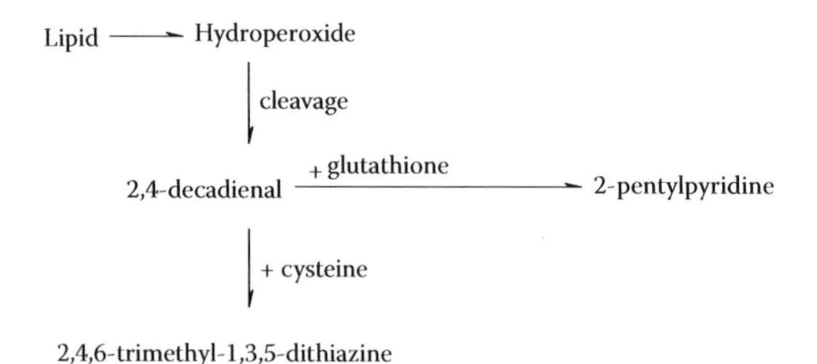

FIGURE 4.7 Reaction of decadienal with sulfur-containing substances.

or microwave heating; only the ratios of volatile products are different. Maillard products are not necessarily active as intermediates. Amino acids are converted into heterocyclic compounds even in the absence of sugars, via interaction with lipid oxidation products (Ohnishi and Shibamoto, 1984). Maillard products, containing glutamic acid, glutamine, and asparagine, were treated under deep-frying conditions, and 29 volatile nitrogen-containing products were isolated (Chun and Ho, 1997).

Sulfur compounds may be found among volatile interaction products as well. Substituted thiazoles, thiazolines, or other heterocycles were observed (Tang et al., 1983).

Sulfur can originate from sources other than amino acids, for example, thiosulfinates and related nonvolatile precursors in garlic. They are decomposed into vinyldithiin and alkyl and alkenyl sulfides under frying conditions. In model experiments simulating frying, more alkylthio and cyclic sulfur-containing compounds (such as thiophenes) were produced from deoxyaliin than from aliin (Yu et al., 1994). Mainly aliphatic sulfur derivatives were found in the volatile fraction of stir-fried garlic, such as diallyl disulfide, diallyl trisulfide, and dithiins. Alkylpyrazines predominated in the reaction mixture, but their composition depended on the amino acid present in the heated mixture.

An example of frying foods with oil interaction is the reaction of 2,4-decadienal (Figure 4.7), a typical flavor compound originating from polyenoic fatty acids, with cysteine and glutathione (Zhang and Ho, 1989). The major product in mixtures with cysteine was 2,4,6-trimethylperhydro-1,3,5-dithiazine. In the case of glutathione, also containing sulfur, it was sulfur-free 2-pentylpyridine.

Even very simple model frying systems result in the formation of many flavor-active compounds under frying conditions. Such complicated mixtures of reactive compounds give, naturally, very complex profiles of flavor notes, typical for particular fried food. Identical food samples give different flavor-active profiles, depending on frying conditions, degree of interaction with the respective frying oil, and the presence of respective degradation products of frying oil.

Sulfur compounds in fried foods are always accompanied by their selenium analogues. Their amounts are extremely small, but selenium compounds have a much more intense flavor than analogous sulfur derivatives. Research in this area requires closer experimental examination.

4.7 CHANGES IN STARCH AND INDIGESTIBLE POLYSACCHARIDES (DIETARY FIBER) DURING FRYING FOOD

The most common polysaccharide in foods is starch. Amylose can form unstable complexes with proteins (Pokorný and Kołakowska, 2003), but several other changes could occur during the frying. Changes in starch are very important (Figure 4.8), when raw potato products are fried. Starch granules are rapidly gelatinized upon contact with hot oil. The rigid structure of raw potatoes is lost in 1–2 min, and the fried chips become soft. On further heating, a firm crispy crust is formed on the surface of fried particles, which is highly appreciated by the consumer. On the surface of chips, where the water content is much lower than in internal layers, the gelatinization is not so intensive, so starch granules partially retain their crystalline structure.

The digestibility of starch (Figure 4.8) after frying is important from the point of view of nutrition. The digestible starch is easily cleaved with amylases, and digested in the small intestine. Another starch fraction is called resistant starch. It is not

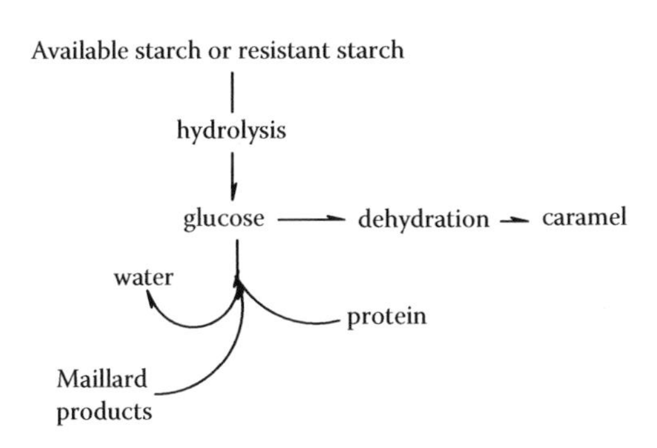

FIGURE 4.8 Changes in starch during frying.

accessible to amylases, and therefore not digested in the small intestine. It is metabolized only by microflora of the large intestine. The nutrition value and utilized energy content are thus decreased.

Resistant starch is present in raw potatoes, too. It is mostly converted into digestible starch on cooking. Frying increases, however, the content of resistant starch. The content of indigestible starch is, therefore, much higher in fried potatoes than in boiled potatoes. Microwave cooking has a similar effect as frying. In the crust, resistant starch decreases, while in the core, it increases. In high-amylose starch (e.g., corn starch), the content of resistant starch increased on cooking at 170°C. Digestible dietary fiber, also present in raw potatoes, becomes insoluble. Because of serious digestive consequences, such as flatulence, the formation of resistant starch is intensively studied (Pinthus and Singh, 1996).

High-molecular-weight nonstarch polysaccharides (dietary fiber) decompose in the course of french frying, especially in the case of low-specific-gravity tubers. Dietary fiber increased, however, in some cases during frying, possibly by formation of melanoidins or other indigestible compounds.

Polysaccharides have another important function in deep-fried products. They already form a compact film on the surface at the beginning of heating. Fat migration into the fried food and water loss from food is thus prevented (Rimac-Brnčić et al., 2004). It is recommended to coat the surface with a slurry of potato starch or other ingredients. A crispy product with a tender interior is obtained after frying. A mixture of insoluble cellulose fiber with other ingredients can be used for this purpose. Addition of 1% cellulose batter improves the color of the fried product, and products with higher moisture content and lower fat content are obtained. Cellulose forms hydrogen bonds with water molecules. Cellulose fibers thus become more hydrophilic than lipophilic, which prevents fat migration during frying. Methyl cellulose and hydroxypropyl cellulose also have satisfactory effect.

Protein-rich additives are also suitable for crust production. They have similar influence as cellulose (Feeney et al., 1992). Pieces of meat may be dipped into frying oil preheated to 200°C–300°C. Meat proteins precipitate on the surface. The prefried samples are then finish-fried in the conventional manner.

4.8 OIL UPTAKE AND MOISTURE LOSS DURING FRYING

The fat content increases the metabolizable energy of fried products. As many consumers try to reduce their energy intake because of the fear of obesity, they prefer low-fat fried products. Various factors influencing fat intake during frying are

1. Frying temperature
2. Duration of frying
3. Moisture content of food, especially in the surface layer
4. Type, size, and shape of food gel strength
5. Porosity of food, particularly of surface layers

6. Prefrying treatments such as predrying or application of batter
7. Type and quality of frying oil, such as the content of polar lipids or surface-active substances (oxidation products).

Average amounts of fat in some fried food are shown in Table 4.5.

The highest fat content was observed in fried vegetables, such as tomato, onion, peeled eggplant, pineapple flesh, or mushrooms. Fried white bread also had high fat content because of porosity higher than that of fried chicken, meat, fish, or prawns. In beef, pork, lamb, or sausages, fat content changes only a little or even decreases in some cases by diffusion of native fat from food into frying oil. Cholesterol and other sterols migrate with fat. In frying egg yolk, yolk phospholipids are transferred to the frying oil (Bennion et al., 1976), causing darker-colored oil.

Potatoes are the most frequently fried food. The loss of water and the oil uptake are interrelated in fried potato slices. They are both linear functions of the square root of frying time and independent of temperature in the range of 145°C–185°C (Pokorný et al., 1993). The effect of drying and the limited content of solids were studied in frying potato slices of different thicknesses (Baumann and Escher, 1995). In their experiments, high initial dry matter content resulted in reduced frying time and higher frying temperature. Oil uptake was negatively correlated with slice thickness as the frying oil mostly remained in the surface layer.

Oil uptake depends on the porosity of restructured potato products, which increased proportionally during frying due to water evaporation. Predrying of blanched potato strips at 80°C diminished the fat content of fried products (Rimac-Brnčić et al., 2004). Treatment with superheated steam reduced the oil layer from the surface, giving chips with not more than 25% oil. Restructured potatoes with higher gel strength took less oil on frying than lower-gel strength products (Pinthus et al., 1992). Porosity influences animal products, too. Deboned breast chicken meat was deep-fried in fat. Pore development depended on water loss, but the relation was exponential, not linear (Kassama and Ngadi, 2004).

Frying oil quality influences the oil uptake. The interfacial tension between frying oil and the potato surface is high in fresh oil. During repeated frying in the same

TABLE 4.5
Fat Content in Fried Food

Fried Food	Range of Fat Content [%]
Potatoes and potato products	15–36
Cereal products, doughnuts	18–30
Bread slices	35–50
Vegetables	35–75
Mushrooms (breaded)	60–80
Beef, pork, meat balls, patties	10–27
Chicken (battered and breaded)	10–30
Fish, prawns (breaded)	20–42
Sausages and other meat products	22–70

oil, the polarity of oil increases by hydrolytic and oxidative processes. In sunflower oil, the polar degradation products rose from the initial 6.2% to a final 18.7% after 15 fryings. The accumulating polar substances decreased the interfacial tension, facilitating oil uptake (Garrido-Polonio et al., 1994). Similar trends were obtained with fried doughnuts.

Doughnuts are another fried product produced in many countries. Similar products with slightly different recipes are encountered in Europe. The fat content in Berliner doughnuts could be minimized similar to in the case of potato fries, by optimizing frying time and temperature. A temperature of 180°C and a frying time of about 3.5 min were recommended for pan frying (Götze et al., 1994). The fat uptake of doughnuts and crullers (Spritzkuchen in German) correlated with moisture loss, and depended on the recipe.

Tortillas and tortilla chips (prepared from corn flour) are popular in America, and are also available in Europe. The fat uptake should be minimized, which is the preference of the majority of consumers. The fat uptake was a first-order exponential function of frying time (Moreira et al., 1995). The rate of moisture loss increased with increasing frying time (3–60 s) and temperature (130°C–190°C), but in the latter case, only after more than 15 s frying time. The initial water content in tortilla chips is important. Starch in chips with higher moisture content was more easily gelatinized, and pores in fried material were smaller and more numerous than in products from low-moisture-content chips. For this reason, the fat content was lower in the former products.

The fat uptake is less crucial in preparing fried animal products such as meat, fish, or sausage than in preparing foods of plant origin. Changes in the weight and shape of beef hamburgers were found to be lower with frying than with other cooking methods. In cooked ground beef, the final fat content was influenced by the initial fat level. Rinsing the samples with hot water reduced the final fat content (Love and Prusa, 1992). In beef patties, the fat content rose with prolonged frying while the temperature had little effect.

4.9 BREADING AND BATTER IN DEEP FRYING OF FOOD

Batter presents a barrier between frying and fried food. The effect of batter formulation on frying frozen squids was reviewed by Llorca et al. (2003). Some examples used for battering and breading are shown in Table 4.6. The batter adhering to fried food interacts with the fried food to form a crust. It is often difficult to separate the

TABLE 4.6
Commercial Products Used for Breading before Frying

Suitable Product	The Most Important Ingredients
Starch	Wheat, maize, cassava
Proteins	Chickpea, soybean
Fiber	Methylcellulose, hydroxypropylcellulose
Synthetic	Tween 80

crust of a battered product from the core. The breading prevents the flow of frying oil in fried product and the release of moisture. The temperature does not reach as high values in battered fried food as in the same product without the batter under identical frying conditions.

Crusts prepared with deep-fried models are similar to those occurring in products prepared in the industry (Visser et al., 2008). Different starches were compared as to their performance in frying chicken breast nuggets. Samples without added starch had lower crispness than a battered product. Pregelatinized tapioca starch was found more suitable than amylomaize, waxy maize starch, or corn in preventing the absorption of frying oil (Altunakar et al., 2004).

The effect of flour type was studied during deep-fat frying at 180°C. Corn flour or wheat flour was partially replaced with chickpea rice and soybean flour, and the microstructure of batter coating was determined. Chickpea batter allowed the formation of larger gas cells on the crust surface, while the addition of soy flour allowed the smallest gas cells (Barutcu at al., 2009).

Coating based on cassava starch and methylcellulose were coated on chicken nuggets. The optimum lipid content of fried nuggets resulted when cassava starch and methylcellulose were in the ratio of 3:1. The hardness was reduced, and the surface had a bright yellow color (Martelli et al., 2008). Coating of potato strips with 1% methylcellulose and 0%–5% sorbitol efficiently reduced the oil uptake and suppressed loss of moisture. Dough disks were affected similarly (García et al., 2004). Blanching combined with frying at 180°C also efficiently reduced the fat content of potato strips.

Batter often contains some egg; egg yolk phospholipids concentrate on the interface between fried food and frying oil, affecting the fat transfer (Bennion et al., 1976). Battered and breaded chicken breasts were fried with edible synthetic surfactant. Various ratios of Tween 80 and a batter mix were tested. Addition of the surfactant had no perceptible effect on moisture content, but fat content increased, due to the lower interfacial tension (Maskat and Kerr, 2004).

In deep-fat-fried breaded chicken thighs, the moisture loss increased with rising frying temperature, as expected. Fat content did not vary with frying temperature (Lane et al., 1982). Batter and breading increased the fat content in deep-fat-fried chicken breasts, thighs, drumsticks, and wing cuts due to fat absorption in the batter layer (Nakai and Chen, 1998). The thermal conductivity of potato crust decreased on frying due to higher moisture loss, compared with inner layers (Zialifar et al., 2009).

When battered fish portions were fried at 144°C–204°C, the necessary cooking time decreased only from 276 to 202 s. Lipid content increased only by 2%–3%, especially at lower frying temperatures. The lipid uptake was moderately influenced by the composition of the shortening used for frying (Flick et al., 1989). When frying fish sticks at 175°C, sticks cooked with breading took up more fat than the uncoated controls. Obviously, the amount of fat absorbed by batter was higher than that absorbed by uncoated controls.

4.10 LIPID OXIDATION IN FRIED FOOD DURING FRYING

Frying oil used for deep frying is used repeatedly for many hours and even days. Its degradation is rapid, so frying oil should be replaced before the polar compounds present exceed about 20%. Lipids present in fried material are oxidized as well. They are exposed to oxygen and to high temperature for only a short time, except the surface layer. Oxidation reactions are limited by the amount of oxygen entering the frying oil, and then entering the fried food by diffusion from the oil.

Oxidation occurs mainly only in the surface layer. Thus, the overall change in food due to oxidation is small. It depends on the amount of oxygen present in food before the frying (only trace amounts) and on oxygen migrating from frying oil. If frying oil contains a higher amount of polyenoic fatty acids, it absorbs more oxygen and, thus, oxygen available for diffusion into fried food is smaller. Data on the amounts of oxidation products are usually available from determination of fluorescence and substances reacting with thiobarbituric acid (Pikul and Niewiarowicz, 1990). Oxidation products of repeatedly used frying oil are easily cleaved with formation of free radicals during frying and even during the subsequent storage. Therefore, low-unsaturated oils, such as olive oil or palm oil, are oxidized only slowly at frying. The oxidation of lipids is then also relatively low (Tabee et al., 2009).

The free radicals in frying oil initiate lipid oxidation in native food. In frying oil with 15% degradation products or less, the off-flavor is similar to frying fresh oil, but frying oil containing 25% degradation products has significantly less acceptable flavor (Hara et al., 2004).

Addition of carrot powder to wheat dough increased the rate of fatty acid oxidation on frying as free radicals were consumed by preferential oxidation of carotenes (Lee et al., 2003). On the contrary, in experiments involving the addition of spinach powder to wheat dough, it was observed that chlorophyll was decomposed on frying. Carotenoids were leached out, and their content increased (Kim et al., 2003).

Fish lipids are far more unsaturated than lipids of land animals (Table 4.7). Therefore, a higher oxidation rate could be expected during the frying, especially in the absence of native antioxidants. Changes in the level of oxidation are difficult to determine in fatty fish species, such as herrings or sardines. When frying such fish, fish lipids migrate into the frying oil during frying. Eicosapentaenoic acid (EPA) and docosahexaenoic acid (DHA), two typical fish fatty acids, were detected in frying

TABLE 4.7
Fatty Acid Composition of Lipids in Some Food Products, Expressed as Percentages

Food	Saturated	Monoenoic	Dienoic and Trienoic	Tetra-, Penta-, and Hexaenoic
Beef	47–80	40–50	1–5	0–0.3
Pork	38–46	48–58	5–10	0–0.6
Chicken	27–30	40–48	20–25	0–0.5
Soybeans	15–20	28–28	52–68	0–0.1
Wheat flour	11–22	10–32	48–72	0.–0.2
Fish	13–30	25–40	2–11	15–55[a]

[a] Substantially lower in freshwater fishes.

oil (Sánchez-Muniz et al., 1992). Their decrease in fried fish could be also due to diffusion into frying oil. When sardines were fried in soybean oil without coating, the content of EPA and DHA decreased because more oil migrated into the frying bath than was absorbed from the bath into the fish. On frying coated fish, oil uptake predominated, so the n-6 to n-3 ratio increased, as EPA and DHA belong to the latter group (Ohgaki et al., 1994). In silver carp (*Hypophalmichtis molitrix* var.), a rather fatty fish, oil losses of 30%–40% were observed on frying (losses of 25%–28% EPA and 23%–40% of DHA, respectively). The more unsaturated DHA is relatively more rapidly destroyed than EPA. When frying Atlantic mackerel (*Scomber scomber* L.), changes in both the foregoing polyunsaturated fatty acids were small, and only traces migrated into the frying oil (Sébédio et al., 1993).

The slightly extended oxidative changes reported in Section 4.11 may be due to rapid transformation of lipid oxidation products, and to nonextractable lipids produced by interaction with protein (see following text). The absolute amount may decrease, but no substantial change is observed in the relative content of different fatty acids in the chromatographic profile.

Sterols and tocopherols are oxidized in fried fish. The more sensitive polyunsaturated fatty acids are oxidized, too. As polyunsaturated fatty acids disappear by oxidation more rapidly than less unsaturated fatty acids, the amount of the latter is relatively increased. For instance, the content of polyunsaturated fatty acids decreased, while that of monoenoic and dienoic acids increased in fried sardines (Castrillon et al., 1997). Polyunsaturated fatty acids, such as EPA and DHA) were rapidly lost by oxidation and by diffusion in frying oil, but their preferential oxidation protected EPA and DHA in enriched margarine (Kitson et al., 2009).

Cholesterol content in channel catfish (*Ictalurus punctatus*) was slowly oxidized, but more during deep-fat frying than when cooking in a microwave oven or oven baking (Wen-Hsin and Lillard, 1998). Phytosterols remained partially absorbed in fried food, and partially migrated to frying oil, so they were distributed in the two fried foods and frying oils in different ratios than before the frying. During eight consecutive fryings, they deteriorated by 9.5%–21.8% and 29.4%–51.2%, respectively. Phytosterol losses during frying were higher than losses of more polyunsaturated oils (Salta et al., 2008). Oxidative changes contribute to a decrease in the nutritional value on frying, especially in fish.

4.11 REACTIONS OF OXIDIZED LIPIDS IN FRIED FOOD WITH FOOD PROTEINS

Oxidized and polar lipids are present in traces even in fresh frying oils, but their amounts gradually increase during frying (Gardner, 1979; Pokorný and Kołakowska, 2003). Oxidized and polar lipids present in frying oil enter the fried material relatively more easily than the original oil. The reason is that they have greater affinity for the hydrophilic surface of fried food. Therefore, lipids extracted from fried food are always more oxidized than the respective frying oil remaining in the fryer.

In fried foods, oxidized lipids react with nonlipidic components, especially with protein. The active functional groups are listed in Table 4.8. Peroxy, epoxy, ketol, and

TABLE 4.8
Functional Groups of Fried Lipids Participating in Interactions

Present in Oxidized Lipids	Present in Proteins
Hydroperoxides, free radicals	Primary and secondary amines
Aldehydes, ketones, ketols	Thiols, sulfides, disulfides
Carboxylic acids	Hydroxyl groups
Hydroxy derivatives	Carboxyl groups

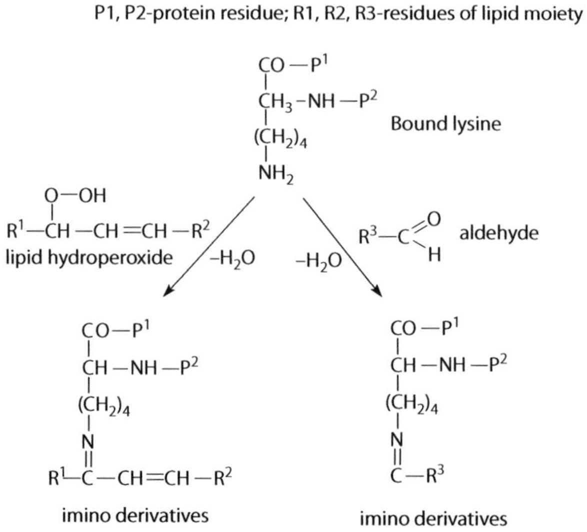

P1, P2-protein residue; R1, R2, R3-residues of lipid moiety

Bound lysine

lipid hydroperoxide

aldehyde

$-H_2O$ $-H_2O$

imino derivatives

imino derivatives

FIGURE 4.9 Interaction of oxidized lipids with protein-bound lysine during frying.

aldehydic groups are the reactive sites in the oxidized lipid fraction, while primary amine groups of lysine; secondary amines such as proline and hydroxyproline; thiol, sulfide, and disulfide groups of sulfur-containing amino acids (cysteine, cystine); and, to a lesser degree, phenol and hydroxyl groups, are the reactive sites of the protein molecule (Pokorný, 1998). Ketols are oxidation products, where the keto group is activated by an adjacent hydroxyl group. Ketols have similar reactivity to aldehydes.

Both hydrogen and covalent bonds are formed by interaction of oxidized lipids with protein. Hydroperoxides or free peroxy radicals present in fried food before the frying may be more available than those penetrating from frying oil. They react readily, for instance, with lysine (Figure 4.9) or with cysteine or methionine (Figure 4.10), but their concentration in hot oil is very low (Gardner, 1979).

Oxidized sterols are very reactive, and tend to form insoluble compounds with protein. Thiol groups of cysteine and amine groups of alanine were found effective in decreasing the level of oxidized sterols in fried fish (Li, 1992). Aldehydes are another group of very reactive lipid oxidation products. They easily react with bound

A = protein-bound methionine; B = lipid hydroperoxide;
C = hydroxylated lipid; D = protein-bound methionine sulfone

P^1 = protein residue; R^1 = lipid residue

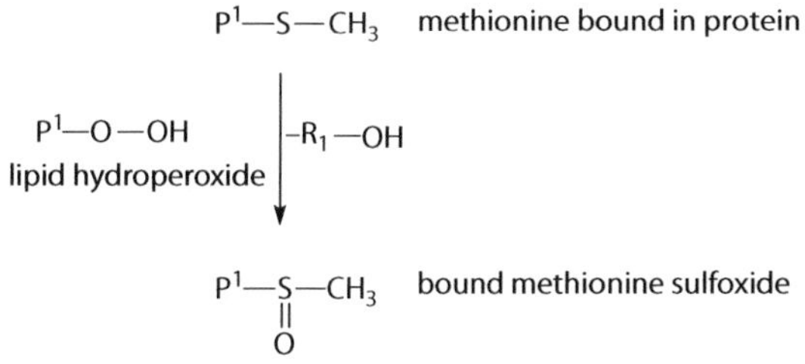

P^1—S—CH_3 methionine bound in protein

P^1—O—OH lipid hydroperoxide → –R_1—OH

P^1—S(=O)—CH_3 bound methionine sulfoxide

FIGURE 4.10 Interaction of oxidized lipids with protein-bound methionine.

lysine, particularly in the presence of water (Pokorný et al., 1993). Another reactive group is the thiol group of cysteine.

The aforementioned oxidized lipid–protein interactions are generally neglected by both researchers and analysts in control laboratories, as the reaction products are insoluble macromolecules (Pokorný and Kołakowska, 2003). These compounds are not easily extracted. Preliminary treatments are necessary to convert the complexes into easier-to-analyze derivatives. For instance, they are boiled with hydrochloric acid for several hours. The low-molecular-weight products can be separated into simple fractions.

Lipid–protein complexes are cleaved with proteases only very slowly, both *in vitro* and *in vivo*, so they may cause digestive problems if present in fried food in larger amounts. The nutritional value of protein is thus decreased. Complaints of some consumers about difficulties after eating fried food may be due, at least partially, to the presence of interaction products rather than to fat. The subject is treated by Pokorný and Kołakowska (2003). These compounds are of particular interest because oxidized lipid–protein interactions are important precursors of atherosclerotic plaques *in vivo*.

Products of fish muscle proteins with oxidized fish lipids deserve attention as they contain highly polyunsaturated fatty acids (EPA and DHA). As fish proteins are rich in basic amino acids and amines, several conjugated double bond systems may be formed by interactions, which are deep brown and contribute to the color of crust in fried fish.

4.12 VITAMIN LOSSES UPON FRYING FOOD

Literature on vitamins during deep frying mostly concerns changes in vitamins in frying oil, especially with tocopherols. Fried food contains several water-soluble vitamins in addition to liposoluble vitamins. Several vitamins are sensitive to higher temperature and dissolved oxygen (Table 4.9). Fortunately, high temperature is reached only in the surface layers of fried food, where their losses are surely

TABLE 4.9
Vitamins Sensitive to Frying

Reaction Type	Examples of Vitamins
Oxidation	Vitamin A and provitamins, tocopherols and tocotrienols, vitamins containing sulfur, vitamin C
Hydrolysis	Thiamin
Pyrolysis	Thiamin, riboflavin, vitamin C

very high. Total losses depend mainly on internal temperature, which usually varies between 70°C and 90°C. This temperature is reached for a very short time, only a few minutes or less. As the core material represents the major part of fried food, vitamin losses depend mostly on the internal temperature. Losses of vitamins are lower with deep-fat frying than with other cooking techniques (Ruiz-Roso, 1998).

Thiamin is one of the most important vitamins of the B group, the intake of which is often deficient. Due to the low oxygen content in fried food, thiamin losses are lower with frying than when the same food is cooked in another way (Kimura et al., 1991; Mašková et al., 1994). Frying caused vitamin losses in the preparation of beef, pork, and chicken meat (Mašková et al., 1994). Compared to boiling, losses of thiamin during frying were low. Stir-fried beef strips retained more thiamin than microwave-heated or broiled foods.

More significant losses of thiamin occur due to irradiation (Thayer et al., 1989). Frying before irradiation substantially increased thiamin retention, but it was not correlated with water loss. Frying after irradiation, on the contrary, increased the destruction of thiamin. When preparing chicken meat by roasting, braising, deep-frying, or microwave cooking, thiamin losses varied between 28% and 64% (Al-Khalifa and Dawood, 1993), but the differences were only moderate. When frying carp (*Cyprinus carpio*), pike-perch (*Lucioper casenetra*), trout (*Trutta facio*), cod (*Gadus callaria*), and silver carp (*Hypophtalmichtis molitrix*), the average thiamin loss was 61% (Duda et al., 1987). In another experiment with fish caught in Ghana, no thiamin was detected in the fried product, probably due to the action of thiaminase before frying. In experiments with several kinds of vegetables, losses of thiamin were lower with stir frying than with boiling or steaming (Pither and Edwards, 1995).

Riboflavin is another vitamin of the B group, often deficient in human diets. The vitamin was better retained in fried chicken meat than thiamin (Al-Khalifa and Dawood, 1993), at least in dark meat. The losses were higher in light-colored meat. Riboflavin losses in frying calves liver and spin liver were 42.5% and 43.5%, respectively (Díaz-Marquínez et al., 1993). Only slight losses were observed when frying various kinds of fish (Duda et al., 1987). Moderate losses were reported during frying of various vegetables (Pither and Edwards, 1995).

Vitamin B_6 (pyridoxamine, pyridoxal, pyridoxol) was analyzed during frying of beef and pork, and losses of 58% and 45%, respectively, were obtained (Bognar, 1993). Low losses occurred during stir frying, particularly when frying chicken muscle.

Niacin is relatively stable on frying. Still, losses of 45% were reported in fried pork muscle, beef, and chicken meat (Mašková et al., 1994). Niacin content even increased in frying peanuts, and the bioavailability was improved, too. Losses of folic acid in vegetables were lower during stir frying in comparison to other cooking methods (Pither and Edwards, 1995).

Losses of vitamin C in calves or swine liver were 34.79% and 37.4%, respectively (Díaz-Marquínez et al., 1993). Losses occurred in soybeans or vegetables, but stir frying in oil resulted in good retention of vitamin C. Frying under vacuum conditions reduces the oxidation of fried food. Vitamin C content increased during vacuum frying of pineapple chips, but moisture losses increased, too (Pérez-Tinoco et al., 2008).

If the frying process is short, losses of vitamin A are low, as with provitamins A (carotenes) and lycopene. Losses of β-carotene during shallow frying are lower than those observed during deep-fat frying, but it is possible that β-carotene partially migrates into the frying oil. In the presence of phenylalanine, the losses are lower, perhaps due to deactivation of free radicals by phenylalanine.

Tocopherols and tocotrienols present in high-fat foods, such as peanuts, may leach into the frying oil and increase its stability. In breaded shrimps or a layer of protein (in fried chicken muscle), tocopherols are protected against free radicals from frying oil (Simonne, 1994). The reverse activity was observed in fried potatoes, where the protein content is very low (Miyagawa et al., 1991). Phospholipids have a similar effect. They are common in fried foods. They accumulate in membranes or partially migrate into the frying oil, where they react with triacylglycerol free radicals. The decomposition of tocopherols is thus inhibited.

4.13 CHANGES IN MINERALS

Mineral components show great changes during cooking operations, such as boiling, because of their solubility in water. However, their losses are much lower during frying in oil, as they are soluble in oil only in small amounts. The availability of some important minerals, such as calcium, magnesium, phosphorus, and especially iron, may decrease (Vaquero, 1998), partially because of their binding in insoluble compounds.

Water migrating from food into frying oil is converted into steam, and lost. Due to water loss, the wet weight of fried food decreases during frying. Most mineral components are nonvolatile; therefore, the content of minerals, expressed as wet weight, would be expected to rise. There occurs, however, another process at the same time, that is, the uptake of frying oil. The dry weight of fried material increases, so a moderate decrease in mineral content would occur. The best way would be to express the content of minerals on the basis of fat-free dry matter. Of course, it would be rather time consuming and inconvenient from the analyst's point of view.

Usually, the content of minerals is reported on a dry basis, without removal of fat. Therefore, the content of iron, zinc, copper, and most other metals is found to decrease slightly after frying because of absorption of some frying fat. The results of iron determination are interesting (Carpenter and Clark, 1995). It seems that the total iron content in the heme fraction is higher than usually assumed. Some iron and copper may migrate into the frying oil, catalyzing its oxidation.

Changes in alkali metal ions are important as they are present in relatively high amounts in food. The amount of sodium increased from 7.5 mg/kg to 43 mg/kg (expressed as sodium oleate) while frying potatoes and other foods (Blumenthal and Stier, 1991). Sodium oleate decreased the interfacial tension, so polar lipids could easily migrate from frying oil into fried food, thus increasing the sodium and oil content. Foaming is stimulated, too, which increases the area of contact between the two phases. It may be assumed that potassium ions have a similar effect as sodium ions.

Volatile mineral components may also be present. Nitrates are mineral components that are found in objectionable amounts in plant foods, such as vegetables and potatoes. Their partial removal during frying could thus be considered positive.

Selenium belongs to other volatile mineral components. Pronounced losses of selenium were observed on frying meat, fish, pulses, cereals, and vegetables (Bratakos et al., 1988).

Mercury, considered a very harmful component, is often present as a toxic contaminant in fish. Some losses of mercury were observed in cooking. No great changes were found, however, during frying of striped bass (Armbruster et al., 1988), but losses were observed in Spanish fish (Hernández-García et al., 1988). Mercury content decreased during frying of tuna without batter, and anchovies during frying. This decrease increases the food safety, even in foods where the original content of mercury does not exceed the limit permitted for organic-bound mercury.

Iodine also is a mineral. Losses of both the mineral-bound and organic-bound content varied between 18% and 30% during frying because of the volatility of iodine (Szymandera-Buszka and Waszkowiak, 2004).

4.14 EFFECT ON AND CHANGES IN PHENOLICS DURING FRYING

Phenolic substances used to be considered rather antinutritional agents as they bind important functional groups of proteins. They are now appreciated as both *in vitro* and *in vivo* antioxidants. They also bind prooxidizing metals into inactive chelates. Flavonoids are particularly valued as nutritionally positive compounds. They are present in most plant foods.

Onion is very rich in flavonoids. During frying of onion (*Allium cepa*), flavonoid glucosides were reduced by 25%, but no hydrolysis of flavonoid glucosides was observed (Price et al., 1997). Similarly, flavonoid glucoside content and free quercetin decreased during commercial frying of tomatoes, celery, or lettuce (Crozier et al., 1997). Oxidation with dissolved oxygen or interaction with protein could be the reason.

If food ingredients or spices containing phenolics or phenolic resins (rosemary and sage are good examples) or lignans are added to meat balls or other animal foods, the oxidation of meat lipids is inhibited during frying. The antioxidant activities of flavonoids and other phenolics in frying oil are discussed in Chapter 8.

TABLE 4.10
Changes of Natural Antinutritional Compounds during Frying

Type of Compounds	Process during Frying
Enzymes	Denaturation
Biogenic amines	Various interactions
Cyanoglycosides	Hydrolysis and evaporation
Glucosinolates	Hydrolysis and interactions
Phytates	Binding in heavy metal salts

4.15 CHANGES IN NATURAL ANTINUTRITIONAL SUBSTANCES DURING FRYING

From the previous sections it is evident that various antinutritional substances are produced during frying, such as Maillard products, oxidized lipids, and lysinoalanine. Many nutritional and even toxic substances are natural components of many foods (Table 4.10). It is interesting and important to learn how they change during frying.

Antinutritional proteins, such as trypsin inhibitors, thiaminases, or lipoxygenase, were discussed earlier. High frying temperature denatures the enzymes, so they are completely deactivated both in shallow and deep frying operations.

Biogenic amines, due to enzymic decarboxylation of amino acids, are produced by microbial action. They are formed in the course of fermentation, for instance, in the production of tempeh from soybeans. Frying in oil significantly decreased levels of putrescine and tyramine (Nout et al., 1993), mainly by reaction with carbonyl groups. High histamine levels are present in fish, especially tuna fish. Deep frying could help to reduce histamine levels in the same way as other amines.

Sorghum (cassava) contains cyanoglycosides, which are rather toxic. If they are not removed during meal preparation, they are hydrolyzed or germinated, and high cyanide levels are released. Toasting decomposes most of the hydrogen cyanide, and frying or grilling reduced hydrogen cyanide levels nearly completely (Padmaja, 1995). In fried cassava chips, some bitter taste persists, similar to that of cyanides. This aftertaste could be, however, attributed to Maillard products.

Glucosinolates are common antinutritional substances in many vegetables of the temperate climate, mainly from the *Brassicaceae* family. They are cleaved to isothiocyanates. Cabbage, a typical representative of this family, was boiled, steamed, and fried for different time intervals. The residual glucosinolate content decreased proportionally to the duration of heating. Boiling was more efficient than frying, as boiling stimulates hydrolysis. The content of thiocyanates, their important degradation products, decreased with increasing frying time, but the final content was higher in fried than boiled product. The effect of different cooking procedures was tested by frying cauliflower, broccoli, and cabbage. Losses of isothiocyanates amounted to 78% (Sultana et al., 2002).

Glycoalkaloids, such as solanine or chaconine, are important antinutritional agents in potatoes. Glycoalkaloids are relatively thermostable under conditions of canning, but high frying temperature decreases their content. The hydrolysis products enable the leaching of alkaloids into frying oil.

Phytates and phytic acid are considered antinutritional agents because they bind bivalent and trivalent cations, such as calcium or iron, into their indigestible salts. The greatest losses were observed during fermentation and boiling, but even frying reduced the phytate content to a significant degree, depending on the food product.

Processing conditions were studied in the preparation of Bengal gram (*Cicer arietinum*) for human consumption (Khan et al., 1988). During soaking in water, the loss of phytates was 12.5%. Losses rose to 25.0% if the grains were boiled. If the boiling was followed by frying, the amount of phytates decreased by 37.5% of the original amount. Losses were still higher with immature brown grams. On the contrary, losses were reduced by presoaking in sodium carbonate solution.

4.16 INACTIVATION OF MYCOTOXINS DURING DEEP FRYING

Products synthesized by microorganisms are dangerous contaminants because of their potential carcinogenicity. Aflatoxins are the most widely known contaminants of this series; they may be partially destroyed on frying.

Fumonisins are mycotoxins produced by *Fusarium verticilliodes* and *F. proliferatum*. They contribute to the incidence of esophageal cancer. They are found mostly in corn or corn products. During frying of corn muffins spiked with 5 mg/kg fumonisin B, 16.3%–27.6% losses were observed. The losses were higher in the crust than in the core (Jackson et al., 1997). The fumonisin degradation was not observed at temperature <170°C, starting only at >180°C. Therefore, fumonisin may be considered heat-stable.

Other examples may be found in the specialized literature.

4.17 DEGRADATION OF ENVIRONMENTAL CONTAMINANTS DURING DEEP FRYING OF FOOD

In addition to natural toxic or antinutritional compounds present in the original food, the fried material may contain some contaminants from the environment (Table 4.11). The most dangerous contaminants are toxic products of microorganisms. Salmonella cells are mostly destroyed during cooking sausages, but a small residual activity of *Salmonella* sp. was detected (Mattick et al., 2002). The data on microbial toxins and pesticide residues are readily available in the literature. For this reason, a few examples of their change on frying will be given here. Experiments with fish muscle contaminated with polychlorinated biphenyls, containing 0.63–6.58 mg/kg congeners, were carried out with several freshwater and marine fish species. Mean losses on the lipid weight basis ranged between 32.1% and 81.1% (Witczak, 2009).

TABLE 4.11
Contaminants Present in Fried Foods

Contaminant	Change during Frying
Nitrates	Partial reduction in nitrites
Iron and other heavy metals	Binding with active groups, proteins
Polychlorobiphenyls	Moderate decrease
Oxytetracycline	Substantial decrease
Sulfonamides	Moderate degradation

In experiments with fish contaminated with oxytetracycline, lower values in muscle and skin than in bone and liver were observed. Baking and frying reduced the residual contaminant by 70%–80% (Maruyama and Uno, 1997).

Oxytetracycline residues were tested in chicken muscle and hamburger. Broiling, stir frying, and deep-fat frying removed 39%–48% contamination in chicken tissues and 8%–34% in chicken hamburgers (Shun-Yao, 1997).

Chicken meat balls contained sulfonamides, such as sulfadiazine, sulfamethazine, sulfamethoxazole, and sulfaquinoxaline. During heating for 3–9 min at 170°C–190°C, sulfonamide losses depended on the moisture losses and internal temperatures, and varied between 27.6% and 37.5% for different sulfonamides (Ismail-Fitry et al., 2008).

The examples show that frying decreases the content of these contaminants, but residual amounts after heating are still quite high. It is necessary to combine cooking with other procedures to reduce the content of contaminants to acceptable levels.

4.18 CONCLUSIONS

The investigation of nutrient changes in fried food during frying is lagging behind research on frying oil. For the evaluation of nutrition value, it should be noted that frying oil is consumed in smaller amounts compared to fried food. Another important factor to be taken into consideration is that many interactions proceed at the same time in fried food. This complexity makes the research more difficult than the study of changes in frying oil. Changes occurring in fried food should be studied in more detail:

1. Determination of differences between changes in the crust and in the core (inner layers) from the standpoint of nutrient losses and the formation of antinutritional substances.
2. Determination of differences between shallow frying and deep frying. During shallow frying, access to oxygen is much better than during deep-fat frying, and overheating is also much more likely. Modern trends of frying without fat should be studied from the standpoint of mutagen increase, nutrient losses, and losses of flavor-active substances.
3. Interactions between substances produced by pyrolysis of proteins, Maillard reactions, vitamins, and oxidized lipids should be studied in more detail.

4. Nutritional value decreases during frying, but sometimes increases because of inactivation of antinutritional components. Which changes predominate and whether frying is preferable to other cooking methods should be studied.
5. Sensory value improves during frying. It should be known, whether the utility of improved sensory value would prevail to the decomposition of important nutrients.

Generally, the progress of knowledge on frying is advancing rapidly, so answers to the foregoing questions would be at least partly known in the near future, at least for specific food products important for human nutrition.

In modern industrial plants, the frying process does not damage fried food to a great extent. If the procedure is strictly controlled, toxic and antinutritional substances are not produced, except in negligible amounts. The main negative change is the increase of metabolizable energy in the product, due to increased fat content. Still, more research is needed to understand the mechanism and kinetics of the formation of antinutritional or, possibly, toxic compounds, even in trace amounts. The optimum frying conditions would then be specified more precisely.

REFERENCES

Abdulkarim, B. G. and Smith, J. S. 1998. Heterocyclic amines in fresh and processed meat products. *J. Agric. Food Chem.,* 46: 4680–4687.

Al-Khalifa, A. S. and Dawood, A. A. 1993. Effect of cooking methods on thiamin and riboflavin contents in chicken meat. *Food Chem.*, 48: 79–74.

Altunakar, B., Sahin, S., and Sumnu, G. 2004. Functionality of batters containing different starch types for deep fat frying of chicken nuggets. *Eur. Food Res. Technol.*, 218: 318–322; FSTA, 2004: Sn1226.

Amin, I. and Khairul Ikram, E. H. 2004. Effects of cooking practices (boiling and frying) on the protein and amino acid contents of four selected fishes. *Nutr. Food Sci.*, 34: 54–59; FSTA: 2004: Rc0596.

Armbruster, G., Gutenmann, W. H., and Lisk, D. J. 1988. The effects of six methods of cooking on residues of mercury in striped bass. *Nutr. Rep. Int.*, 37: 193–196; FSTA, 1989: R0033.

Baixauli, R, Salvador, A., Fiszman, S. M., and Calvo, C. 2002. Effect of oil degradation during frying on the colour of fried, battered squid rings. *J. Am. Oil Chem. Soc.*, 79: 1127–1131.

Barutcu, I., Sahin, S., and Sumnnu, G. 2009. Effects of microwave frying and different flour types addition on the microstructure of batter coatings. *J. Food Eng.*, 95: 684–692.

Baumann, B. and Escher, F. 1995. Mass and heat transfer during deep-fat frying of potato slices. I. Rate of drying and oil uptake. *Lebensm.-Wissensch.-Technol.*, 28: 395–403.

Bennion, B., Stirk, K. S., and Ball, B. H., 1976. Changes in frying fats with batters containing egg. *J. Am. Diet. Assoc.*, 68: 234–236.

Blumenthal, M. M. and Stier, F. F. 1991. Optimization of deep-fat frying operations. *Trends Food Sci. Technol.*, 2: 144–148; FSTA, 1991:E0004.

Bognar, A. 1993. Studies on the influence of cooking on the vitamin B_6 content of food. *Proc. Symp., Nutr. Bioavaiability.*, 93: 346–351; FSTA, 1993: A0072.

Bognar, A. 1998. Comparative study of frying to other cooking technique influence on the nutritive value. *Grasas Aceites*, 49: 250–260; FSTA, 1999: Ag1985.

Borowski, I., Kozikowski, W., Rotkiewicz, W., and Amarowicz, R. 1986. Influence of cooking methods on the nutritive value of turkey meat. *Nahrung*, 30: 987–993.

Bratakos, M. S., Zafiropoulos, T. T., Siskos, P. A., and Joannou, P. V. 1988. Selenium losses on cooking Greek foods. *Int. J. Food Sci. Technol.*, 2: 585–590.

Brockstedt, U. and Pfau, W. 1998. Formation of 2-amino-α-carbolines in pan-fried poultry and 32-postlabelling analysis of DNA adducts. *Z. Lebensm.-Unters.-Forsch.*, 207: 472–476.

Buckley, D. J., Zabik, J. M., Gray, J. I., Crackel, R. L., and Beatty, E. 1989. *N*-Nitrosamine formation in Irish and US bacon as influenced by sodium nitrite concentration and the presence of α-tocopherol. *Irish J. Food Sci. Technol.*, 13: 109–117; FSTA, 1991: S0055.

Carpenter, C. E. and Clark, E. 1995. Evaluation of methods used in meat iron analysis and iron content of raw and cooked meats. *J. Agric. Food Chem.*, 43: 1824–1827.

Castrillon, A. M., Navarro, P., and Alvárez-Pontes, E. 1997. Changes in chemical composition and nutritional quality of fried sardine (*Clupea pilchardus*) produced by frozen storage and microwave heating. *J. Sci. Food Agric.*, 75: 125–132.

Chen, C., Pearson, A. M., and Gray, J. I. 1990. Meat mutagens. *Adv. Food Nutr. Res.*, 34: 387–394; FSTA, 1991: 50027.

Chun, H.-K. and Ho, C. T. 1997. Volatile nitrogen-containing compounds generated from Maillard reactions under simulated deep-fat frying conditions. *J. Food Lipids*, 4: 239–244.

Crozier, A., Lean, M. E. J., McDonald, M. S., and Black, C. 1997. Quantitative analysis of the flavonoid content of commercial tomatoes, onions, lettuce, and celery. *J. Agric. Food Chem.* 45: 590–595.

Díaz-Marquínez, A., Orzáez-Villenueva, M. T., and Palomino-Fernández, P. 1993. Variations in water-soluble vitamins in liver caused by cooking. *Alimentaria*, 239: 73–76; FSTA, 1994: S0029.

Duda, G, Maruszewska, M., Kulesza, C., Gertig, H., and Szajkowski, Z. 1987. Losses of B-vitamins in selected fish species during cooking. *Bromatol., Chem. Toksykol.*, 20: 89–95; FSTA, 1988: R0029.

Feeney, R. D., Haralampu, S. G., and Gron, A. 1992. Method of coating foods with an edible oil barrier film and product thereof. U.S. Pat. 5126152; FSTA, 1992: N0043.

Fiddler, W., Pensalene, J. W., Gates, R. A., Hale, M., and Jahncke, M. 1992, Nitrosodimethylamine formation in cooked frankfurters containing Alaska pollock mince and surimi. *J. Food Sci.,* 57: 569–571, 595.

Fiselier, K., Grob, K., and Pfefferle, A. 2004. Brown potato croquettes low in acrylamide by coating with egg/breadcumbs. *Eur. Food Res. Technol.*, 219: 111–115.

Flick Jr., G. J., Gwo, Y. Y., Ory, R. L., Sasiela, R. J., Boling, J., Vinnett, C. H., Martin, R. E., and Arganosa, G. C. 1989. Effects of cooking conditions and postpreparation procedures on the quality of battered fish portions. *J. Food Qual.*, 12: 227–242; FSTA, 1990: J0006.

Fukuda, M., Kumisada, Y., and Toyoswa, I. 1989. Some properties and *in vitro* digestibility of fried and roasted soybean proteins. *Nihon Eiyo Shokuryo Gakkaishi,* 42: 305–311; FSTA, 1991: J0126.

García-Arias, M. T., García-Linares, M. C., Capita, R., García-Fernández, M. C., and Sánchez-Muniz, F. J. 2003. Deep frying of chicken meat and chicken-based products. Changes in the proximate and fatty acid compositions. *Ital. J. Food Sci.*, 15: 225–239; FSTA: 2003: Sn2015.

García, M. A., Ferrero, C., Campana, A., Bertola, N., Martino, M., and Zaritzky, N. 2004. Methyl cellulose coatings applied to reduce oil uptake in fried products. *Food Sci. Technol. Int.*, 10: 339–346.

Gardner, H. W. 1979. Lipid hydroperoxide reactivity with proteins and amino acids. *J. Agric. Food Chem.* 27: 220–229.

Garrido-Polonio, M. C., Sánchez-Muniz, F. J., Arroyo, R., and Cuesta, C. 1994. Small scale frying of potatoes in sunflower oil: Thermoxidative alteration of the fat content in fried product. *Z. Ernährungswissensch.*, 33: 267–276; FSTA. 1995: J0142.

Gertz, C. and Klostermann, S. 2002. Analysis of acrylamide and mechanisms of its formation in deep-fried products. *Eur. J. Lipid Sci. Technol.*, 104: 762–771.

Gertz, C., Klostermann, S., and Kochhar, S. P. 2003. Deep frying: The role of water from being fried and acrylamide formation. *OCL*, 10: 297–303; FSTA. 2004: Cf0712.

Giese, J. 2002. Acrylamide in foods. *Food Technol.*, 56: 71–72.

Gloria, M. B. A. Barbour, J. F., and Scanlan, R. A. 1997. Volatile nitrosamines in fried bacon. *J. Agric. Food Chem.*, 15: 1816–1818.

Götze, S., Tscheuschner, H. D., and Bindrich, U. 1994. Can fat uptake during frying of Berliner doughnuts be reduced? *Zucker-Süßwaren-Wirtsch.*, 47: 268–274; FSTA, 1994: M0110.

Granada, C., Moreira, R. G., and Tiehy, S. E. 2004. Reduction of acrylamide formation in potato chips by low-temperature vacuum frying. *J. Food Sci.*, 69: E405–E411.

Hamza, A. M., El-Tabey-Shebata, A. M., and Stegemann, H. 1987. Effect of traditional methods of processing on the electrophoretic patterns of faba-bean water-soluble proteins. *Qualitas Plantarum*, 36: 253–262; FSTA, 1987: J 0111.

Han-Seung, S., Rogers, W. J., Gomaa, E. A., Strasburg, G. M., and Gray, J. I. 2002. Inhibition of heterocyclic aromatic amine formation in fried ground beef patties by garlic and selected garlic-related sulfur compounds. *J. Food Protection*, 65: 1766–1770; FSTA, 2003: Sg0759.

Hara, T., Ando, M., Fujimura-Ito, T., Inoue, S., Otsuka, K., Ono, Y., Okamura, Y, Shiranasam H., and Takamura, H. 2004. A relationship between the amount of polar compounds and the sensory scores of fried foods and deep-frying oil. *Nippon Shokuhin Kagaku Kogaku Kaishi*, 51: 23–27; FSTA, 2004: Na0620.

Hatch, F. T., Knize, M. G., and Felton, J. J. 1991. Quantitative structure-activity relationship of heterocyclic amine mutagens formed during the cooking of food. *Environ. Mol. Mutagenesis*, 17: 4–19; FSTA, 1991: L0060.

He, X.-L.,Tan, X.-He, Wang, Y., Xiong, X.-Y., Wu, W.-G., Zhang, Y., Zeng, Y., Zeng, M., Jiang, M., and Wang, J.-H. 2007. Study of influence factor on acrylamide formation during frying of potato crisps. *Food Sci. Technol.*, 3: 54–58.

Hernández-García, M. T., Martínez-Para, M. C., and Massoud, T. A. 1988. Changes in the content of mercury in fish samples subjected to different cooking. *Anal. Bromatol.*, 40(2): 291–297; FSTA, 1989: R0037.

Ibrahim, N. and Unclesbay, N. F. 1986. Comparison of heat processing by rotating hot air and deep fat frying. *J. Can. Diet. Assoc.*, 47: 26–31.

Ismail-Fitry, M. R., Jinap, S., Jamilah, B., and Saleha, A. A. 2008. Effect of deep frying at different temperature and time on sulfonamide residue in chicken meat-balls. *J. Food Drug Anal.*, 16: 81–86.

Jackson, L. S., Katta, S. K., Fingerhutt, D. D., DeVries, J. W., and Bullerman, L. B. 1997. Effects of baking and frying on the fumonisin B1 content of corn-based foods. *J. Agric. Food Chem.*, 45: 4800–4805.

Johansson, M. and Jägerstad, M. 1993. Influence of oxidized deep-frying fat and iron on the formation of food mutagens in a model system. *Food Chem.Toxicol.*, 31: 971–979.

Kassama, L. S. and Ngadi, M. O. 2004. Pore development in chicken meat during deep-fat frying. *Lebensm.-Wiss.-Technol.*, 37: 841–847.

Khan, N., Zaman, R., and Elahi, M. 1988. Effect of processing on the phytic acid content of Bengal grams (*Cicer arietinum*) products. *J. Agric. Food Chem.*, 36: 1274–1276.

Kim, M., Lee, J., and Choe, E. 2003. Effects of the number of fryings on pigment stability in frying oil and fried dough containing spinach powder. *J. Food Sci.*, 68: 866–869.

Kimura, M., Itokawa, Y., and Fujiwara, M. 1991. Cooking losses of thiamin in food and its nutritional significance. *J. Nutr. Sci. Vitaminol.,* Suppl. 36: 17–24; FSTA, 1991: A0105.

Kita, A., Brathen, E., Knutsen, S. H., and Wicklund, T. 2004. Effective ways of decreasing acrylamide content in potato crisps during processing. *J. Agric. Food Chem.*, 52: 7011–7016.

Kitson, A. P., Patterson, A. C., Izadi, H., and Stork, K. D. 2009. Pan-frying salmon in an eicosapentaenoic (EPA) and docosahexaenoic acid (DHA) enriched margarine prevents EPA and DHA loss. *Food Chem.*, 114, 927–932.

Krbavčič, I. P. and Barič, I. C. 2004. Influence of deep fat frying on some nutritional parameters of novel food based on mushroom and fresh soft cheese. *Food Chem.*, 84: 417–419.

Lane, R. H., Nguyen, H., Jones, S. W., and Midkiff, V. C. 1982. The effect of fryer temperature and raw weight on yield and composition of deep fat fried chicken thighs. *Poultry Sci.*, 61: 286–291.

Lee, J., Kim, M., Park, K. and Choe, E. 2003. Lipid oxidation and carotenoid content in frying oil and fried dough containing carrot powder. *J. Food Sci.*, 68: 1248–1253.

Leszkowiat, M. J., Barichello, V., Yada, R. Y., Coffin, R. H., Loghheed, E. C., and Stanley, D. W. 1990. Contribution of sucrose to non-enzymatic browning of potato chips. *J. Food Sci.*, 55: 281–282, 284.

Li, Y.-J. 1992. Processing induced changes in fish lipids with emphasis on cholesterol. *Dissert. Abstr. Int.*, B: 52.

Llorca, E., Hernando, I., Pérez-Munvera, I., Quiles, A., Fiszman, S. M., and Lluch, M. A. 2003. Effect of batter formulation on lipid uptake during frying and lipid fraction of frozen battered squid. *Eur. Food Res. Technol.*, 216: 297–302.

Love, J. A. and Prusa, K. J. 1992. Nutrient composition and sensory attributes of cooked ground beef: Effect of fat content, cooking method, and water rinsing. *J. Am. Diet. Assoc.*, 92: 1367–1371.

Maruyama, R. and Uno, K. 1997. Oxytetracycline residues in tissues of cultured eel and ayu and the effect of cooking procedure on the residues. *Shokuhin Eiseigaku Zasshi,* 38: 425–429; FSTA, 1998: Rc0493.

Maskat, M. Y. and Kerr, W. L. 2004. Effect of surfactant and batter mix ratio on the properties of coated poultry product. *Int. J. Food Properties*, 7: 341–342; FSTA, 2004: Sn2491.

Martelli, M. R., Carvalho, R. A., Sobral, P. J. A., and Santus, J. S. 2008. Reduction of oil uptake in deep fat fried chicken nuggets using edible coatings based on cassava starch and methylcellulose. *Ital. J. Food Sci.*, 20: 111–118; FSTA, 2009: Sn0420.

Mašková, E., Rysová, J., Fiedlerová, V., and Holasová, M. 1994. Vitamin and mineral retention in meat cooked by various methods. *Potrav. Vědy*, 12: 407–416; FSTA, 1996: S007.

Matthäus B., Haase, N. U., aand Vosmann, K. 2004. Factors affecting the concentration of acrylamide during deep-fat frying of potatoes. *Eur. J. Lipid Sci. Technol.*, 106: 793–801.

Mattick, K. L., Bailey, R. A., Jørgensen, F., and Humphrey, T. J. 2002. The prevalence and number of *Salmonella* in sausages and their destruction by frying or barbecuing. *J. Appl. Microbiol.*, 93: 541–547.

Mitka, M. 2002. Fear of frying. Is acrylamide in foods a cancer risk? *J. Am. Med. Assoc.*, 288: 2105–2106.

Miyagawa, K., Hirai, K., and Takezoe, R. 1991. Tocopherol and fluorescence levels in deep-frying oil and their measurement for oil assessment. *J. Am. Oil. Chem. Soc.*, 68: 163–166.

Moreira, R., Palau, J., and Sun, X. 1995. Simultaneous heat and mass transfer during the deep fat frying of tortilla chips. *J. Food Process. Eng.*, 18: 307–320.

Moyero, P. C., Rioseco, V. K., and González, P. A. 2002. Kinetics of crust colour changes during deep-fat frying of impregnated french fries. *J. Food Eng.*, 54: 249–255.

Murkovič, M., Fridrich, M., and Pfannhauser, W. 1997. Heterocyclic aromatic amines in fried poultry meat. *Z. Lebensm,-Unters.-Forsch.*, 205: 347–350.

Nakai, Y. and Chen, T. C. 1998. Effects of coating preparation methods on yields and composition of deep-fat fried chicken parts. *Poultry Sci.*, 65: 307–313.

Nielsen, P. A., Vahl, M., and Gry, J. 1988. HPLC profiles of mutagens in lean ground pork fried at different temperatures. *Z. Lebenms.-Unters.-Forsch.*, 187: 451–456.

Nout, M. J. R., Ruikes, M. M. W., Bouwmeester, H. M., and Beljars, P. R. 1993. Effect of processing conditions on the formation of biogenic amines and ethyl carbamate in soybean tempe. *J. Food Safety*, 13: 293–303.

Ohgaki, S., Kamei, M., and Morita, S. 1994. Quantitative and qualitative changes in sardine lipids by cooking. *Annu. Rep. Osaka City Inst. Publ. Health Environm. Sci.*, 56: 24–31; FSTA, 1995: R0021.

Ohnishi, S. and Shibamoto, T. 1984. Volatile compounds from heated beef fat and beef fat with glycine. *J. Agric. Food Chem.*, 32: 987–992.

Olsson, V., Solyakov, A., Skog, K., Lundstrom, K., and Jägerstad, M. 2002. Natural variations of precursors in pig meat affect the yield of heterocyclic amines—effect of RN genotype, feeding regime, and sex. *J. Agric. Food Chem.*, 50: 2962–2969.

Övervik, E., Kleman, M., Berg, I., and Gustafsson, J. A. 1989. Influence of creatine, amino acids and water on the formation of the mutagenic heterocyclic amines found in cooked meat. *Carcinogenesis*, 10: 2293–2301.

Padmaja, G. 1995. Cyanide detoxification for food and feed uses. *Crit. Rev. Food Sci. Nutr.*, 35: 299–339.

Pedreschi, F., Kaack, K., and Granby, K. 2004. Reduction of acrylamide formation in potato slices during frying. *Lebensm.-Wissensch.-Technol.*, 37: 679–685.

Pérez-Tinoco, M. R., Pérez, A., Salgado-Cervantes, M., Reynes, M., and Vaillant, F. 2008. Effect of vacuum frying on main physicochemical and nutritional quality parameters of pineapple chips. *J. Sci. Food Agric.*, 88: 945–953.

Persson, E., Oroszvári, B. K., Tornberg, E., Sjoholm, I., and Skog, K. 2003. Heterocyclic amine formation during frying of frozen beef burgers. *Int. J. Food Sci. Technol.*, 43: 62–68.

Persson, E., Graziani, G., Ferracane, R., Fogliano, V., and Skog, K. 2003a. Influence of antioxidants in virgin olive oil on the formation of heterocyclic amines in fried beefburgers. *Food Chem. Toxicol.*, 41: 1587–1597; FSTA, 2004: Sg0192.

Persson, E., Sjoholm, I., Nyman, M., and Skog, K. 2004. Addition of various carbohydrates to beef burgers affects the formation of heterocyclic amines during frying. *J. Agric. Food Chem.*, 52: 7561–7566.

Pikul, J. and Niewiarowicz, A. 1990. Effects of deep fat frying and refrigerated storage of main broiler parts on lipid oxidation in broiler muscles and skin. *Archiv Geflügelkunde*, 54: 85–93; FSTA, 1990: S0117.

Pinthus, E. J. and Singh, R. P. 1996. Resistant starch formation during deep-fat frying of restructured starch product. *U.S. Inst. Food Technol. Annu. Meeting*, Abstract p. 142; FSTA, 1997: L0021.

Pinthus, E. J., Weinberg, P., and Saguy, I. S. 1992. Gel strength in restructured potato products affects oil uptake during deep-fat frying. *J. Food Sci.*, 57: 1359–1360.

Pither, R. J. and Edwards, M. C. 1995. The effect of domestic cooking and preparation techniques. *Nutr. Compos. Vegetables, Rept.*, Camden Choleywood Food Res. Assoc., 18: 1–50.

Pokorný, J. 1998. Substrate influence on the frying process. *Grasas Aceites*, 49: 265–270; FSTA, 1999: Ea0439.

Pokorný, J. and Kołakowska, A. 2003. Lipid–protein and lipid–saccharide interactions. In: *Chemical and Functional Properties of Food Lipids.* Z. E. Sikorski and A. Kołakowska (eds.). CRC Press, Boca Raton, FL, pp. 345–362.

Pokorný, J., Réblová, Z., Kouřimská, L., Pudil, F., and Kwiatkowska, A. 1993. Effect of interactions with oxidized lipids on structure change and properties of food proteins. In: *Food Proteins: Structure and Functionality.* K. D. Schwenke and R. Mothes (eds.). Verlag Chemie, Weinheim, Germany, pp. 232–235.

Price, K. R., Bacon J. R., and Rhodes, M. J. C. 1997. Effect of storage and domestic processing on the content and composition of flavonoid glycosides in onion. *J. Agric. Food Chem.*, 45: 938–942.
Ribarova, F., Yurukov, K., and Shishkov, S. 1994. Stability of tryptophan during heat treatment of meat products. *Khranit. Prom.*, 43: 9–11; FSTA, 1994: S0052.
Rimac-Brnčić, S., Lelas, V., Rade, D., and Šimundić, B. 2004. Decreasing of oil absorption in potato strips during deep fat frying. *J. Food Eng.*, 64: 237–241.
Robbana-Bornat, S., Rabache, M., Rialland, E., and Fradin, J. 1996. Heterocyclic amines: Occurrence and prevention in cooked food. *Environ. Health Perspectives*, 104: 280–288; FSTA, 1997: C200195.
Romani, S., Bacchiocca, M., Rocculi, P., and Dalla Rosa, M. 2008. Effect of frying time on acrylamide content and quality aspects of french fries. *Eur. Food Res. Technol.*, 226: 555–560.
Ruiz-Roso, B. 1998. Vitamins. *Grasas Aceites*, 49: 347–351; FSTA, 1990: Ay1980.
Salta, F. N., Kalogeropoulos, N., Karavanou, N., and Andrikopoulos, N. K. 2008. Distribution and retention of phytosterols in frying oils and fried potatoes during repeated deep and pan frying. *Eur. Food Res. Technol.*, 227: 391–400.
Sánchez-Muniz, F. J., Viejo, J. M., and Medina, R. 1992. Deep frying of sardinesin different culinary fats. Changes in the fatty acid composition of sardines in fryin fats. *J. Agric. Food Chem.*, 40: 2252–2256.
Sébédio, J. I., Ratnayake, W. M. N., Ackman, R. G., and Prévost, J. 1993. Stability of polyunsaturated ω-3 fatty acids during deep fat frying of Atlantic mackerel. *Food Res. Int.*, 26: 163–172.
Shashikala, M. and Prakash, J. 1995. In vitro digestibility of protein in black gram and green gram papads. *Nahrung*, 39: 42–47.
Shih, F. F., Boue, S. M., Daigle, K. W., and Shih, B. Y. 2004. Effects of flour sources on acrylamide formation and oil uptake in fried batters. *J. Am. Oil Chem. Soc.*, 81: 265–268.
Shun-Yao, H. 1997. Effects of cooking and processings on oxytetracycline residues in chicken bones and muscle tissues. *J. Chinese Agric. Chem. Soc.*, 35: 573–579; FSTA. 1998: Sn0609.
Simonne, A. H. 1994. Stability of vitamin E and added retinyl palmitate in frying oils during frying. *Dissert. Abstr. Int., B*, 54: 4461.
Steiner-Asiedu, M., Julshamn, K., and Lie, O. 1991. Effect of local processing method on three fish species from Ghana, I. *Food Chem.*, 40: 309–321.
Sultana, T., Savage, G. P., and Porter, N. G. 2002. Allyl isothiocyanate content of three common raw and cooked *Cruciferae*. *Proc. Nutr. Soc. N. Zealand*, 27: 86–91; FSTA, 2004: 3q3134.
Szymandera-Buszka, K. and Waszkowiak, K. 2004. Iodine retention in ground pork burgers fried in fat free conditions. *Acta Sci. Pol. Technol. Alim.*, 3: 157–162; FSTA, 2005: Sj2082.
Tabee, E., Jägerstad, M., and Dutta, P. C. 2009. Frying quality characteristics of french fries prepared in refined olive oil and palm olein. *J. Am. Oil Chem. Soc.*, 86: 885–893.
Tang, J., Jin, Q. Z., Sher, G.-H., Ho, C. T., and Chang, S. S. 1983. Isolation and identification of volatile compounds from fried chicken. *J. Agric. Food Chem.*, 31: 1287–1292.
Thayer D. W., Shieh, J. S., Jenkins, R. K., Phillips, J. G., Wierbicki, E., and Ackerman, S. A. 1989. Effect of gamma-ray irradiation and frying on the thiamin content of bacon. *J. Food Qual.*, 12: 115–134.
Thomson, B. M., Lake, R. J., Cressey, P. J., and Knize, M. G. 1996. Estimated cancer risk from heterocyclic amines in cooked meat—a New Zealand perspective. *Proc. Nutr. Soc. New Zeal.*, 21: 106–115; FSTA, 1997: Sz0027.
Vaquero, M. 1998. Minerals. *Grasas Aceites*, 49: 352–358; FSTA, 1989: Aj1960.
Visser, J. E., De Beukelaer, H., Hamer, R. J., and Van Vliet, T. 2008. A new device for studying deep-frying behavior of batters and resulting crust properties. *Cereal Chem.*, 85: 417–424.

Vivas-Rodríguez, N. E., Saldivar, S. O., Waniska, R. D., and Rooney, L. W. 1990. Effect of tortilla chip preparation on the protein fractions of quality protein maize, regular maize and sorghum. *J. Cereal Sci.*, 12: 289–296.

Weisburger, J. H., Veliath, E., Larios, E., Pittman, B., Zang, E., and Hara, Y. 2002. Tea polyphenols inhibit the formation of mutagens during the cooking of meat. *Mutation Res.*, 516: 19–22.

Wen-Hsin, W. and Lillard, D. A. 1998. Cholesterol and proximate composition of channel catfish (*Ictalurus punctatus*) fillets—changes during cooking by microwave heating, deep-fat frying, and oven baking. *J. Food Quality*, 21: 41–51.

Witczak, A. 2009. Effect of frying on polychlorinated biphenyls content in muscle meat on selected fish species. *Polish J. Food Nutr. Sci.*, 59: 157–161; FSTA, 2009: Rc1457.

Yi, Y.-H. and Chen, T.-C. 1987. Yields, color, moisture and microbial contents of chicken patties as affected by frying and internal temperature. *J. Food Sci.,* 52: 1183–1185.

Yu, T.-H., Chung, M.-W., and Ho, C.-T. 1994. Volatile compounds generated from thermal interaction of glucose and alliin or deoxyalliin in propylene glycol. *Food Chem.*, 51: 281–286.

Zhang, Y. and Ho, C.-T. 1989. Volatile compounds formed from thermal interaction of 2,4-decadienal with cysteine and glutathione. *J. Agric. Food Chem.*, 37: 1016–1020.

Zialifar, A. M., Heyd, B., and Courtois, F. 2009. Investigation of effective thermal conductivity kinetics of crust and core regions of potato during deep-fat frying using a modified Lees method. *J. Food Eng.*, 95: 373–378.

5 Enzymes and Thermally Oxidized Oils and Fats

Francisco J. Sánchez-Muniz, Juana Benedi, Sara Bastida, Raul Olivero-David, and Maria J. González-Muñoz

CONTENTS

5.1 INTRODUCTION

The physiology of all the body's cells and tissues may be affected by the amount and quality of the fats and oils in the diet (Chow, 2008; Elmadfa and Wagner, 1999; Mahungu et al., 1999) that supply essential fatty acids and other lipids with key metabolic roles. The dietary unsaturated fatty acids—oleic (C18:1, n-9), linoleic

(18:2, n-6), and linolenic (18:3, n-3) acids—are important metabolic energy substrates and may also be transformed into long-chain fatty acids consisting of 20-C and 22-C atoms. After their inclusion in membrane phospholipids or sphingomyelin-enriched areas of cell membranes (Jump, 2004), these long-chain fatty acids regulate cellular gene expression and production of certain bioactive molecules via complex metabolic pathways. In addition, the consumption of diet with a correct content of fatty acids may maintain the cholesterolemia between normal ranges, promote lipoprotein oxidation, and help to balance lipogenesis and lipolysis and thrombogenesis among other functions (Massaro et al., 2002; Massaro et al., 2006; Mata et al., 2002; Mensink and Katan, 1992). An adequate dietary fat intake is thus clearly important in maintaining good health. Most nutrition associations agree that total fat intake should represent no more than 30% of the diet, and that less than 10% of the total energy intake should consist of saturated fatty acids (Kris-Etherton, 1999; Ros, 2001). However, some nutritional guidelines recommend lowering saturated fatty acid consumption even below 7% of the total energy intake (Arbones et al., 2003; WHO, 2003; WHO, 2009; Sánchez-Muniz and Nus, 2008). Due to the significant role of dietary fats and oils in human health, great emphasis has been given to the study of their composition and their susceptibily to oxidation and hydrolysis (Dobarganes et al., 2000; Sánchez-Muniz et al., 2008), as well as the association of atherosclerosis and other degenerative diseases with peroxidation of cell membrane and serum lipoproteins (Chow, 2008; Vemuri and Kelley, 2008).

The Mediterranean diet is considered to be one of the healthiest diets in the world (Keys et al., 1986; De Lorgeril et al., 1999). This traditional diet is based on the consumption of vegetables, cereals, fruits, and olive oil (Trichopoulou et al., 2003); the frequent intake of fried foods is one of its characteristic features (Varela and Ruiz Roso, 1988; Sánchez-Muniz and Bastida, 2006). Fried foods are often considered to be unhealthy, but many investigators (Bognár, 1998; Henry, 1998; Marquez-Ruiz and Dobarganes, 1996; Ringseis et al., 2007; Sánchez-Muniz, 2006; Sánchez-Muniz and Bastida, 2006) report that these foods provide important nutrients and are not harmful when eaten in moderation. Arbitrary recommendations to reduce the consumption of fried foods may thus actually have a negative effect on the dietary habits of populations that have traditionally consumed foods prepared in this manner (Bastida and Sánchez-Muniz, 2001).

Frying represents a traditional method of food preparation that employs oil or fat as a fluid heat vector, and it has to be performed using adequate oils from a stability and nutritional point of view (Blumenthal, 1991; Choe and Min, 2007; Sánchez-Muniz and Bastida, 2006).

This culinary technique alters the physicochemical and organoleptic properties of foodstuffs (Varela, 1988), producing pleasurable food textures (such as crispness), flavors, and aromas (Sánchez-Muniz and Bastida, 2006). An advantageous feature of this method is that it requires no more than 10 min to prepare fried foods (Bastida et al., 2003; Blumental, 1991; Cuesta and Sánchez-Muniz, 2001; Gertz, 2000; Márquez Ruiz et al., 1998; Varela, 1988).

The two major frying techniques are *shallow* and *deep-fat* frying. In contrast to deep-fat frying, in which the entire product is immersed in large quantities of hot oil in domestic or industrial fryers or deep pots, shallow frying involves the use of relatively flat pots or pans and small amounts of oil that do not completely cover the foodstuffs. In shallow frying, only the food in contact with the oil is actually fried, while the rest simply cooks. Bognár (1998) reports that deep-fat-fried meat and fish absorb less oil than the same products prepared using the shallow frying method.

The frying process may include *frequent*, *slow*, or *null* fresh oil replenishment. Oil alteration is minimal when fresh oil is added frequently during frying (Friedman, 1991; Romero et al., 2000a; Romero et al., 2000b; Romero et al., 2007; Sánchez-Muniz et al., 2008).

Discontinuous frying refers to the procedure in which oil is allowed to cool off (normally to room temperature) between uses, in contrast to *continuous* frying, in which foods are uninterruptedly added to the hot oil. Under laboratory conditions, potatoes fried using the continuous method exhibited less oil alteration than those fried employing the discontinuous technique (Jorge et al., 1996).

As a result of the complex reactions and numerous compounds formed in frying, it is difficult to evaluate all the changes that take place in the fatty acids of fats and oils during the frying process. However, considerable information regarding the nature and concentrations of the modified fatty acids is now available, as a result of the development of chromatographic techniques over the past 20 years. The concentration of modified fatty acids in frying oils and fats with a polar compound content of about 20%–25%, close to the upper limit permitted by present legislation, is approximately 10%–12% (DGF, 2000; Sanchez-Muniz et al., 2008). Most of these compounds are fatty acid dimers and higher oligomers, although a sizable portion consists of oxidized fatty acid monomers. The latter include primarily epoxy, keto, and hydroxyl function compounds, while much smaller quantities of short-chain fatty acids are also found. In this context, further investigation is needed to better understand the specific structures of the modified fatty acids produced during frying, and more work must be done to evaluate their impact on human nutritional status when consumed at the levels normally present in fried foods and in fats and oils used for frying.

The aim of the present review is to describe how enzymes have been used (1) to study the alteration and/or degradation of oils and fats heated or used in frying and (2) as biomarkers in the study of fresh and altered fat and oil intake.

5.2 PHYSIOLOGICAL IMPORTANCE OF ENZYMES

The enzymes modulate important physiological processes such as digestion, absorption, and metabolism of fats and oils (Welch and Borlak, 2008). Enterocytes of the intestinal wall absorb the products of the hydrolysis of ingested fats and oils (free fatty acids, monoacylglycerols, lysophospholipids, and free cholesterol molecules) (Welch and Borlak, 2008). Between 95% and 99% of the fatty acids in the diet are consumed in the form of acyl esters, normally triacylglycerols, but also as mono- or diacylglycerols. To a lesser degree, dietary fatty acids are also found in the form of free fatty acids, phospholipids, and other complex lipids that exist in plant and animal tissues (Gunstone and Norris, 1983). These enzymes are biomarkers of ingested altered fat, and the study of their physiological roles will aid in the investigation of oil/fat alteration.

5.2.1 Lipases

Lipases (triacylglycerol acyl hydrolases, E.C.3.1.1.3), found in most living organisms, have essential physiological functions in addition to important industrial applications. These enzymes play a key role in the digestion, absorption, and reconstitution of dietary lipids and in the lipoprotein metabolism of eukaryotic organisms, and are also found in the food-reserve tissues of plants.

The term *triglycerides* refers to acylglycerols in which the glycerol molecule is esterified with three fatty acids. Fatty acids may vary according to the length of the chain (short, medium, or long), the ramification (branched or unbranched molecular structure), degree of saturation (saturated, monounsaturated, or polyunsaturated), the location of the double bonds (Δ, when the relative position of the double bounds is the acid end; *n* or ω, when the relative position is the methyl end), and the existence of geometric isomerization (*cis* or *trans*).

In humans, the most important lipases, such as the gastric, pancreatic, and bile salt-stimulated lipases, participate in the digestion and assimilation of dietary fats, while others, including the hepatic, lipoprotein, and endothelial lipases, participate in lipoprotein metabolism (Mukherjee, 2003). The pancreatic, hepatic, lipoprotein, and endothelial lipases are members of the lipase gene family. Digestion of dietary fats in the intestine is made possible by pancreatic lipase, an enzyme produced by exocrine acinar cells and present in pancreatic juice. Rather than using a single molecule substrate, this enzyme uses a nonaqueous phase of aggregated lipids consisting of aggregates of ester molecules, micelles, or monolayers interfacing with an aqueous medium as its substrate (Esposito et al., 1973; Lagocki et al., 1973). The enzymatic activity of pancreatic lipase depends on its cofactor, colipase, which helps anchor the lipase to the surface and stabilizes it in the "open" active conformation (Lowe, 1997; Brockman, 2000).

Lipases do not induce substrates dissolved in bulk fluids to hydrolysis and are only activated when adsorbed at an oil–water interface (Martinelle et al., 1995). Emulsified triolein and tripalmitin, and other long-chain fatty acid esters are hydrolyzed by true lipases. Soluble substrates in aqueous solutions are little affected by lipases, which are members of the serine hydrolase superfamily. Interfacial interaction between lipids and lipases is little understood to date, and important research on the subject is currently underway (Aloulou and Carrière, 2008).

Investigation of lipases frequently centers on the structure, mechanisms of action, kinetics, and overall performance of these enzymes, as well as on lipase gene cloning and sequence analysis (Alberghina et al., 1991; Bornscheuer, 2000). Numerous industries, including those specializing in organic chemical processes, oleo-chemical applications, dairy products, agrochemical materials, paper, cosmetics, biosurfactants, detergents, nutrition, and pharmaceuticals currently use lipases (Nus et al., 2006b).

Technologies employing lipases to produce new compounds for industrial use are rapidly being developed, increasing the potential applications of these enzymes (Liese, 2001). Use of enzymatic interesterification has increased markedely in recent years due to the search for more thermostable dietary oils with a healthy fatty acid composition (Nus et al., 2006b). In contrast to the chemical interesterification process, enzymatic interesterification does not require the use of numerous chemical catalysts and high temperatures. Enzymatic interesterification is less costly and more efficient, and reduces the formation of a number of undesirable molecules. In addition, one of the most significant novel uses of immobilized enzymes, including lipases, is the conversion of waste oils into biodiesel (Watanabe et al., 2001).

The oleochemical industry uses lipases in its chemical transformations (Bornscheuer, 2000; Soumanou and Bornscheuer, 2003), including the processing of the polyunsaturated fatty acid (PUFA) γ-linolenic acid, the food colorant astaxanthin, the methyl ketones responsible for the characteristic flavor of blue cheeses, the fruit flavor precursor γ-decalactone, dicarboxylic acid prepolymers, and the inexpensive acylglycerols for different uses (e.g., cocoa butter replacers) (Undurraga et al., 2001). Human milk fat mimetics are now produced for use in baby formulas, using lipases to modify the 2-position of vegetable oil triacylglycerols (Schmid et al., 1999; Adamczak, 2004).

Under various conditions, Blasi et al. (2009) studied the production of structured triacylglycerols using an enzymatic reaction between *sn*-1,3-diacylglyerol and conjugated linoleic acid (CLA), employing immobilized lipase from *Rhizomucor miehei.* These authors observed a direct correlation between the addition of CLA and the enzymatic load and reaction time, while reporting an inverse correlation between CLA incorporation and the temperature and the 1,3-diacylglyerol/CLA substrate ratio variables in total triacylglycerols and in the *sn*-2 position.

Applications for lipases also include synthesis of lipid esters (Bornscheuer et al., 2002), specifically that of isopropyl myristate, used in the cosmetics industry, and production of the monoacylglycerols used in the food and pharmaceutical industries as emulsifiers. Demand for more efficient industrial production of pure enantiomers, including chiral anti-inflammatory drugs such as naproxen (Xin et al., 2001) and ibuprofen (Arroyo et al., 1999; Chen and Tsai, 2000), antihypertensive agents (including the angiotensin-converting enzyme inhibitors captopril, enalapril, ceranopril, zofenapril, and lisinopril), and calcium-blocking drugs such as diltiazem has increased due to the recognition of the importance of chirality in biological activity. Lipases are employed in the production of these drugs (Berglund and Hutt, 2000).

5.2.2 Phospholipases

Phospholipases (PL) (EC 3.1.1.4) are enzymes that hydrolyze phospholipids, forming free fatty acids and their lysophospholipids counterparts (Vemuri and Kelley, 2008). Although phospholipids may also be susceptible to the action of pancreatic lipase at position 1 (1,3 acylglycerol ester hydrolases), most of these compounds are usually hydrolyzed by PL A2. Pancreatic PL A2 produces 1-lyso compounds through its action at the *sn*-2 position of phospholipids, which are further absorbed by enterocytes.

Despite the fact that PLs have an essential function in the digestion and absorption of fats, few studies appear to have been carried out recently on the use of these enzymes to study oils or on the effect of fat and oil intake on intestinal enzyme gene expression and activity, and virtually no studies using altered oils/lipids have been published. The aim of certain investigations has been to study the use of PLs to eliminate phosphorus from refined oils and fats and improve their organoleptic characteristics (Clausen et al., 2000).

Ample information is available on PLs regarding the function of these enzymes in metabolic pathways associated with eicosanoid and second messenger synthesis (Lee and Hwang, 2008). Thus, *in vivo* fatty acid composition of membrane phospholipids can be modified by dietary fat and by alteration of cellular desaturation and acylation reactions. Cellular lipases and PLs participate in the release of fatty acids from membrane phospholipids. Endogenous fatty acids that act as second messengers are apparently released by agonists that activate these lipases and/or PLs (Bruckner, 2008). Classification of the cellular PLs depends on the phospholipid bonds they cleave. Thus, fatty acids from *sn*-1 and *sn*-2 positions in phospholipid molecules are specifically targeted by PLs A1 and A2. PL B cleaves both the 1- and 2-lysophospholipids, produced by PL A1 and PL A2, respectively. The bond between glycerol and a phosphate is broken by PL C, and the phospholipid amino alcohol moiety undergoes hydrolysis due to the action of PL D (Figure 5.1). The principal function of PL A2 is to release fatty acids with 20-C atoms, including arachidonic acid, which undergoes most of its esterification in the *sn*-2 position of the phospholipids in cell membranes. Other activated PLs, notably PL C, followed by diacylglycerol and monoacylglycerol lipases, also release 20-C fatty acids.

5.2.3 Arylesterase

Differing from other enzymes that participate in the metabolism of fat and cholesterol, the esterases act in a homogenous polar phase rather than at interfaces.

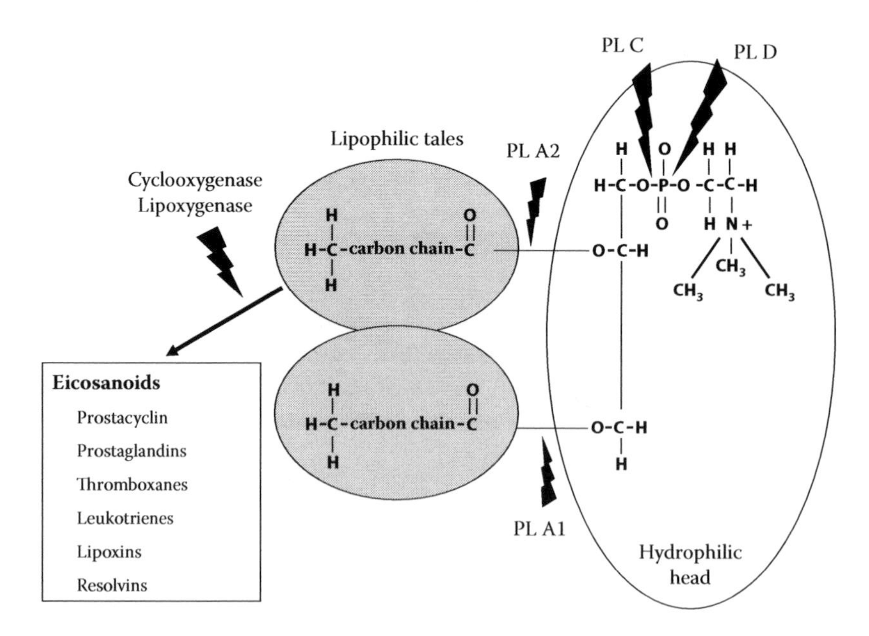

FIGURE 5.1 Hydrolytic point of different phospholipases for the study of phospholipids. The figure also shows the formation of eicosanoids. PL: Phospholipase.

Arylesterase (E.C. 3.1.1.2) is an esterase that displays one of the three enzymatic activities of paraoxonase-1 (PON-1) (Canales and Sánchez-Muniz, 2003). According to different studies, arylesterase may be a high-density lipoprotein (HDL)-bound enzyme found in the same locations as apolipoprotein (apo) A1 and apo J, also called *clusterin* (Blatter et al., 1993; Kelso et al., 1994). Although it is known that arylesterase hydrolyzes aromatic esters, including phenylacetate (Figure 5.2), no information is available regarding its native substrate (Nus et al., 2007).

An important role of arylesterase is to facilitate reverse cholesterol transport and safeguard LDL and HDL from oxidation (Parthasarathy et al., 1990; Aviram et al., 1998). The capacity of the enzyme to cleave cholesteryl linoleate hydroperoxides (Aviram et al., 1998) and certain oxidized phospholipids (Navab et al., 1996) in oxidized LDL probably enables it to fulfill this function. However, arylesterase displays both antioxidant and the arylesterase itself activities. Using sulfhydryl-blocking agents such as *p*-hydroxymercurybenzoate (PHMB), Aviram et al. (1999) have demonstrated that arylesterase requires a free sulfhydryl group on cysteine-284 in order to hinder oxidation of LDL-cholesterol. The replacement of cysteine-284 by serine or alanine does not alter the arylesterase activity of PON-1, which does, however, depend on the presence of calcium. This mineral, on the other hand, is not needed by arylesterase for its role in protecting LDL-cholesterol from oxidation.

Arylesterase activity varies between individuals due to mutations in the PON-1 encoding gene found on the long arm of chromosome 7 (Canales and Sánchez-Muniz, 2003; Nus et al., 2007). These variant forms utilize, similarly, phenylacetate as substrate, but the antioxidant capacity of the different PON-1 polymorphisms may vary. It is as yet undetermined which allozyme displays the greatest protection against LDL-cholesterol oxidation.

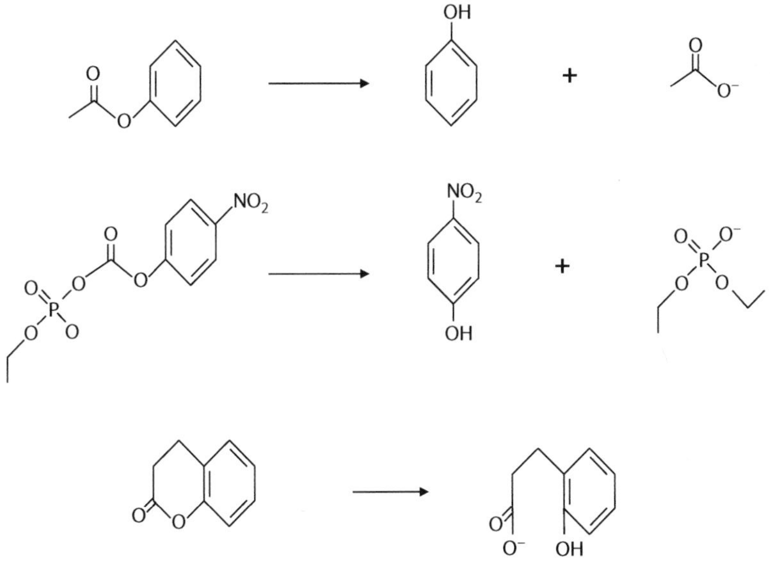

FIGURE 5.2 Types of reactions catalyzed by the PON-1 enzyme.

The arylesterase activity of PON-1 is reduced in hypercholesterolemia, type-2 diabetes, and cardiovascular disease (Sutherland et al., 2001; Azarsiz et al., 2003), while certain dietary customs (Sarandol et al., 2003; Wallace et al., 2001) and lifestyle habits such as smoking (Nishio and Watanabe, 1997) affect both its antioxidant and arylesterase activities. Free calcium must be available for proper arylesterase activity (Eckerson et al., 1983; Nus et al., 2006a), and chloride in the medium increases this capacity. The activity of this enzyme decreases, on the other hand, in the presence of high concentrations of bicarbonate (Nus et al., 2006a).

5.2.4 Lipoxygenases and Cyclooxygenases

A common class of enzymes that uses molecular oxygen to catalyze the oxidation of lipids with a *cis-cis*-1,4-pentadiene group acts in plant systems. The principal function of these enzymes in animal lifeforms is the oxidative transformation of arachidonic acid and other 20-C PUFAs into multifunctional compounds such as prostaglandins, thromboxanes, and leukotrienes. This process occurs in all mammalian tissues. Cyclooxygenase participates in the formation of prostaglandin G2 (PGG2) or G3 (PGG3), the first product of arachidonic acid or eicosapentaenoic acid, respectively. The cyclic endoperoxides PGG2 or PGG3 are the precursors of prostaglandins and thromboxanes derived from omega-6 and omega-3 PUFAs, respectively. Arachidonic and eicosapentaenoic acids can also be oxidized by lipoxygenases, thus being precursors of different hydroperoxides and epoxides, which in turn are precursors of 4- and 5-series of leukotrienes (Padley et al., 1986; Lee and Hwang, 2008). A summary of the implicated reactions is shown in Figure 5.3.

As an iron carrier, lipoxygenase (linoleate: oxygen oxidoreductase, EC 1.13.11.12) is one of many metalloproteins with important biological functions. Soy lipoxygenase is one of the best-known enzymes in this group. Legumes, peas, soy, and other beans represent the primary plant sources of lipoxygenase, although to a lesser degree, this enzyme is also found in potatoes, wheat, radishes, and peanuts. This enzyme is also present in virgin olive oil (Georgalaki et al., 1998). Oxidation of the unsaturated fatty acids present in seeds stored for lengthy periods of time is primarily due to the action of lipoxygenases (Suzuki et al., 1996). As heat-labile proteins, however, these enzymes display a propensity toward denaturation at high temperatures, and are thus not present in refined oils. For this reason, enzyme-catalyzed oxidation of fats and the majority of oils during storage and utilization is not a problem.

There are two types of lipoxygenases present in plants. Type I lipoxygenase is only responsible for the peroxidation of free fatty acids with high stereo region selectivity, and acts on them after lipase has freed them from triacylglycerols. Type II lipoxygenase, less specific for free linoleic acid, catalyzes autoxidation in general, and acts directly on triacylglycerols.

Gardner (1996) and Andreou and Feussner (2009) have reviewed the mode of action of lipoxygenase. The reaction catalyzed by lipoxygenases is highly specific for the *cis-cis* methylene-interrupted sequence and the basis of the analytical method to measure essential fatty acids (Chism, 1985).

Initiation and/or acceleration of autoxidation during fat/oil processing, refining, and storage may be triggered by a number of factors, such as the presence of heat,

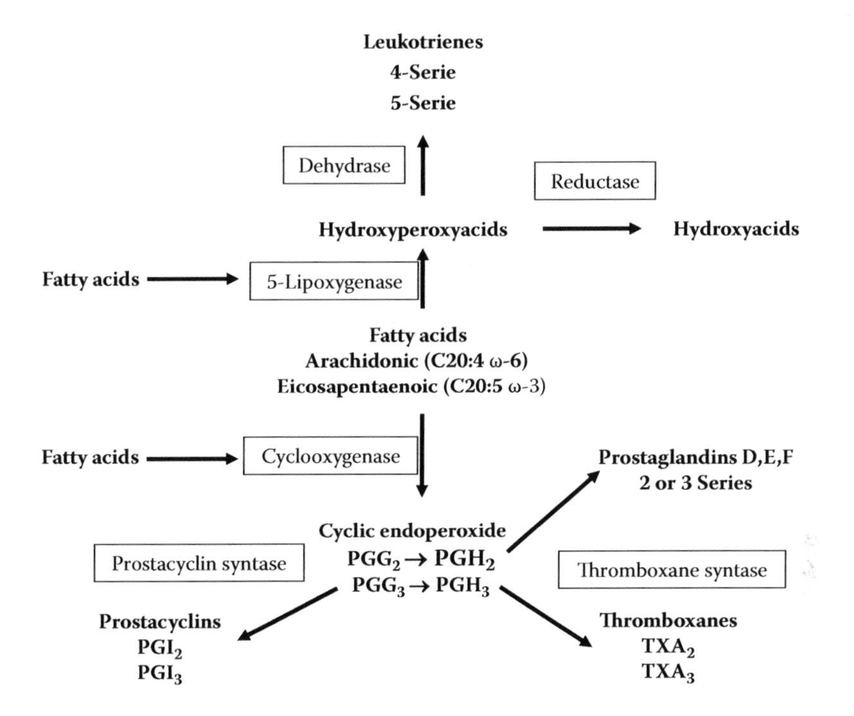

FIGURE 5.3 Formation of different eicosanoids and related compound formed through the lipoxygenase and cycloxygenase pathways. *Note*: PG, prostaglandin; PGI, prostacyclin; TX, thromboxane.

light, metals, oxygen, free fatty acids, and lipoxygenases. Lipoxygenases, which catalyze polyunsaturated fatty acid oxidation, exist in all plant and animal tissues (Engeseth et al., 1987; Patterson, 1989; Sherwin, 1978).

Lipid peroxidation arising from altered fat consumption affects eicosanoid production. Lipid peroxidation plays an important role in certain noncytotoxic natural biological processes, such as prostaglandin production, but may accelerate free-radical chain reactions that occur when lipid peroxidation conditions are uncontrolled. According to Giani et al. (1985), accumulation of peroxides may affect homeostatic control of eicosanoids, allowing synthesis of thromboxanes and leukotrienes but halting that of PGE2 and prostacyclin and decreasing concentrations of tocopherol *in vivo*. A vitamin E supplement restored the prostacyclin/thromboxane balance to normal again.

5.2.5 Glutathione Enzyme System

Molecules or atoms with unpaired electrons in their outermost shell and aggressive oxidative behavior are continually being produced in body tissues. Enzyme-mediated electron-transfer reactions often produce free radicals in cells. One of the most important of such reactions is the mitochondrial respiratory chain that releases the energy of nutrients and also produces superoxide radicals.

Proliferation of free-radical reactions when endogenous defence mechanisms are insufficient may result in co-oxidation of such nucleophilic macromolecules as enzymes, DNA, and certain membrane components (Yuan and Kitts, 1997). *In vivo* peroxidation modifies the fluidity and permeability characteristics of biological membranes and is related to cellular toxic phenomena and tissue damage. While there is considerable controversy regarding the role and significance of exogenous antioxidants in organic antioxidant defence, the important function of endogenous antioxidants is universally accepted. The enzymes primarily responsible for detoxifying reactive oxygen species (ROS) and lipid hydroperoxides are Cu,Zn-superoxide dismutase (SOD), found in cytoplasm; Mn-SOD, located in mitochondria; catalase, present in peroxisomes; the two glutathione redox cycling enzymes, selenium-dependent glutathione peroxidase (GPx) and glutathione reductase (GR), found in cytoplasm and mitochondria; and nonselenium-dependent glutathione-*S*-transferase, which is present in cytoplasm.

The actions and interactions of these enzymes are represented in Figure 5.4.

The two principal enzymes of the glutathione redox system are GR and GPx. GR converts GSSG into GSH, while GPx produces primary alcohol from hydroperoxides. A third component of the glutathione redox system is glutathione-*S*-transferase, an enzyme that produces glutathione conjugates by catalyzing the conjugation reaction between GSH and electrophilic, xenobiotic, and carcinogenic compounds.

Hepatic activity of the glutathione enzyme system in rats was studied *in vivo* by Saka et al. (2003). The glutathione defence system in living cells detoxifies and eliminates certain endogenous materials and xenobiotic substances by forming water-soluble compounds. Enzymes of this system interact with others to reduce or eliminate the toxicity of pesticides, medications, hormones, and other substances in the body.

Purushothama et al. (2003) studied the effect of long-term feeding with fried peanut, rice bran, and palm oil added in the diet on the hepatic antioxidant enzyme

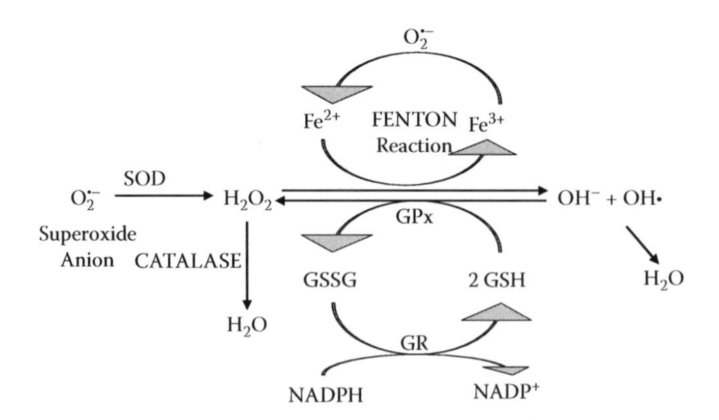

FIGURE 5.4 Relationship of glutathione and antioxidant enzymes; mechanisms involving neutralizing free radicals and reactive species. (SOD, superoxide dismutase; GPx, glutathione peroxidase; GR, glutathione reductase; GSH, reduced glutathione; GSSG, oxidized glutathione.)

status in laboratory rats. The rats were fed oils fried at 20% level in the diet for 18 weeks. Although some significant changes for the GPx and GST activity were found in frying, the results indicated no appreciable damage with respect to these antioxidant enzymes.

Certain pathological conditions have been associated with modified antioxidant enzyme behavior. For example, concentrations of lipid peroxides in the aorta and other tissues may decrease as a result of increased GPx activity (Yuan and Kitts, 1997; Huang et al., 2002). However, it is difficult to interpret the meaning of the tissue-specific and antioxidant enzyme-specific modifications associated with various morbid conditions, due to the complex adaptive or compensatory organic metabolic processes taking place. Due to the tissue-specific nature of antioxidant enzyme activity, it is often very difficult to ascertain the cause–effect relationship between significant enzymatic modifications and particular pathological conditions.

Many alterations in metabolism, growth, and cell differentiation can be attributed to the substantial effects of dietary fats on gene expression. These effects represent the adaptive response of body tissues to changes in the amount and quality of ingested fats (Jump and Clarke, 1999). Major specific fatty acid-regulated transcription factors in mammalian tissues include the peroxisome proliferator-activated receptors (PPARS) HNF4α, NFkB, and SREBP1c. These transcription factors can be regulated by (1) direct binding of fatty acids, fatty acyl-coA, or oxidized fatty acids; (2) oxidized fatty acids (as eicosanoids) regulation of G-protein-linked cell surface receptors and activation of signaling cascades that target the nucleus; and (3) oxidized fatty acid regulation of intracellular calcium levels that affect cell signaling cascades targeting the nucleus (Jump and Clarke, 1999).

5.3 ENZYME METHODS FOR FAT COMPOSITION ASSESSMENT

This section deals with some major enzymatic methods used to study various characteristics related to lipid and fat composition (type, structure, and alteration/oxidation status) that may aid in understanding the stability and shelf-life of fats and oils. Estimated antioxidant enzyme activity data and information regarding the genetic expression of these enzymes in tissues as a consequence of the intake of altered fats and foods are also included.

Enzymatic changes are important for olives and olive oil quality. Information about the degradation reactions by enzymes can be obtained in the books edited by Boskou (2006). The first step in the process of oil degradation is the release of fatty acids by lipolytic enzymes. Acyl hydrolases, for example, are active in olives during their maturation, selection, and oil extraction. The crushing of the cells of the pulp liberates enzymes that degrade the lipids present. Morales and Przybylski (2000) have summarized the sequence of enzymatic actions that may contribute to the deterioration of oils such as olive oil. During the olive ripening and storage stages, and particularly in the malaxation stage of oil extraction, enzymes such as lipase, lipoxygenase, PL, and hydroperoxide lyase/isomerase or dehydrogenase are active (Sciancalepore and Longone, 1984; Goupy et al., 1991; Morales and Aparicio, 1999; Salas et al., 1999). In this view, undesirable enzymatic activity can be confirmed when good manufacturing practice rules for virgin olive oil production are not followed. Lipases can be active even during the short time between the grinding of the olives and the separation of the oil from the solid material. Free fatty acids are usually released during the cellular degradation that occurs throughout the storage period of harvested olives and in the oil extraction process. During this phase, unsaturated fatty acids, pigments, and phenols also undergo oxidation. These alterations are related to lipoxygenase and hydroperoxide lyase activities (Boskou, 2006). Lipoxygenase activity leads to the rapid deterioration of low-quality extracted oils (such as olive pomace oils), which are produced by grinding the pits with part of the pulp. Milling liberates the enzymes present in the pits, accelerating the formation of oxidation compounds (Boskou, 1996).

Enzymes of bacterial origin may also be present in veiled olive oil (Ciafardini and Zullo, 2002; Koidis et al., 2008). According to Ciafardini and Zullo (2002), their most abundant source in olive oil is yeast. These authors report that microbiological glycosidases and other hydrolytic enzymes in stored olive oil may also contribute to a reduction of bitterness. Pomace oil, produced by grinding olive pits and pulp (known as "orujo" in Spain), contains enzymes that contribute to deterioration during storage (Boskou, 1996). Enzymatic activity, however, is necessary, as it is essential for the production of the volatiles that produce the desirable green odor of olive oil (Olias et al., 1993; Morales and Aparicio, 1999; Aparicio and Morales, 1998; Angerosa et al., 2000; Ridolfi et al., 2002). There is generally a plethora of publications relating the sensory quality of olive oil to the activity of lipoxygenase and other enzymes in olives (Williams et al., 2000; Briante et al., 2002; Angerosa, 2000; Angerosa et al., 2000). Proteins and oxidative enzymes (lipoxygenase and phenoloxidase) may be present and active in virgin olive oil, although this information is not conclusive (Georgalaki et al., 1998).

5.3.1 Pancreatic Lipase

Enzymatic hydrolysis was first used in a systematic manner to demonstrate the different distributions of fatty acids in the three possible positions of the triacylglycerol moiety (Christie, 1986). Fatty acids at the *sn*-2 position can be analyzed specifically using pancreatic lipase. It is possible to more thoroughly analyze the positional distributions of fatty acids using more complicated stereospecific hydrolysis methods (Christie, 1986).

As commented in the previous section, the oil and fat industries rely heavily on the use of lipases. Hydrolysis of triacylglycerols and acyl and aryl esters (Christie, 1986; Faber, 2004) is activated by these enzymes, many of which can also catalyze other organic reactions in nonaqueous media (Klibanov, 1986; Sánchez-Montero et al., 1991; Hernáiz et al., 1994a).

Due to its low cost and availability, pancreatic lipase is widely used under experimental conditions as a digestive tract lipase. This enzyme hydrolyzes or synthesizes triacylglycerols that display positional and fatty acid specificities. Although human pancreatic lipase is clearly the enzyme of choice to study the nutritional repercussions of fat hydrolysis and digestion in humans, porcine pancreatic lipase, which is 86% homologous with the human enzyme, is used instead (Winkler et al., 1990).

The three stages of lipase hydrolysis include interface enzyme adsorption, interfacial activation, and catalysis. Pancreatic lipase, present in aqueous medium, penetrates between the lipophilic substrate and water into the interface. The enzyme–substrate complex, formed when both elements are at the interface, leads to the formation of the catalytic product, on the one hand, and enzyme regeneration on the other (Faber, 2004; Hermoso et al., 1996).

Triacylglycerol molecules that include short-chain fatty acids are more rapidly hydrolyzed by pancreatic lipase than those with only long-chain fatty acids. Hydrolysis of ester bonds with proximal substituent groups may be slow as the result of steric hindrance. This is the case in some especially long PUFAs (e.g., docosahexaenoic acid), certain *trans* fatty acids (e.g., *trans*-3-hexadecenoid acid), and the phytanic acid ester of glycerol (Faber, 2004; Bottino et al., 1967). The fact that pancreatic lipase does not efficiently hydrolyze triacylglycerols with behenic acid is of importance to the food industry, in the manufacturing of foods with low amounts of available energy (Bastida and Sánchez-Muniz, 2001; Arishima et al., 2009).

The presence of calcium ions and bile salts is necessary for the structural analysis of triacylglycerols. In addition, as the enzyme only reacts to the micellar form of fats, these must undergo vigorous shaking before analysis can be carried out. To increase the solubility of saturated fats that have high melting points, carriers such as methyl oleate (Barford et al., 1966) or hexane (Brockerhoff, 1965) are sometimes used to facilitate the reaction. An alternative to the use of these carriers is preincubation at 42°C (Rossell et al., 1983). When microbial lipases are used to analyze triacylglycerols, the recommended carriers are isooctane and cyclohexane (Kim et al., 1984; Hernáiz et al., 1999). To rapidly achieve an acceptable degree of hydrolysis (50%–60%), it is necessary to adjust enzyme, cation, and bile salt concentrations and determine the optimal temperature and pH of the buffer (Hernáiz et al., 1994b).

The most highly recommended practical method of studying triacylglycerol structure is the semi-micro technique of Luddy et al. (1964), the principle of which is described in the following text.

Previous to the addition of pancreatic lipase formulation, triacylglycerols are mixed with tris(hydroxymethyl)methylamine (Tris) buffer, calcium chloride solution, and a solution of bile salts in a stoppered test tube that is left for 1 min in a 40°C water bath. The necessary degree of hydrolysis is achieved by vigorous shaking via a mechanical shaker, and the reaction is terminated with the addition of ethanol, followed by the addition of hydrochloric acid. The solution is centrifuged to break any emulsions present, if necessary, and then extracted three times with diethyl ether. The resultant solvent layer is washed two times with portions of distilled water and then dried over anhydrous sodium sulfate. Thin layer chromatography (TLC) and high-performance liquid chromatography (HPLC) are two of the methods available to study the compounds formed through hydrolysis. Gas chromatography (GC) can be used to determine triacylglycerol composition if the hydrolysis products are resuspended with *N*-methyl-*N*-trimethylsilylheptafluorbutyramide to transform the free fatty acids into their corresponding silyl ether (Schmitt et al., 2002). TLC is usually employed to isolate the 2-monoacyl-*sn*-glycerols formed by hydrolysis, and these compounds are transesterified to fatty acid methyl esters before GC is performed. HPLC is preferable to TLC separation, as retention factor (Rf) values of 2-monoacyl-

sn-glycerols are similar to those of many compounds present in altered oils. Other nutrients present, such as proteins, can influence the digestibility of lipids.

Pancreatic lipase is used to study the effect of globular interfacial protein cross-linking on *in vitro* digestibility of emulsified corn oil. Sandra et al. (2008) stabilized emulsions of corn oil in water using lecithin or beta-lactoglobulin (pH 7). Heat (85°C, 20 min) was used to cross-link the adsorbed globular proteins of a quantity of the β-lactoglobulin-stabilized emulsions. Changes in particle charge, size, appearance, and liberated free fatty acids were measured during 2 h after pancreatic lipase and bile extract was added to each portion of emulsion at 37°C. Few differences were observed between the rate and extent of lipid digestion in lecithin and β-lactoglobulin-stabilized emulsions or between unheated (BLG-U) or heated (BLG-H) β-lactoglobulin-stabilized emulsions. For example, the initial rate of lipid digestion was found to be 3.1, 3.4, and 2.3 mM fatty acids s^{-1} m^{-2} of lipid surface for droplets stabilized by BLG-U, BLG-H, and lecithin, respectively. Neither the initial composition of the interface nor the fact that a part of the original emulsifier stayed on the oil–water interface during digestion altered the ability of pancreatic lipase to adsorb onto the droplet surfaces and reach the emulsified lipids; this partially explains the efficiency displayed by the human body in the digestion of dietary triacylglycerols.

p-Nitrophenyl esters with short (acetate or butyrate) or long (palmitate or oleate) aliphatic acyl chains are frequently used to study esterase and lipase activities, respectively. The release of *p*-nitrophenol is measured spectrophotometrically (Gilham and Lehner, 2005). To date, no known lipolytic enzyme is able to differentiate between triacyl-*sn*-glycerol positions 1 and 3. Investigators have developed several original stereospecific analytic procedures to establish the composition of positions *sn*-1, *sn*-2, and *sn*-3 (Christie, 1986; Breckenridge, 1978).

Electrospray ionization mass spectrometry (ESI-MS), chiral-phase HPLC, and GC have successfully been used to identify regioisomers (reverse isomers) of 1,2-diacyl-*sn*-glycerols that possess various pairs of saturated and unsaturated acyl groups. The isomers were prepared by partial Grignard degradation of the corresponding triacylglycerol and converted into 3,5-dinitrophenylurethanes (DNPU), 3,5-dinitrobenzoates, nicotinates, and 1- and 2-anthrylurethanes (Mu et al., 2001; Itabashi et al., 2000).

Both hydrolysis and autoxidation produce structural modifications of fats and oils (Dobarganes et al., 2000; Sánchez-Muniz and Bastida, 2006). Unsaturated fats are most susceptible to fat oxidation (e.g., linoleic acid, which is 40 times more susceptible than oleic acid). Repeated frying produces thermally oxidized oils that contain a complicated mixture of compounds including oxidized triacylglycerol monomers, triacylglycerol dimers, and triacylglycerol polymers, which are primarily the result of changes in the physical and chemical properties of fats (Gutiérrez González-Quijano and Dobarganes, 1988; López-Varela et al., 1995; Gere, 1982; Sánchez-Muniz et al., 2008). Consequently, the oil–water interface could vary, and the lipolytic activity of the lipase could be altered. To date, hydrolysis of oxidized or polymerized oils has only been studied using pancreatic lipase.

As previously mentioned, triacylglycerols with more hydrophilic fatty acids, such as C10 and C12, react better to pancreatic lipase than those containing C16 and C18

(Linscheer and Vergroesen, 1994). Fats and oils normally contain both linoleic and linolenic acids. Under prooxidant conditions, these fatty acids are converted into their respective hydroperoxide forms. Altered oils appear to undergo less hydrolyzation than unaltered oils using pancreatic lipase (Yoshida and Alexander, 1983; Miyashita et al., 1990; Sánchez-Muniz et al., 1999). Nevertheless, several factors such as the presence of surfactants in the oil can improve the hydrolysis (Arroyo et al., 1997) that will be discussed in the following text. Thus, pancreatic lipase may display a preference for hydrolyzing hydroperoxy linoleoyl and linolenoyl groups due to their higher polarity.

Miyashita et al. (1990) proposed a mechanism to explain the hydrolysis of triacylglycerides monohydroperoxides. The resulting products are schematized in Figure 5.5. This author also suggested that hydroperoxides do not inhibit lipase activity, although lipid peroxidation products react with the enzyme protein, reducing its biological activity. Protein-centered free-radical intermediates may decrease enzyme activity as a result of structural alterations in the proteins (Funes and Karel, 1981; Funes et al., 1982; Leake and Karel, 1982).

Enzymatic hydrolysis was used to study acylglycerol products from thermally oxidized corn, sunflower, and soybean oils (Yoshida and Alexander, 1983). The

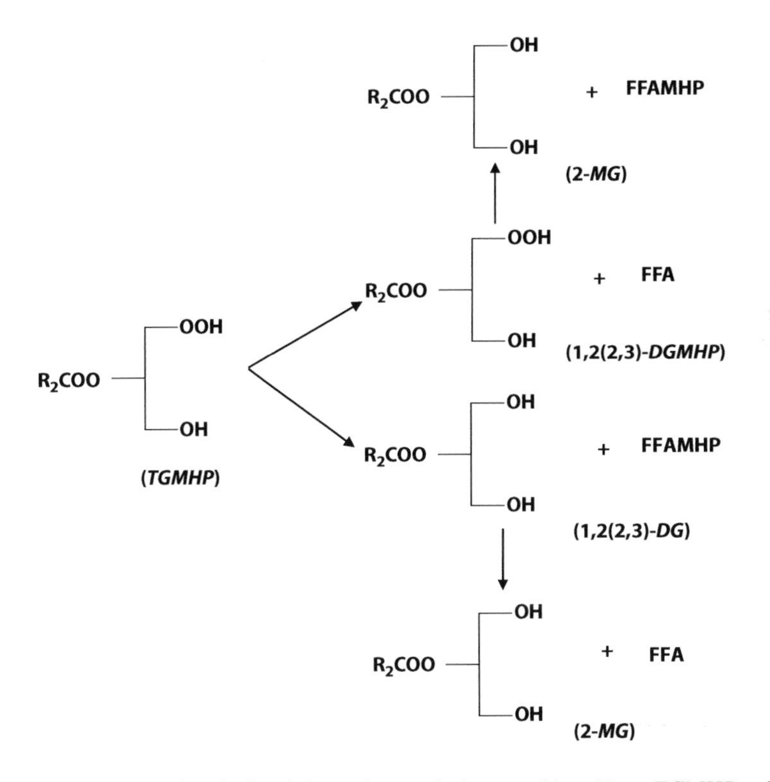

FIGURE 5.5 Hydrolysis of triacylglycerol monohydroperoxides. *Note*: TGMHP, triacylglycerol monohydroperoxides; DGMHP, diacylglycerol monohydroperoxides; FFAMHP, free fatty acid monohydroperoxides. (Modified scheme, from Miyashita et al. 1990. *Biochim. Biophys. Acta* 1045:233–8.)

oils studied were first heated at 180°C for 50, 70, and 100 h with aeration. The three fractions eluted from the oils included a nonpolar fraction with monomeric compounds, a somewhat polar fraction with dimeric compounds, and a polar fraction with polymeric compounds. Pancreatic lipase hydrolyzed the monomers in the nonpolar fraction as quickly as in the corresponding unheated oils. With the same enzyme, it took much more time to hydrolyze the dimers in the polar fraction, and the polymers of the polar fraction were not hydrolyzed at all. Pancreatic lipase hydrolyzed heated corn, sunflower, and soybean oils to almost the same degree. Yoshida and Alexander (1983) offer an explanation of the findings of Ohfuji et al. (1972) regarding the greater hydrolysis of monomers than that of thermally oxidized dimers. The section on *in vivo* methods contains further relevant information on this topic.

Márquez-Ruiz et al. (1992a) tested the *in vitro* action of pancreatic lipase on complex acylglycerols from thermally oxidized oils. These authors used olive oil heated at 180°C for 150 h. Pancreatic lipase was used to hydrolyze samples of non-heated olive oil, heated olive oil, a 1:1 mixture of both oils, and the polar fraction isolated from the heated oil for 2 min and 30 min under standardized conditions. Considerably more free fatty acids were released when hydrolysis time was longer, but the amount decreased as the degree of oil alteration increased. In the olive oil heated for 150 h, triacylglycerol dimers underwent much greater pancreatic lipase hydrolysis than triacylglycerol polymers. Hydrolysis of the polar fraction of the olive oil heated for 150 h displayed similar results. The fact that no significant losses of triacylglycerol polymers or release of fatty acids occurred as a result of hydrolysis after the 15 min reaction period indicates the difficulty of hydrolyzing these high molecular mass compounds.

Pancreatic lipase has been used by Márquez-Ruiz et al. (1998) to study two aspects of *in vitro* hydrolysis of abused oils. These authors investigated (a) the susceptibility to hydrolysis of altered compounds in the oils used for frying and (b) the result of the degree of frying oil degradation on hydrolysis of nonoxidized triacylglycerols. Combined adsorption chromatography, high-performance size-exclusion chromatography (HPSEC), TLC-FID, and *in vitro* hydrolysis with pancreatic lipase (2 min and for 20 min) were used to study used frying oils with a 3.1%–61.4% polar compound content. Enzymatic hydrolysis was performed on the following three types of substrates: (1) polar fractions isolated from trilinolein heated at 180°C, (2) polar fractions isolated from thermally oxidized frying oils, and (3) oils used in frying, without preliminary separation of nonoxidized triacylglycerol fractions. The relative hydrolysis rates of the two polar fraction substrates were similar. The type of altered compounds that were hydrolyzed most (80%–89%) after 20 min of hydrolysis were oxidized triacylglycerol monomers. Hydrolysis values of polymers were low (11%–42%), while those of dimers were somewhat higher. Hydrolysis of oils used in frying revealed that the most degraded oils produce a significantly reduced level of total hydrolytic products (diacylglycerols, monoacylglycerols, and fatty acid monomers). The hydrolysis rate of nonoxidized triacylglycerols is highly influenced by the number of polymers and dimers in the oil. Hydrolysis of these triacylglycerols is about 95% in somewhat degraded oil but is as low as 52% in degraded oils with a 47.6% dimer and polymer content.

Enzymatic hydrolysis of triacylglycerols may be influenced by the position of the oxygenated group. Differentiation between the oxidized acyls in positions 1 and 3 of the triacyl-*sn*-glycerol is possible through the classical method of Brockerhoff (Turon et al., 2002). *In vitro* hydrolysis of thermally oxidized triacylglycerols by pancreatic lipase has also been investigated by our group (Arroyo et al., 1995; Arroyo et al., 1997) and others (Henderson et al., 1993; Arzoglou, 1994; Márquez-Ruíz et al., 2006) using both the "final point" and the "dynamic hydrolysis" techniques. The former method was used to analyze total hydrolysis of thermally oxidized oils, while the latter was applied to study the enzymatic hydrolysis kinetics of altered oils.

5.3.1.1 Pancreatic Lipase—Final Point Technique

The final point technique, which our group has used extensively to study both used and nonused oils, facilitates the quantification and analysis of the compounds produced in altered oil during a given hydrolysis time. Using porcine pancreatic lipase, our research group compared the results of enzymatic hydrolysis of palm olein after 40 or 90 fryings of potatoes. The food-to-oil ratio in the fryers was maintained at 500 g/3 L by emptying the contents of one fryer into the others after each 10 uses. Five hundred gram portions of sliced potatoes were fried without replenishing the used palm olein with fresh oil (Arroyo et al., 1995; Arroyo et al., 1997). Total polar content in the palm olein rose significantly after 90 successive potato-frying uses, from 9.3 ± 0.1 mg/100 mg of oil to 26.4 ± 0.3 mg/100 mg of oil (Figure 5.6). Combined use of column chromatography and HPSEC (Dobarganes et al., 1988) revealed that polymer, dimer, and oxidized triacylglycerol contents of this used palm olein were also several times higher than those of nonused oil (Figure 5.6).

Employing a slight variation of the method of Luddy et al. (1964), porcine pancreatic lipase (E.C. 3.1.1.3) was used for 20 min to hydrolyze fresh palm olein and palm oleins from the 40th and 90th frying operation, as well as polar fractions from the used and nonused palm oleins that had been isolated by applying a slight modification of the column chromatographic method of Waltking and Wessels (1981).

The nonused palm olein displayed a degree of hydrolysis similar to that of the palm olein used 40 times to fry potatoes (Figure 5.7), while hydrolysis of palm olein used 90 times was significantly lower. These results may be explained by the presence of nonoxidized triacylglycerols in the oil, which are the natural substrates for pancreatic lipase.

Fewer intact triacylglycerols were found in palm olein that had been used 90 times than in fresh palm olein or in palm olein used 40 times (Figure 5.7). Fresh palm olein and palm olein used 40 times contained a much higher net amount of hydrolyzed triacylglycerols (83.8% and 87.1%, respectively) than palm olein from the 90th frying use (65.8%), clearly indicating that pancreatic lipase hydrolysis of triacylglycerols decreases in highly altered oil but is only slightly affected, or not at all, by moderate thermal alteration.

According to the results of a previous study, a balance between factors that improve or impair pancreatic lipase hydrolysis may exist (Arroyo et al., 1996). The

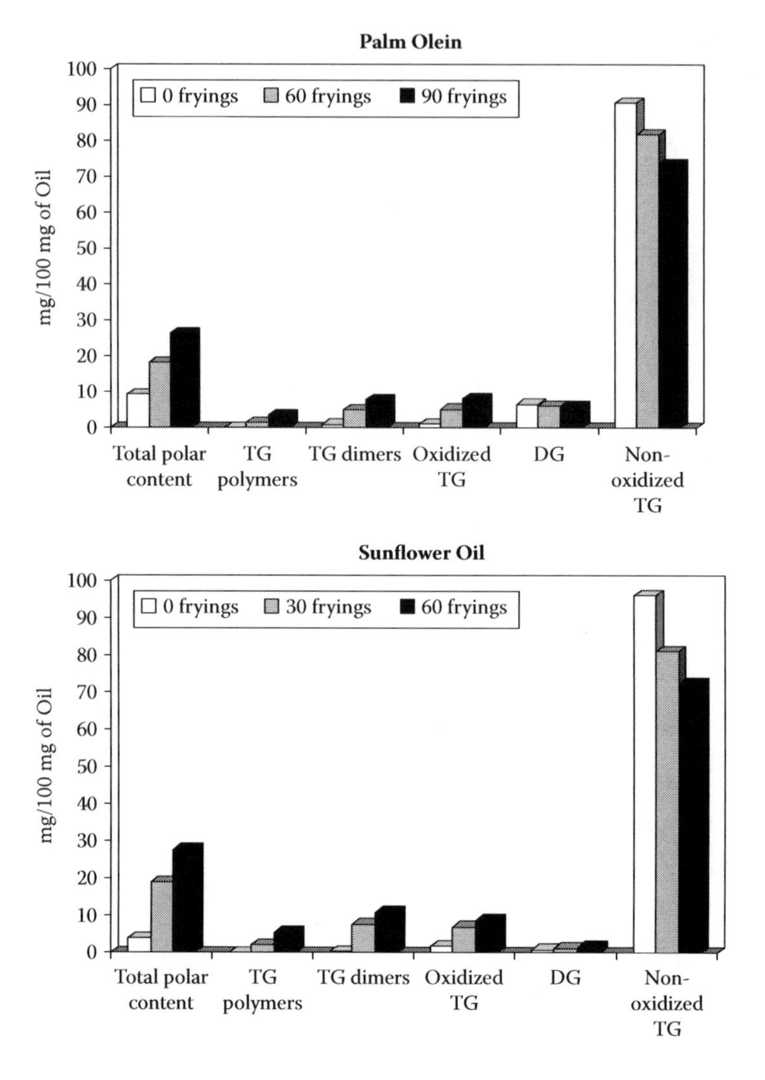

FIGURE 5.6 Composition of palm olein and sunflower oil before and after being used in potato frying. (Modified from Arroyo et al. 1996. *Lipids*. 31:1133-9.)

following section includes a discussion of this balance. In agreement with Márquez-Ruiz et al. (1998), the study of Arroyo et al. (1995) showed that oils with the higher level of oligomers underwent less hydrolysis of intact triacylglycerols.

Figure 5.8 contains information regarding the polymers hydrolyzed from palm oleins and from their respective polar fractions. It appears that some compounds formed during frying influence the hydrolytic action of pancreatic lipase, as the polar fractions of the oleins contained considerably more nonhydrolyzed polymers.

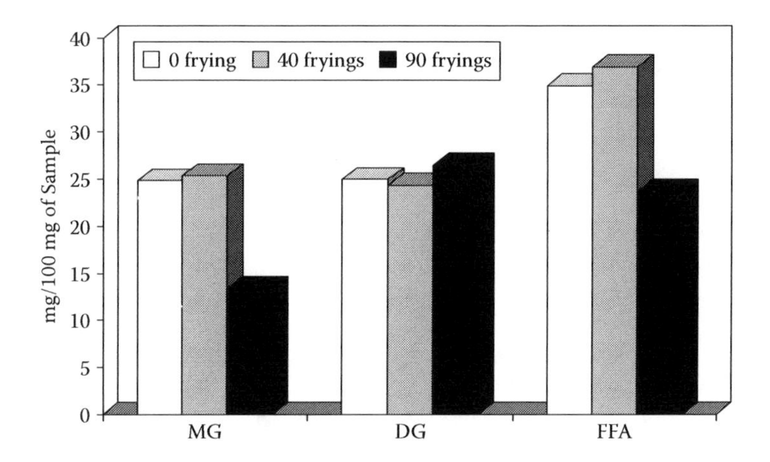

FIGURE 5.7 Diacylglycerol (DG), monoacylglycerol (MG), and free fatty acid (FFA) contents in nonused and used palm oleins 40 and 90 times for frying potatoes after a 20 min pancreatic lipase enzymatic hydrolysis. (Modified from Arroyo, R. et al., 1996. *Lipids,* 31:1133-9.)

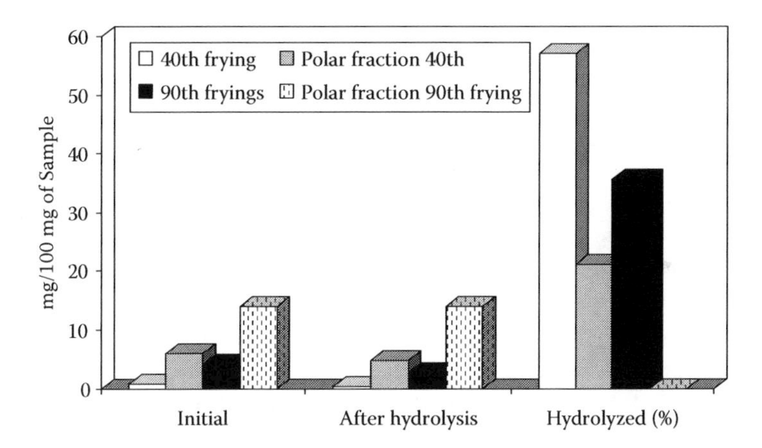

FIGURE 5.8 *In vitro* enzymatic hydrolysis of polymers from palm olein and from the polar fraction of fresh palm olein and that used 40 and 90 times for frying potatoes. (Modified from Arroyo 1996. *Lipids*, 31:1133–9.)

5.3.1.2 Pancreatic Lipase—Dynamic Hydrolysis Technique

Lipolytic activity was determined by means of a titrimetric assay (Arzoglou, 1994). Continuous titration of the acid released during hydrolysis was performed at 37°C with NaOH solutions at pH 8.3. Reaction time was 10 min. The initial reaction rates were used to measure lipase activity in order to avoid any possible inhibitions due to the presence of reaction products. The specific lipase activity was defined as the μmol of free fatty acids released per minute and per milligram of crude enzyme. Substrate concentration in the reactor was determined taking into account its molecular mass, calculated by HPSEC (Husain et al., 1988), using acylglycerol standards and polyethylene glycol in different aggregation states as MW markers (Arroyo et al., 1996).

Porcine pancreatic lipase was also used to analyze the kinetics of enzymatic hydrolysis of vegetable oils with various degrees of alteration due to frying (Arroyo et al., 1996). The aim of the study was to understand the effect of the thermally oxidized products that appear during frying on *in vitro* lipolysis by pancreatic lipase. In addition, the authors studied the effect of thermally oxidized acylglycerol models on pancreatic lipase activity to better understand lipase hydrolysis of thermally oxidized oils (Arroyo et al., 1997).

The successive use of palm olein and sunflower oil to fry potatoes without oil renewal resulted in a significant increase in their total polar compound contents (Arroyo et al., 1996). Palm olein used in 90 fryings had 37.0 times more triacylglycerol polymers, 7.9 times more triacylglycerol dimers, and 7.5 times more oxidized triacylglycerols; sunflower oil used in 60 fryings presented 56.0 times more triacylglycerol polymers, 22.0 times more dimers, and 4.7 times more oxidized triacylglycerols in comparison with nonused oils (Figure 5.6).

The presence of complex compounds, such as triacylglycerol dimers and polymers, in the substrate, and of other alteration products such as fatty acids that are formed during hydrolysis, can influence the activity of pancreatic lipase. For this reason, Arroyo et al. (1995) measured the relationship between reaction velocity and substrate concentration and calculated the corresponding kinetic parameters.

The activity of pancreatic porcine lipase in palm olein and sunflower oil containing approximately 18 mg of polar compounds/100 mg of oil is shown in Figure 5.9. Hydrolysis of both oils using pancreatic lipase followed Michaelis–Menten saturation kinetic behavior. Each set of data was fitted to the Michaelis–Menten equation by nonlinear regression, and the corresponding apparent V_{max} and apparent K_m were estimated.

The hydrolytic efficiency of pancreatic lipase (evaluated by the apparent K_m or V_{max}) was evidently greater in palm olein than in sunflower oil, according to the results of pancreatic lipase activity on these two oils (used or nonused for frying) (Figure 5.9). However, the kinetic parameters for palm olein and sunflower oil presented in Table 5.1 do not indicate a clear-cut relationship with the number of fryings, and thus with the content of altered compounds in the oils. The palmitic and oleic acid contents of palm olein are higher than that of sunflower oil, while sunflower oil contains more linoleic acid than palm olein. Pancreatic lipase is known to release oleic acid from oils with the highest relative speed (Yang et al., 1990), and the used sunflower oils contained more polymers and dimers but fewer diacylglycerols than palm olein. Both these factors explain the preceding results. Minor constituents present in the oils can also affect the enzymatic reaction. The amount of tocotrienols in palm olein is high, while sunflower oil contains elevated level of tocopherols. Moreover, tensioactive substances formed in these two oils during frying (Sánchez-Muniz et al., 1998; Sánchez-Muniz et al., 2003) are other factors possibly influencing the enzymatic activity and thus the reaction rate of the enzymatic process.

The degree of oil alteration was not responsible for the changes observed in the apparent K_m and apparent V_{max}. Compounds with high molecular mass and polarity do not appear to have an important effect on the enzymatic hydrolysis

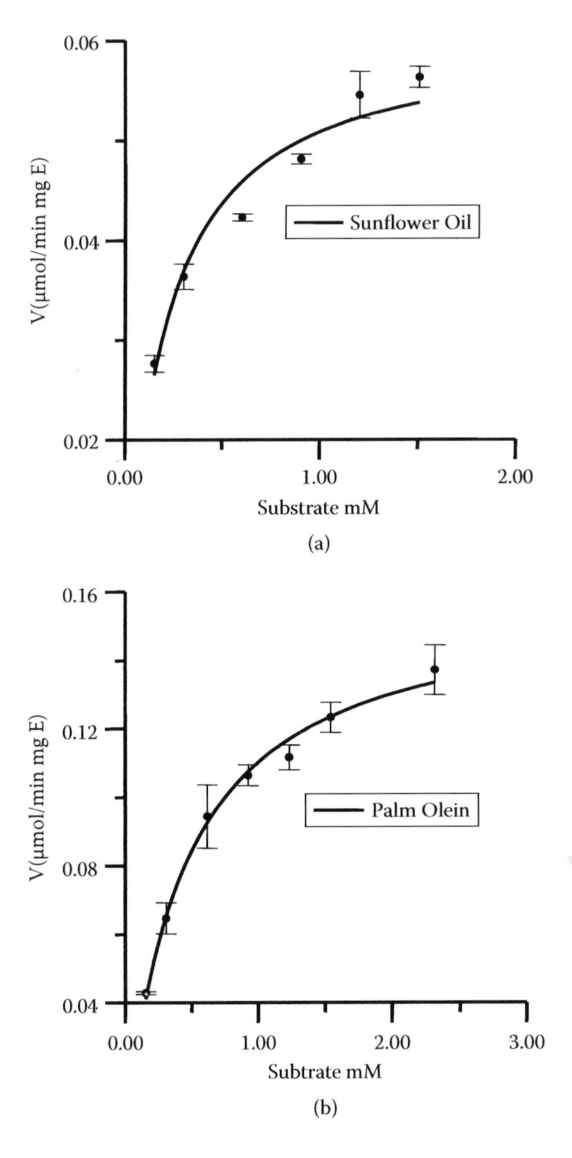

FIGURE 5.9 Activity of porcine pancreatic lipase toward sunflower oil and palm olein both with approximately 18% in polar content. Error bars refer to 95% confidence limits using *t*-Student. (Adapted from Arroyo et al., 1996. *Lipids* 31:1133–9, and Nus et al. 2006b. *Food Technol. Biotechnol.* 44:1–15.)

activated by porcine pancreatic lipase. However, these data must be evaluated taking into account the process itself, because the lipase dynamic hydrolysis occurs in the oil–water interface. Certain factors (e.g., sample complexity, tocopherol loss) hinder the enzymatic reaction, while others (e.g., appropriate interface substrates, sample polarity, surfactant concentration) appear to facilitate it. Triacylglycerols and diacylglycerols are always the preferred interface

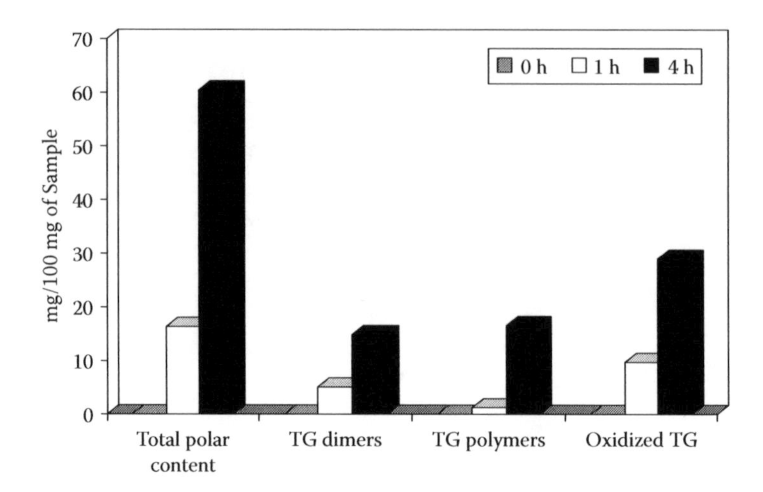

FIGURE 5.10 Composition of nonpolar triacylglycerols of palm olein (NPTPO) before and after thermal oxidation at 180°C. (Modified from Arroyo, R. et al., 1997. *J. Am. Oil Chem. Soc.* 74:1509–16.)

substrates of pancreatic lipase, regardless of the composition of the sample. In addition, heating is known to decrease tocopherol levels significantly (Jorge et al., 1996; Choe and Min, 2007).

According to Yoshida and Alexander (1983) and Ohfuji et al. (1972), enzymatic hydrolysis decreases as the molecular mass of the acylglycerol sample increases. As cited earlier, monohydroperoxides do not inactivate porcine pancreatic lipase, which prefers molecules with esterified monohydroperoxide fatty acids to those with nonoxidized fatty acids as a substrate (Miyashita et al., 1990). According to Henderson et al. (1993), pancreatic lipase hydrolyzes the high-molecular-mass polymers of triacylglycerols in oxidized fish oils *in vivo.* The frying process produces polymers with surfactant characteristics in the frying oil (Blumenthal, 1991). Surfactants appear to increase enzymatic hydrolysis, and emulsions that contain more surfactants display greater stability and have a larger interface surface than emulsions with lower surfactant content. Pancreatic lipase tends to hydrolyze palm olein (expressed by the V_{max}) more efficiently in substrates with a high concentration of altered compounds (Arroyo et al., 1996). The enzyme always has the same affinity for palm olein (expressed by the apparent K_m), regardless of its content in altered compounds. In used palm olein containing a polar content of about 18%, the kinetic parameter apparent V_{max}/apparent K_m was higher than that of nonused oils or that of palm olein with a polar content of about 27%. This effect was also found in the hydrolysis of the nonpolar triacylglycerols isolated from palm olein (NPTPO) that were heated to 180°C for 1–4 h (Arroyo et al., 1997). The kinetic behavior demonstrates that pancreatic lipase acts better in the whole fresh palm olein than in the unheated NPTPO (Table 5.1).

The differences between the two substrates reside in their content in major and minor compounds. The only compounds present in NPTPO are nonoxidized triacylglycerols, while these compounds as well as others with higher polarity (e.g., oxidized triacylglycerols, oligomers) are found in nonused palm olein. Pancreatic lipase

TABLE 5.1
Kinetic Parameters for Hydrolysis by Porcine Pancreatic Lipase of: Sunflower Oil, Palm Olein. Sunflower Oil Used in Successive Fryings of Potatoes. Palm Olein Used in Successive Fryings of Potatoes. Non-Oxidized Triacylglycerols Isolated from Palm Olein (NPTPO)

	Frying/Heating	Apparent K_m*	Apparent V_{max}**	Apparent K_m/V_{max}***
Palm olein	Non-heated	1	1	1
	60th frying	1.16	0.96	1.19
	90th frying	1.04	1.36	0.77
Sunflower oil	Non-used	0.43	0.37	1.16
	30th frying	0.76	0.61	1.23
	60th frying	0.41	0.46	0.90
NPTPO	Non-heated	0.63	0.70	1.16
	Heated 1 hour 180°C	0.29	0.64	2.26
	Heated 2 hours 180°C	0.24	0.54	2.39
	Heated 4 hours 180°C	0.49	0.56	1.19

Notes: Values are *multiples of the Apparent K_m of unused palm olein; **multiples of Apparent V_{max} of unused palm olein; ***apparent (K_m/V_{max}) of unused palm olein. Modified from Arroyo et al. (1996, 1997).

behavior may be conditioned by the dissimilarities of these two substrates. The loss of natural emulsifier in NPTPO may account for the drop in the apparent K_m and apparent V_{max} after 2 h, while increased production of polar surfactants would explain the increase in the K_m after 4 h, in relation to the NPTPO samples heated for 2 h.

5.3.1.3 *In Vivo* Studies

A number of interactions involving complex physicochemical reactions occur between lipolytic products, phospholipids, bile salts, proteins, and carbohydrates during the lipid digestion process. In this process, the alimentary canal enzymes, of which pancreatic lipase is the most significant, transform the water-insoluble triacylglycerols into more hydrophilic molecules, such as diacylglycerols, monoacylglycerols, and free fatty acids, thus preparing them for absorption and transport through the enterocyte membrane (Chapus et al., 1988).

The triacylglycerol content of edible fats and oils is generally above 95%. Acyl groups of triacylglycerols are characterized by their number of C-atoms (chain length) and the number and position of their double bonds (unsaturation and positional isomerization). Conversion of triacylglycerols into free fatty acids and monoacylglycerols must occur before absorption can take place. These triacylglycerol-derived compounds are placed into intestinal micelles and absorbed. A reesterification process later occurs within the enterocytes, and the resynthetized triacylglycerols, cholesterol esters, and phospholipids are incorporated to chylomicrons (Sánchez-Muniz and Sánchez-Montero, 1999). Chylomicrons and other lipoproteins participate in the very complex lipoprotein metabolism carrying cholesterol, fatty acids, and other fat-soluble compounds to/between the different body tissues (Elmadfa and Wagner, 1999).

Little research has been done to date on the digestion and absorption of thermally oxidized and polymerized oils (including micelle formation and transport through cell membranes). While heated fats are generally thought to be less digestible and more difficult to absorb than unheated fats (Márquez-Ruiz and Dobarganes, 1996), some authors believe that frying does not significantly affect oil digestion (Lanteaume et al., 1966; Le Floch et al., 1968; Varela et al., 1986).

According to Deuel (1955), the degree of polymerization of frying fat largely affects its digestibility. Crampton et al. (1953) reported that the presence of dimers of fatty acids was the main reason for the poor digestion of diets containing linseed oil heated to 275°C. Nonetheless, it has to be taken into account that this high temperature does not correspond to the nomal range (160°C–200°C) at which fryings are performed. Similarly, Potteau et al. (1970, 1977) found that digestion of oil with a high degree of polymerization was less efficient than that of oil with a low polymer content. However, differences in the hydrolysis and digestion of dimers have been suggested. Thus, lymph studies indicate that hydrolysis and absorption of dimers and polymers is higher than that of nonpolar dimers (Combe et al., 1981).

Digestibility values of oxidized dimers and polymers were high (Márquez-Ruiz et al., 1992b). Kajimoto and Mukai (1970) considered the dimer digestibility values reported by Bottino et al. (1967) (30%–70%) too high. According to Ohfuji et al. (1972), rats can absorb the dimers of thermally oxidized oils. However, while hydrolysis of two-thirds of the ester bonds in triacylglycerol monomers takes place without any difficulty, the larger triacylglycerol dimers contain numerous chemical entities and C-C bonds that prevent hydrolysis of the internal ester groups (Ohfuji and Kaneda, 1973). Paulose and Chang (1973) report that the coexistence of intramolecular and intermolecular bonds complicates the dimeric and polymeric structures of heated oils, and this causes a hindrance of hydrolysis of dimers and polymers. Ohfuji and Kaneda (1973) reported that polymers in oxidized oils underwent no apparent enzymatic hydrolysis.

Radiolabeled markers, whose composition may differ from that of the altered compounds in the oils used for frying, are often employed in altered fat digestion assays. These markers, in addition to being expensive, are potentially unsafe. Our group has studied *in vivo* digestibility and absorption coefficients of oils over the past few years without the use of these markers (González-Muñoz et al., 1996; González-Muñoz et al., 1998; González-Muñoz et al., 2003; Sánchez-Muniz et al., 1999; Olivero-David et al., 2010a). Young adult Wistar rats given 1 g unheated olive oil/100 g body mass were tested 2, 4, 6, and 7 h after oil administration to determine the true digestibility of the oil (Figure 5.11, right). Isotonic saline solution was given to control rats at the rate of 1 mL/100 g body weight. To obtain luminal fat, 50 mL of isotonic saline solution were slowly passed through the section of the digestive tract between the distal esophagus and the distal ileum (Figure 5.11, left). A linear but inverse relationship ($r = -0.99$; $p < 0.001$) between the remaining gastrointestinal luminal fat and the length of the experiment was seen (Figure 5.11, right) (Sánchez-Muniz et al., 1999). A 4 h test was found to be adequate for this assay, as after this time approximately half of the oil administered remained in the lumen; as a result, an accurate determination

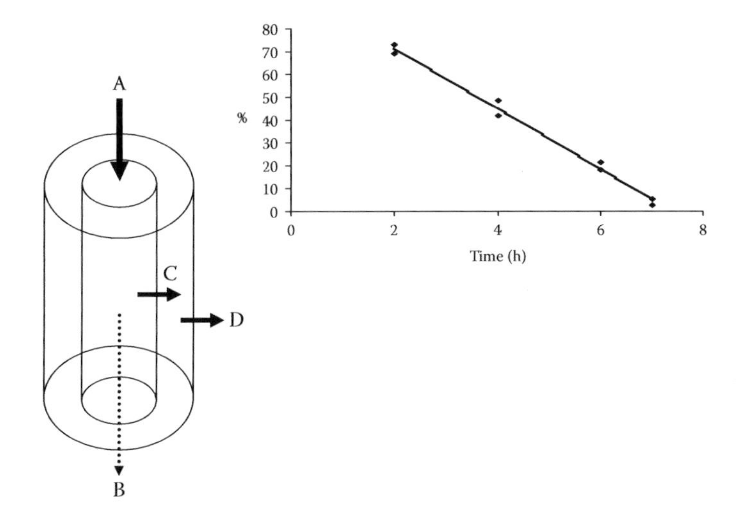

FIGURE 5.11 *In vivo* approximation for the study of thermal oxidized compounds of oil. Right: Relationship between length of experimental period (h) and luminal fat (recovery ratio) of administered fat (%). Left: Scheme for digestibility. (A = administrated fat; B = remaining fat; C = fat in the intestinal mucosa; D = absorbed fat.)

of the various nondigested and/or nonabsorbed thermally oxidized compounds was possible (Sánchez-Muniz et al., 1999).

Thermal oxidation levels of palm oleins used to fry potatoes were related to their *in vivo* digestibility and absorption coefficients (González-Muñoz et al., 1996). The palm oleins discussed in the *in vitro* studies section were the same ones used in this *in vivo* study (Figure 5.6). True digestibility and true absorption coefficients of these used palm oleins decreased significantly, but intracellular gastrointestinal fat content remained unaltered. Total polar content and triacylglycerol polymer, triacylglycerol dimer, and oxidized triacylglycerol concentrations in the oil were highly correlated with the changes observed in fat digestibility and absorption (González-Muñoz et al., 1996).

In agreement with Carey et al. (1983), the results of González-Muñoz et al., (1996) indicate that the intestinal lumen was the site of the most important reactions that took place during the experiment, as no significant changes in the clearance of intracellular gastrointestinal fat were observed. Investigation of *in vivo* hydrolysis of thermally oxidized and polymerized products of palm oleins used in repeated potato-frying procedures was carried out in a 4 h long experiment (González-Muñoz et al., 1998). This study combined column chromatography, HPSEC techniques, and *in vivo* short-term fat digestion. The authors also tested the possibility that certain altered products, such as oligomers, exert an inhibitory role on fat digestibility of nonthermally oxidized triacylglycerols. To this end, palm oleins containing different concentrations of altered compounds were subjected to hydrolysis and the results compared. There was a sharp increase in the percentage of oligomers and oxidized triacylglycerols, and a drop in the percentage of diacylglycerols and free fatty acids in the polar fraction of the luminal fat after palm oleins used in frying were given to the rats.

Nonoxidized triacylglycerols exhibited higher true digestibility values than oligomers. Nevertheless, the digestibility ratio of nonoxidized triacylglycerols decreased 12% and 37% with palm olein used for potato frying 40 and 90 times, respectively (Figure 5.12). These findings indicate that thermally oxidized compounds in the oil inhibit or slow down hydrolysis of nonoxidized triacylglycerols by pancreatic lipase under the experimental conditions of the study. Henderson et al. (1993) found that triacylglycerols and polymers in oils with low concentrations of polymers (<4%) underwent almost complete *in vitro* hydrolysis with pancreatic lipase in 1 h, but some triacylglycerols were left intact when highly oxidized oils (with a 20%–30% triacylglycerol polymer content) were tested. The complex balance between the formation of oxidized triacylglycerols from polymers and dimers and their elimination through hydrolysis makes it difficult to explain the information available, concerning oxidized triacylglycerol digestibility coefficients.

The reduced digestibility of these compounds contrast with data reported by other authors (Funes and Karel, 1981; Matsuchita, 1975) who suggest that oxidized triacylglycerols are well-absorbed and appear to be adequately hydrolyzed by pancreatic lipase as they have higher polarity and similar molecular mass of nonaltered triacylglycerols. Carey et al. (1983) believe that the absorption that occurs during the first phase of digestion depends on lipolytic enzyme activity. In the following step, lipids enter into the micellar phase greatly aided by molecular polarity; molecular mass reduces the luminal uptake of fat in the final stage of digestion.

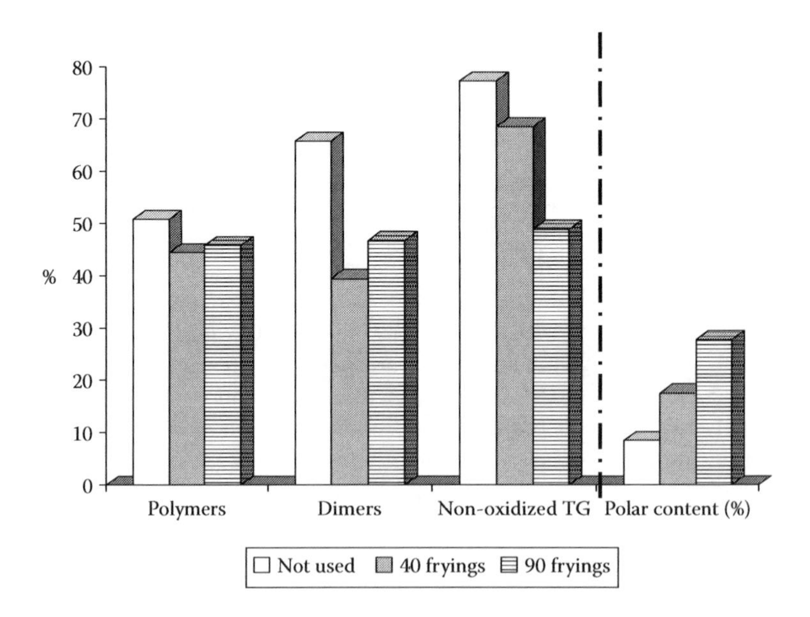

FIGURE 5.12 True digestibility (% of ingested fat) of triacylglycerol polymers (polymers), triacylglycerol dimers (dimers), and nonoxidized triacylglycerols (nonoxidized TG) from nonused palm olein and from palm olein used 40 and 90 times for frying of potatoes after 4 h *in vivo* experiment. The polar content (mg/100 mg oil, %) is also included. (Modified from González-Muñoz et al. 1998. *J. Agric. Food Chem.* 46:5188–93.)

The percentages and amounts of monoacylglycerols and free fatty acids in luminal fat decreased as the number of frying uses of the palm olein increased, possibly as a result of the reduction in pancreatic lipase activity (González-Muñoz et al., 1998). However, the alteration in the oil would mean additional polar groups from oxidation in monoacylglycerols and free fatty acids that increases their absorption. Sánchez-Muniz et al. (1999) reported similar results following the short-term digestion of olive oil heated for 50 h at 180°C. The amount of polar material in the heated oil was significantly greater than that in unheated oil (46 g/100 g heated oil versus 3.6 g/100 g unheated oil). The quantity of remaining lumen fat (nonhydrolyzed and/or nonabsorbed) with the heated olive oil samples tended to be higher than that observed with the unheated oil after 4 h. The true digestibility coefficient of the heated oil was 24% lower than that of the unheated olive oil, as a consequence of its lower degree of polymer hydrolysis and absorption. Thermally oxidized products from the heated oils displayed a true digestibility of 30%–40%, much lower than that of the compounds from the unheated oils (about 80%). The presence of large amounts of thermally oxidized compounds decreased nonoxidized triacylglycerol hydrolysis. In summary, results of this 4 h study indicate that active hydrolysis of thermally oxidized compounds from highly altered olive oil takes place under these *in vivo* experimental conditions.

A more recent short-term, *in vivo* digestibility study on sunflower oil heated for 50 h at 180°C (González-Muñoz et al., 2003) followed the experimental conditions of esophageal cannulation and recovery of the remaining luminal fat previously cited (González-Muñoz et al., 1998; Sánchez-Muniz et al., 1999). True digestibility coefficients were calculated after correcting data with values observed in a control group cannulated with saline solution. A nonsignificant decrease of 38% for the true digestibility coefficient of the heated oil was observed in this study. True digestibility data of the heated sunflower oil tended to be higher than that reported for heated olive oil in another study (Sánchez-Muniz et al., 1999). These data suggest that, although sunflower oil was much more altered than olive oil, according to the polymer plus dimer contents (69.4 versus 22.6 g/100 g oil), and theoretically more resistant to hydrolysis by pancreatic lipase, some compounds in this heated oil (e.g., surfactants) should enhance the dynamic hydrolysis that occurs at the oil–water interface in comparison with the heated olive oil.

True digestibility of polymers (30%), dimers (59%), and total oligomers (dimers plus polymers) (38%) of heated sunflower oil (González-Muñoz et al., 2003) was much lower than that of the corresponding compounds in unheated oil (95%, 90%, and 91%, respectively). The true digestibility of triacylglycerol monomers (oxidized plus nonoxidized) from the sunflower oil used in frying was not significantly different than that of the nonused oil. The high alteration levels of the oil used in this study make it difficult to definitively explain the finding; these may be due in part to a balance between the formation of triacylglycerol monomers (both oxidized and nonoxidized) from polymers and dimers, and their disappearance due to hydrolysis and absorption.

Our group has recently carried out further *in vivo* digestibility studies (Olivero-David 2010a). Hydrolysis of thermally oxidized and polymerized compounds of a sunflower oils used in 40 frying operations of various frozen and nonfrozen foods was performed in a 4 h *in vivo* experiment. The used sunflower oil in this study contained a polar content of over 30%, while that of the unused (control) sunflower oil had less than 5%. The study used a combination of *in vivo* short-term fat digestibility, minicolumn chromatography, and HPSEC techniques (Dobarganes et al., 2000).

As the presence of nutrients of any kind may modify pancreatic lipase activity and fat absorption, short-term *in vivo* digestibility was tested in rats with ad libitum access to food before the experiment as well as in those fasted for 15 h. The remaining fat in the stomach and small intestine was removed and analyzed separately, since altered fats are known to be poorly digested and delay gastric emptying. Figure 5.13 contains data of the different amounts of fat in the stomach and intestines of rats administered used and control sunflower oils under fasting and nonfasting conditions. As expected, rats given oil used in frying had more fat in the intestine than those fed the nonused oil. It is interesting to note, however, that there were no differences between the amount of fat remaining in the stomachs of fasting and nonfasting rats given unaltered (control) oil, while fasting rats to which altered oil was administered had much more gastric fat than their nonfasting counterparts (Olivero-David et al., 2010a).

There is general evidence that a large quantity of undigested/unabsorbed food in the duodenum slows down gastric emptying time. The difficult hydrolysis of altered triacylglycerols—in the form of polymers, dimers, or monomers—discussed in the following text appears to be the reason that certain material remains undigested or unabsorbed in the duodenum. Thus, administration of altered fats slowed down the fat moving from the stomach to the intestine, especially in the case of rats in fasting conditions administered fat via cannulation.

Rats given sunflower oil used in frying, and especially fasting rats, displayed a notably higher percentage of polymerization compounds in the luminal fat and a lower hydrolyzed compound content than rats fed nonused sunflower oil (data not shown). As previously commented, Henderson et al. (1993) reported that pancreatic lipase almost completely hydrolyzed the triacylglycerols and polymers of oils with low amounts (<4%) of polymer substrates after 1 h *in vitro*, but some triacylglycerols

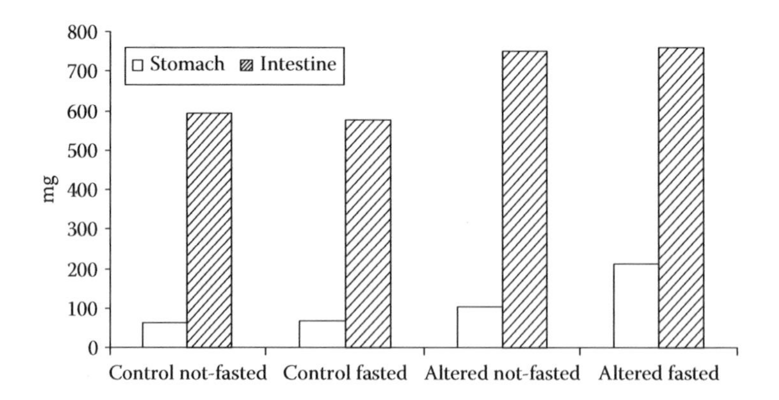

FIGURE 5.13 Fat (mg) not absorbed in the stomach and intestine after 4 h of administration. Rats received 1 g/100 g body weight (Modified from Olivero-David et al. 2010a. *J. Agric. Food Chem.* doi 10.102/jf 101715g).

remained nonhydrolyzed when highly oxidized oils (containing 20% or 30% of triacylglycerols polymers) were tested.

5.3.2 Phospholipases

According to Christie (1986), by using small variations of the classical method of Brockerhoff (1965), it is possible to obtain phospholipid derivatives (e.g., phosphatidylcholines) from diacylglycerols. PLs can further hydrolyze the phospholipid derivatives formed in this process. The structure of the original triacylglycerol can be studied further by analyzing the diacylglycerols produced.

Although the level of phospholipids in foods is much lower than that of triglycerides, their study has scientific and economical importance. Elimination of phospholipids from some fats may be advisable under certain conditions, as these compounds contain a high level of polyunsaturated fatty acids and are thus highly susceptible to oxidation. The various classes of PLs that can be analyzed include PL A, PL B, PL C, and PL D, as previously mentioned (Figure 5.1).

Fatty acids are eliminated from the *sn*-1 and *sn*-2 positions by PL A1 and PL A2, respectively. Joining these hydrolyses with PL B can be used to know the composition of both the 1- and 2-lysophospholipids produced by PL A1 and PL A2, respectively. PL C and PL D can be further employed to determine the complete composition of a specific phospholipid because the function of PL C is to break the bond that unites glycerol and phosphates, while that of PL D is to hydrolyze the ester bond in the moiety containing the amino alcohol.

The main features of the various PL techniques are reviewed in the following text.

5.3.2.1 Phospholipase A2

PL A2 is one of the principal phospholipid enzymes. Extracellular PL A2 is released by activated platelets or secreted into the bloodstream by the pancreas as a type I digestive enzyme. Various methods using spectrophotometry, colorimetry, and titration are available to measure the fatty acids released as a result of PL A action on phosphatidylcholine. Some of these techniques include a pH indicator and measure the changes in absorbance (e.g., phenol red absorbance values at 558 nm) (Lobo de Araujo and Radvanyi, 1987). As can be easily deduced, the free fatty acids released by PL A2 would be titrated using a similar method to the one described in Section 5.3.1.2.

5.3.2.2 Phospholipase B

The method described by Ottolenghi et al. (1966, 1973) is used to determine PL B activity. The enzyme attacks lysolecithin with liberation of 1 mol of glycerylphosphorylcholine and 1 mol of fatty acid. The fatty acid liberated by enzymatic hydrolysis can be measured by different methods.

5.3.2.3 Phospholipase C

This enzyme activates the hydrolysis of glycerophosphatide at the diacylglycerol-phosphate bond and that of sphingomyelin at the *N*-acylsphingosine phosphate linkage.

5.3.2.4 Phospholipase D

PL D is used to liberate choline from phospholipids. The choline released from phospholipids through the action of PL D is oxidized to betaine and hydrogen peroxide by the enzymatic action of choline oxidase. In contact with horseradish peroxidase, the mixture of the hydrogen peroxide with phenol and 4-aminoantipyrene produces a complex whose color can be measured spectrophotometrically at 500 nm. This method is an accurate, relatively easy, and fast technique that uses less sample than others, but whose results are influenced by blood, ascorbic acid, and bilirubin. Due to a discrepancy between previously published results, Tosi et al. (2007) used two methods to determine the soybean PL D activity. One method is based on the extraction of the enzyme from whole soybean flour, quantifying the enzyme activity on the extract. The other method quantifies the enzymatic activity on whole soybean flour without enzyme extraction. In the extraction-based-method, both the extraction time and the number of extractions were optimized. The highest PL D activity values were obtained from the method without enzyme extraction. This method is less complex, requires less running time, and the conditions of the medium in which PL D acts resemble the conditions found in the oil industry.

The content of phosphorus compounds can be reduced in edible oils through the use of PLs when the minimum concentration of nonhydratable phosphorus is 50 ppm. Although there are several possibilities, Clausen et al. (2000) selected a procedure where the oil is placed in contact with an aqueous solution of a PL A1, A2, or B at pH 1.5–8. The phosphorus content of the oil in which the enzyme is emulsified decreases to below 11 ppm; the aqueous phase and the treated oil are subsequently separated. PLs used for the patented methods of Clausen et al. (2000) were obtained from a strain of the genus *Fusarium*, including *F. culmorum*, *F. heterosporum*, *F. solani*, or from a particular strain of *F. oxysporum*. These enzymes can be used for hydrolysis of fatty acyl groups of phospholipids or lysophospholipids.

5.3.3 Enzymes as Biomarkers of Altered Oils Ingestion and Metabolism

In this section, the use of enzymes as biomarkers of altered oils consumption is reviewed. Enzymes can be used to split up tissue lipids and study their fatty acid constituents. Triacylglycerols can be studied quantitatively and qualitatively. Most recent methods used in clinics and research measure the total triacylglycerol amount in plasma using sequential action of lipase, glycerol kinase, glycerol phosphate oxidase, and peroxidase (Bucolo and David, 1973). The enzyme method is associated with the Trinder color reaction (Trinder, 1969). Measurements can be performed in a few minutes. Moderate hypertriglyceridemia is indicated by triacylglycerol levels between 150 and 200 mg/dL but present recommendations suggest <110 mg/dL (1.24 mmol/L) for children and <130 mg/dL (1.46 mmol/L) for adults as desirable values.

Another important aspect is derived from the fatty acid composition of plasma triacylglycerol, which gives information of the recent adherence to diet and the nutritional status (e.g., the triene/tetraene ratio (20:3/20:4) is a biomarker of malnutrition). Similar results can be obtained by studying the composition of phospholipids and cholesterol ester plasma fractions. The study of erythrocyte membrane phospholipid composition gives accurate information about the quality of fat consumed over the preceding weeks (Sánchez-Muniz, 2003), while the fatty acid composition of fat tissue is a valid indicator of the fatty acid composition of the diet over the preceding 1 to 3 years (Beynen and Katan, 1985). These results are related to the presence of fatty acids into pools with very low/medium/high turnover or, with the physiological mode, phospholipids were included in cell membrane or triacylglycerols or cholesterol are included in serum lipoproteins or certain tissues (e.g., liver). For instance, human platelets have a life span of 9 days, while with erythrocytes it is 120 days; thus, fatty acid analysis of membrane phospholipids informs about the moment dietary fatty acids were incorporated to these cells. However, tissue fatty acid concentrations are also the result of lipolysis and lipogenesis, which are very active in some tissues and actively modified by diet, nutritional, physiological, and hormonal status (Sánchez-Muniz, 2003). In fact, lipogenesis is normally decreased but lipolysis activated when oxidized fatty acids are available (Kersten, 2001).

As thermally oxidized fatty acids are absorbed and can be incorporated to cells, increased ingestion of thermally oxidized oils or food cooked with these oils may, in turn, alter tissue lipid composition. An increase of palmitic, palmitoleic, stearic, oleic, and linoleic acids and a decrease of arachidonic acid in liver phospholipids were observed in rats administrated 15% thermally oxidized sunflower oil for 45 days (Kode et al., 2005). An increase of geometrical and positional isomers in meat was observed in broilers that had received oils heated at 190°C–195°C (Bou et al., 2005).

5.3.3.1 Studies with Arylesterase

The interest in arylesterase has grown in the last few decades, as previously mentioned. Still, the available information about the uses of this enzyme is poorer in comparison with that of pancreatic lipase. This enzyme is used to predict cardiovascular risk and gives information about the oxidative modification of lipoproteins (Parthasarathy et al., 1990; Aviram et al., 1998). Arylesterase has only recently been purified and added to the Protein Data Bank (Harel et al., 2004).

Eckerson et al. (1983) first proposed a method based on the spectrophotometric measurement of the rate of phenol formation as the product of the reaction catalyzed by the arylesterase (Figure 5.2). The initial rates of hydrolysis were determined at 270 nm. The assay mixture included phenyl acetate and $CaCl_2$ in Tris/HCl, pH = 8.0 at 25°C. This seems to be the elective method for many authors.

Our group has developed a new method using simulated body fluid (SBF) as buffer instead of Tris/HCl, at 37°C and pH = 7.34–7.4 (Nus et al., 2006a; Nus et al., 2008). Higher precision was achieved using this method.

Experiments using the method proposed by Eckerson et al. (1983) as well as that by Nus et al. (2006) show that arylesterase exhibits Michaelis–Menten kinetics (Figure 5.14), with the apparent K_m and V_{max} values shown in Table 5.2. As this enzyme has not been commercialized, the results are given per microliters of serum diluted in ratio 1:40 used for the assay.

Another method was proposed by Lorentz et al. (2001). It is an authomatized assay based on the diminution of the hexacyanoferrate-III as a consequence of the coupled reaction of the thiophenyl acetate hydrolysis catalyzed by the arylesterase.

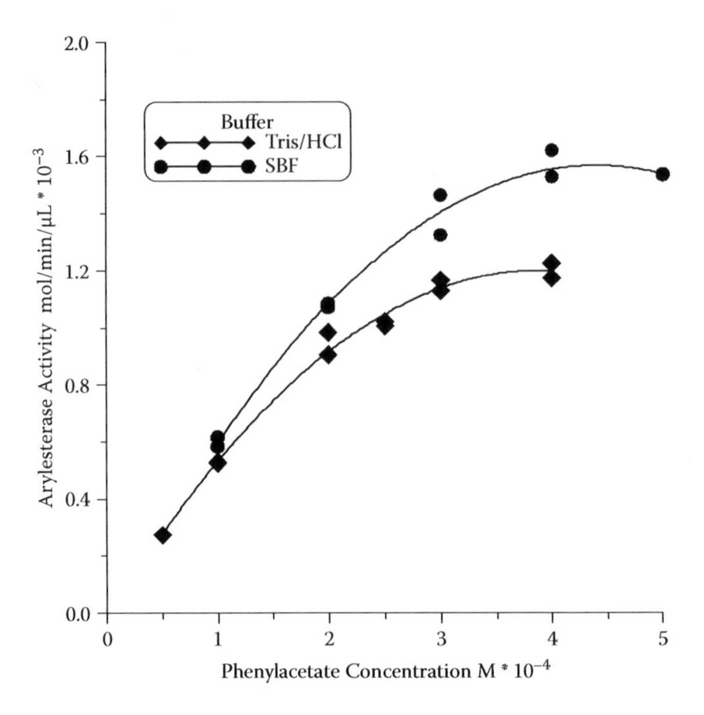

FIGURE 5.14 Michaelis–Menten kinetic adjustment of human serum arylesterase using simulated body fluid (SBF) and Tris/HCl. Curves were obtained by changing phenylacetate (PA) concentration (1·10–4 to 5·10–4 M) and maintaining a fixed serum volume (50 µL). (From Nus et al. 2006a. *Atherosclerosis* 188:155–9. With permission from Elsevier.)

TABLE 5.2
Apparent Kinetic Parameters of Arylesterase Obtained Using Tris/HCl and Using Simulated Body Fluid (SBF) Buffer

Buffer	*V*max/U	*V*max 95% CL/U	*K*m/M	*K*m 95% CL/M
Tris/HCl	5.299×10^{-5}	4.18×10^{-5}–6.42×10^{-5}	2.765×10^{-4}	1.62×10^{-4}–3.91×10^{-4}
SBF	6.305×10^{-5}	4.62×10^{-5}–7.99×10^{-5}	2.643×10^{-4}	1.13×10^{-4}–4.16×10^{-4}

Source: Tris/HCl data from Eckerson et al. 1983. *Am. J. Hum. Genet.* 35:214–27. SBF data from Nus, M., F.J. Sánchez-Muniz, and J.M. Sánchez-Montero. 2006b. *Food. Technol. Biotechnol.* 44:1–15. With permission from *Food Technol. Biotechnol.*

Note: CL, confidence limits.

As previously discussed, hydrolysis of oxidized phospholipids by paraoxonase destroys the biologically active lipids in mildly oxidized LDL (Sutherland et al., 1999). This action of paraoxonase could potentially attenuate the development of atherosclerosis. Low paraoxonase activity is present in subjects at high risk of coronary artery disease, including those with hypercholesterolaemia (Mackness et al., 1991) and diabetes (Abbott et al., 1995). Also, mice lacking serum paraoxonase activity are susceptible to atherosclerosis (Shih et al., 1998). High levels of lipid oxidation products in the diet increases levels of lipid oxidation products in human chylomicrons (Staprans et al., 1993). In mice that are susceptible to arterial lesion development, an atherogenic diet and injection of mildly oxidized LDL and oxidized lipids into the circulation reduces serum paraoxonase activity (Sutherland et al., 1999). Whether a meal rich in lipid oxidation products alters postprandial serum paraoxonase activity and susceptibility of LDL to oxidation in humans is unknown. (Wallace et al., 2001).

According to Aviram et al. (1998), the arylesterase works as a suicide enzyme, which is used when required. Thus, it seems that it would be better to have high arylesterase levels when nonantioxidant or low levels of antioxidant are available.

Very recent data from our group (Vázquez-Velasco et al., 2009) also indicated that the consumption of a sunflower oil enriched in hydroxytyrosol with respect to the control sunflower oil increased the activity of arylesterase and decreased the LDL oxidation level in volunteers.

5.3.3.2 The Glutathione Enzyme System

Although there are several enzymes implicated in the elimination of free radicals, among the most important enzymes implicated and used in laboratory are the glutathione system enzymes.

The total amount of glutathione can be determined by different methods (e.g., Griffith, 1980) using the enzyme recycling method in which the GSH is oxidized by the 5,5′-dithiobis-2-nitrobenzoic acid (DTNB) and reduced by NADPH in presence of the GR enzyme. The formation of 2-nitro-5-thiobenzoic acid (TNB) can be measured at 412 nm. The total glutathione amount in the sample is determined by extrapolating the observed value to a standard curve generated using known GSH concentrations.

The GSSG level is determined by the recycling methods previously described. The samples have to be treated with 4-vinylpyridine to eliminate the from GSH the sample, leaving the GSSG as the only substrate assayable (Afzal et al., 2002; Timur et al., 2008). The GSH is calculated by subtracting the obtained GSSG from the total glutathione. Results are expressed in nmol/g liver (Afzal et al., 2002).

Glutathione reductase (GR; EC 1.6.4.2) catalyzes the reduction of GSSG to GSH by NADPH and is an essential part of the glutathione redox cycle, which maintains adequate levels of cellular GSH. The GR is a homodimeric enzyme integrated by two identical subunits that are bound by a disulfur bridge. The GR activity can be determined by monitoring the oxidation of NADPH to $NADP^+$ during the GSSG reduction. A classic method of determining GR activity involves measuring the rate of NADPH oxidation spectrophotometrically at 340 nm (Tietze, 1969). This method, although still widely used, is relatively insensitive and is susceptible to interference from other NADPH-consuming enzymes present in cellular homogenates, as well as from any species that absorb UV light. An improved method (Rahman et al.,, 2007) based on the increase in absorbance at 412 nm when DTNB is reduced to 5-mercapto-2-nitrobenzoic acid by GSH overcomes some of the interference problems inherent in the classical assay. A fluorometric GR assay, based on the formation of a fluorescent compound when GSH reacts with *N*-(9-acridinyl)maleimide (NAM), was found to be 50 times more sensitive than the classical method. However, the fluorescent compound formed has an emission maximum of 435 nm, and hence, the assay is susceptible to interference from autofluorescence and from the inherent fluorescence of NADPH, which is an integral part of the assay (Kamata et al., 1993). This problem was overcome by Maeda et al. (2005), who employed a 2,4-dinitrobenzenesulfonyl fluorescein probe for thiol quantification. Cleavage of this latent fluorophore by GSH releases fluorescein, which has an emission maximum at 520 nm and hence is less susceptible to autofluorescence than NAM.

The glutathione peroxidase (GPx) (EC. 1.11.1.9) enzyme activity can be measured following a modification of the Paglia and Valentine (1967) method. Thus, the GPx catalyzes the GSH oxidation by the cumene hydroperoxide. GSSG in the presence of GR and NADPH is immediately transformed in GSH with concomitant transformation of NADPH to $NADP^+$. The absorbance decrease is measured at 340 nm. The GPx activity can be also measured by other methods, such as the Flohe and Gunzler (1984) one. The GSSG formed by the GPx enzyme action is instantaneously and continuously reduced by an excess activity of GR, giving rise to a constant level of GSH.

Pure fatty acid hydroperoxides are very toxic to experimental animals when administered intravenously (Olcott and Dolev, 1963; Findley et al., 1970). The 24 h lethal intravenously dose of a high-purity preparation of methyl linoleate hydroperoxides in adult male rats is approximately 0.07 mmol/100 g body weight (Cortesi and Privett, 1972). However, lipid hydroperoxides are not that toxic when given orally (Dubouloz et al., 1951), and only very small amounts of hydroperoxides are detected in some tissues and plasma, indicating that most hydroperoxides are destroyed/metabolized after their digestion/absorption.

Iritani et al. (1980) fail to detect the presence of hydroperoxides per se in the lymph or tissues in the next 24 h after giving oxidized corn oil orally to rats with thoracic lymph fistula. The finding suggests that peroxide lipids absorbed are metabolized. Subsequent studies show that the metalloprotein (suggested to metabolizing peroxidized fatty acids) was the seleno-enzyme GPx that utilizes the reducing equivalent of glutathione (GSH) to reduce hydrogen peroxide as well as lipid hydroperoxides (Little and O'Brien, 1968; Ursini et al., 1982). Figure 5.15 illustrates dietary compounds and possible body pathways for eliminating altered products.

Vilas et al. (1976), after feeding weanling rats for 14 weeks 7% oxidized corn oil with a peroxide value of 1000 Eq/kg in the framework of a diet with adequate Se and vitamin E, observed increases in the GPx activity in stomach mucosa but not in the plasma or in adipose tissue. Ingestion of heated, at 98°C for 48 h, sunflower oil increased activities of GPx and GR in rat liver (Ammouche et al., 2002). As the activity of GPx and metabolic-related enzymes are increased in tissues of animals fed oxidized oils (Chow, and Tappel, 1972; Chow, et al., 1973; Negishi et al., 1980), the enzymic system plays an important role in the protection against the deleterious effects of hydroperoxides (Chow, 1979; Chow, 1988a). Similarly, significantly higher lipase activities observed in the exocrine pancreatic juice of growing pigs fed oxidized canola oils suggest that exocrine pancreas is able to adapt to variations in the level and quantity of oxidized lipids in the diet (Ozimek et al., 2005). In addition, the metabolic response to the consumption of oxidized oils is supported by the findings that the activities of hepatic aminopyridine *N*-demethylase,

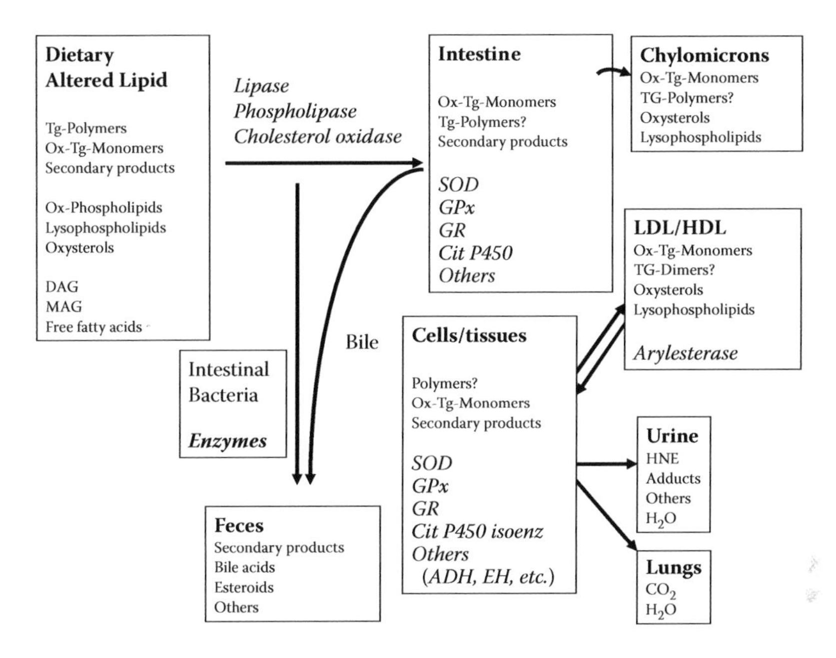

FIGURE 5.15 Possible metabolic fate of dietary oxidized and polymerized fats/oils. The scheme illustrates dietary compounds and possible body pathways for eliminating those observed altered products. (*Note*: Tg = triacylglycerols; O = oxidized; secondary products (short- or medium-chain alkyl group, carbonyl, hydroxyl, aldehyde, ester, epoxy, carboxyl, etc.); GPx = glutathione peroxidase; GR = glutathione reductase, ADH = aldehyde dehydrogenase, EH = epoxy hydrase; HNE, hydroxynonenals, adducts, HNE adduct, and HNE-conjugates.)

aniline hydrolase, NADPH-cytochrome-c-reductase, UDP-glucuronyl transferase, and glutathione-*S*-transferase, as well as cytochrome P450 content are significantly increased in rats fed a diet containing 15% deteriorated soybean frying oil (Huang et al., 1988).

Rancid fats have been shown to inactivate certain enzymes such as succinic oxidase, cytochrome oxidase, and choline oxidase (Bernheim et al., 1952, quoted by Purushothama et al., 2003). On the contrary, no harmful effect either due to commercial frying of foods or compounds derived from heated fats has been reported (Lang, 1978). Feeding peroxidized groundnut oil (pv-90 meq/kg) resulted in suppression of Na^+-K^+ ATPase and acetylcholine esterase activities in erythrocytes (Behniwal et al., 1991). The effect of oxidized soybean oil on the activities of different enzymes such as SOD, catalase, GPx, GR, and glucose-6-phosphate dehydrogenase in both erythrocytes and liver have been reported (Hayam et al., 1995). Reduction in the activities of several enzymes such as isocitric dehydrogenase, carnitine palmitoyl transferase-1, and glucose-6-phosphate dehydrogenase and increase in the activity of cytochromes P450 and b5 in rats fed heated hydrogenated soybean oil have been reported (Lamboni and Perkins, 1996). Ingestion of heated fats is also reported to affect a variety of physiological functions resulting in poor appetite and growth retardation. However, Narasimhamurthy and Raina

(1999) have reported no adverse effect either in reduced food intake or excess fecal fat excretion upon feeding heated oil at 20% levels for 20 weeks. The safety of heated/fried oils for consumption has been questioned for a long time, and the literature is abundant with studies carried out on heated fats under a variety conditions. Heating fats at higher temperatures apart from resulting in loss of unsaturation also affects metabolic functions.

Hepatic aldehyde dehydrogenase is the major enzymic system responsible for malonyldialdehyde (MDA) metabolism (Chow, 1988b; Esterbauer, 1993) (Figure 5.15). Multiple forms of aldehyde dehydrogenase isoenzymes have been found in mitochondria, cytosol, and microsomes (Mitchell and Petersen, 1989). According to Esterbauer et al. (1991), alcohol dehydrogenase and GSH-transferases play a role eliminating MDA. Rats fed a diet containing 10% oxidized linoleic acid for 4 weeks showed increases in the activities of microsomal aldehyde dehydrogenase and NADH cytochrome c reductase, suggesting metabolic adaptations of rats to metabolize absorbed aldehydes from the diet (Hochgraf et al., 1997).

The importance of decreasing hydroxynonenals (HNE) in biological systems has been recognized for years. HNE are eliminated or metabolized rapidly in hepatocytes under normal conditions (Esterbauer, 1993) as an important part of the antioxidant defense against protein modification by the aldehydic group. Two major pathways have been proposed: (a) reduction/oxidation of the aldehyde group, and (b) conjugation to endogenous GSH leading to mercapturic acid conjugated in urine (Chow, 2008).

Metabolism by the alcohol/ADH pathways account for about the 10% of the HNE elimination, whereas bioconversion by GSH-*S*-transferase represents 50%–60% of the total HNE removal by hepatocytes (Hartley et al., 1995). At present, very active research on glutathione-*S*-transferases aims at studying the locus for recognizing and binding the HNE to the active center of glutathione-*S*-transferases (Chow, 2008).

Although research on the effect of fatty acids in cell signaling and gene regulation is growing, very few papers have focused on the effect of oxidized fat ingestion and gene regulation of enzymes related to the antioxidant defence. Gene expression modulation of enzymes occurs within hours of feeding animals diets rich in PUFAs (Jump et al., 1994; Kim and Edsall, 1999; Sampath and Ntambi, 2008). Thus, it can be accepted that the oxidized fatty acid intake, directly or through their body metabolites or second messengers, can affect the gene expression of enzymes implicated in the antioxidant defense.

Oxidized LDL (oxLDL) and its component hydroxy fatty acids have been shown to activate peroxisome proliferator-activating receptor alpha (PPARα) and gamma (PPARγ). Chao et al. (2001) tested the hypothesis that lipid oxidation products in oxidized frying oil obtained by frying wheat dough sheets in soybean oil at 205°C ± 5°C for 24 h can activate PPARα and up-regulate its target genes on Sprague-Dawley male weanling rats. Oxidized frying oil dose dependently and significantly increased mRNA of acyl-CoA oxidase and cytochrome P450 4A1 (CYP4A1) in liver of rats. Dietary-oxidized frying oil also dose-dependently increased liver microsomal CYP4A protein. The activity of hepatic acyl-CoA oxidase of the group consuming a high percentage of oxidized oil was sixfold that of the group consuming a high percentage of fresh oil. The results of Chao et al. (2001) support the hypothesis that dietary-oxidized frying oil, by activating PPARα, up-regulates the expression of PPARα downstream genes.

Trying to obtain information on this subject, the effect of thermally oxidized sunflower oil ingestion on antioxidant levels, enzyme activities, and expressions in the small intestine of fed and fasted rats has been studied by our group (Olivero-David et al. 2010b). For three consecutive days, 12 male Wistar rats received 0.5 g unused sunflower oil/100 g body weight (controls, C), while another 12 were given 0.5 g thermally oxidized sunflower oil/100 g (test group, T). On the night of day 3, 6 rats from each group were fasted (FC and FT, respectively), while the other 6 animals from each group were given free access to food (NFC and NFT, respectively). On day 4, FC and NFC rats received 1 g of unused oil/100 g body weight, while FT and NFT rats were given 1 g of frying used oil/100 g body weight. Small intestines were extracted after 4 h exposure to the oils. Altered oil ingestion under fasting conditions increased intestinal TBARS content and decreased SOD, Se-dependent GPx, and non-Se-dependent GPx activities, increased catalase activity. SOD, GPx, and TNFα expressions increased significantly in FT rats. Most of these effects were partially diminished in fed rats, suggesting that long fasting and consumption of food containing oxidized fat should both be avoided to prevent intestinal oxidative stress.

5.4 CONCLUSIONS AND FUTURE STUDIES

Lipids containing polyunsaturated fatty acids are susceptible to oxidation and polymerization during processing that involves high temperatures. The adverse effects of oxidized fats and oils differ considerably, depending on the conditions of their oxidation/heating and the extent to which they are present in the diet. Although peroxidized fatty acids and their derived metabolites are toxic, they can be readily metabolized by GPx and other related enzymes once ingested or absorbed. Aldehydic compounds such as MDA and hydroxynonenal (HNE) can be rapidly metabolized by alcohol dehydrogenase or eliminated through urine, following conjugation with GSH and other compounds.

Pancreatic lipase, which has been extensively employed to study oils and fats, is used to determine the oxidized and nonoxidized fatty acids located at the 1,3-positions of the glycerol moiety. Moreover, this enzyme can be used to analyze thermally oxidized fats. Future studies must be conducted to investigate the kinetic behavior of isolated altered compounds (such as polymers, dimers, etc.), and to establish the influence of minor compounds (e.g., sterols, polyphenols) present in some oils on the hydrolytic behavior of this digestive enzyme. The possible *in vitro* balance between factors that improve and impair lipase hydrolysis has to be tested extensively. Finally, an effort must be made to find relationships between the results of *in vitro* and *in vivo* studies to better comprehend the major stages of the digestion and absorption of thermally oxidized oils, and to assess the degree of oil alteration when the enzymatic capacity of lipase is completely inhibited. Moreover, studies have to be addressed to analyze the effect of processed foods with different alteration degree on the activity and gene expression of pancreatic lipase, PLs, and other digestive enzymes, taking into account the physiological conditions (hormone concentrations, sex, age, etc.)

of the consumers Arylesterase is a little-known enzyme that seems to play a very important role against LDL oxidation. Its mechanism of action has not yet been described, so further studies must be carried out to better understand the process of LDL oxidation and the manner in which it can be inhibited. Research into the effect of fried food consumption on arylesterase activity and the protective action of this enzyme on LDL oxidation are urgently demanded. Modulation of its activity by prooxidant and antioxidant compounds and/or other substances *in vitro* as well as *in vivo* should also be investigated.

GPx, GR, and glutathione-*S*-transferase play critical roles in the detoxification of reactive oxygen species. Several authors have indicated that modifications in the activity and gene expression of these enzymes observed in animals may be related to consumption of abused oils. Few studies, however, have been performed in animals or human volunteers to test the effects of diets including fats with low/medium levels of alteration. The role of these enzymes in the endogenous antioxidant enzymatic system must be determined within the framework of diets containing different types and concentrations of digestible carbohydrates, proteins, fiber, minerals, vitamins, antioxidants, and minor bioactive compounds and considering different physical activity status. Because dietary fats/lipids have profound effects on gene expression related to enzymes and metabolism, the effect of different fatty acids oxidized, nonoxidized, alone, or in combination with other macro and micronutrients on gene expression has to be investigated. Knowledge of the effects of dietary fatty acids on the genome may provide new insight into how dietary fat might play a role in health and disease. Furthermore, because differences in response to diet depend on existing gene polymorphisms, future work on the effect of diets containing altered fatty acids should focus on volunteers with different gene–enzyme polymorphisms. Finally, extensive studies in individuals, used to consuming high levels of fats and fried foods, and who present major polymorphisms for the enzymes involved in antioxidant and detoxification defense systems, must be carried out following strict ethical standards.

In conclusion, although adverse effects are observed after the consumption of highly altered fats, the intake of foods prepared with oils that have been heated for limited periods of time, at controlled temperatures and under regulated oil-turnover conditions, is generally considered safe. These oils are deemed nontoxic because of their low level of reactive oxidation products. Consumption of such oils presents no potential health risk when the nutritional status of the individual is adequate and the heated oil consumed is not the unique source of dietary lipids.

ACKNOWLEDGMENTS

Thanks are due to Dr. C. Cuesta, A. Romero, R. Arroyo, S. López-Varela, C. García Polonio, and to Ph.D. students D. Zulim-Botega, L. Di Lorenzo, A. Schultz, and L. Gonzalez-Torres. The study was supported by AGL-2005-07204-C02-01/ALI and AGL-2008-04892-C03-02 del Ministerio de Educación y Ciencia de España and by the Project Consolider-Ingenio 2010, Reference CSD2007-00016. We also thank Fundación Gran Mariscal de Ayacucho (FUNDAYACUCHO) from Venezuela fellowship to Raul Olivero-David.

REFERENCES

Abbott, C.A., M.I. Mackness, S. Kumar, A.J. Boulton, and P.N. Durrington. 1995. Serum paraoxonase activity, concentration, and phenotype distribution in diabetes mellitus and its relationship to serum lipids and lipoproteins. *Arterioscler Thromb. Vasc. Biol.* 15:1812–8.

Adamczak, M. 2004. The application of lipases in modifying the composition, structure and properties of lipids—a review. *Pol. J. Food Nutr. Sci.* 13:PJ3–10.

Afzal, M., A. Afzal, A. Jones, and D. Armstrong. 2002. A rapid method for the quantification of GSH and GSSG in biological samples. *Methods Mol. Biol.* 186:117–22.

Alberghina, L., R. Gnandori, S. Longhi, and M. Loti. 1991. Molecular cloning of a lipase and of a lipase related gene from *Candida cylindricea.* In *Lipases: Structure, Mechanism and Genetic Engineering*, ed. L. Alberghina, R.D. Schmid, R. Verger, 231–35. Weinheim: Wiley-VCH.

Aloulou, A. and F. Carrière. 2008. Gastric lipase: An extremophilic interfacial enzyme with medical applications. *Cell Mol. Life Sci.* 65:851–4.

Ammouche, A., F. Rouaki, A. Bitam, and M.M. Bellal. 2002. Effect of ingestion of thermally oxidized sunflower oil on the fatty acid composition and antioxidant enzymes of rat liver and brain in developments. *Ann. Nutr. Metab.* 46:268–75.

Andreou, A. and I. Feussner. 2009. Lipoxygenases—Structure and reaction mechanism. *Phytochemistry* 70:1504–10.

Angerosa, F.R., C. Mostallino Basti, and R. Vito. 2000. Virgin olive oil odour notes: Their relationships with volatile compounds from the lipoxygenase pathway and secoiridoid compounds. *Food Chem.* 68:283–7.

Angerosa, F. 2000. Sensory quality of olive oils. In J. Harwood and R. Aparicio, Eds., *Handbook of Olive Oil: Analysis and Properties*. 355–72. Gaithersburg, MD: Aspen Publications, Inc.

Aparicio, R. and M.T. Morales. 1998. Characterization of olive ripeness by green aroma compounds of virgin olive oil. *J. Agric. Food Chem.* 46:1116–22.

Arbones, G., A. Carbajal, B. Gonzalvo, M. González-Gross, M. Joyanes, I. Marques-Lopes et al. 2003. Nutrición y recomendaciones dietéticas para personas mayores. Grupo de trabajo "Salud pública" de la Sociedad Española de Nutrición (SEN). *Nutr. Hosp.* 18:113–41.

Arishima, T., N. Tachibana, M. Kojima, K. Takamatsu, and K. Imaizumi. 2009. Screening of resistant triacylglycerols to the pancreatic lipase and their potentialities as a digestive retardant. *J. Food Lipids* 16:72–88.

Arroyo, R. 1995. Comportamiento de aceite de girasol y oleína de palma en fritura de patatas. Estudio *in vitro* de la actividad hidrolítica de la lipasa pancreática porcina sobre sustratos termooxidados. Ph.D Thesis. Facultad de Farmacia. Universidad Complutense. Madrid, Spain.

Arroyo, R., C. Cuesta, J.M. Sánchez-Montero, and F.J. Sánchez-Muniz. 1995. High performance size exclusion chromatography of palm olein used for frying. *Fett. Wiss. Technol.* 97:292–6.

Arroyo, R., F.J. Sánchez-Muniz, C. Cuesta, F.J. Burguillo, and J.M. Sánchez-Montero. 1996. Hydrolysis of used palm oil and sunflower oil catalysed by porcine pancreatic lipase. *Lipids* 31:1133–9.

Arroyo, R., F.J. Sánchez-Muniz, C. Cuesta, J.V. Sinisterra, and J.M. Sánchez-Montero. 1997. Thermoxidation of substrate models and their behaviour during hydrolysis by porcine pancreatic lipase. *J. Am. Oil Chem. Soc.* 74:1509–16.

Arroyo, M., J.M. Sanchez-Montero, and J.V. Sinisterra. 1999. Thermal stabilization of immobilized lipase B from *Candida antarctica* on different supports: Effect of water activity on enzymatic activity in organic media. *Enzyme Microb. Technol.* 24:3–12.

Arzoglou, P. 1994. Titrimetric assay of pancreatic lipase: State-of-the-art. *Ann. Biol. Clin.* 52:165–70.

Aviram, M., M. Rosenblat, C.L. Bisgaier, R.S. Newton, S.L. Primo-Parmo, and B.N. La Du. 1998. Paraoxonase inhibits high density lipoprotein (HDL) oxidation and preserves its functions: A possible peroxidative role for paraoxonase. *J. Clin. Invest.* 101:1581–90.

Aviram, M., M. Rosenblat, S. Billecke, S. Mahmood, S. Milo, A. Hoffman et al. 1999. Human serum paraoxonase (PON1) is inactivated by oxidized low density lipoprotein and preserved by antioxidants. *Free Radic. Biol. Med.* 26:892–904.

Azarsiz, E., M. Kayikcioglu, S. Payzin, and E. Yildirim Sozmen. 2003. PON1 Activities and oxidative markers of LDL in patients with angiographically proven coronary artery disease. *Int. J. Cardiol.* 91:43–51.

Barford, R.A., F.E. Luddy, and P. Magidman. 1966. The hydrolysis of long-chain triglycerides by pancreatic lipase. *Lipids* 1:287–96.

Bastida, S. and F.J. Sánchez-Muniz. 2001. Selección del aceite en la fritura de los alimentos. Desde el aceite de oliva a los nuevos aceites. *Rev. Nutr. Pract.* 5:67–79.

Bastida, S., F.J. Sánchez-Muniz, and G. Trigueros. 2003. Aplicación de un test colorimétrico al estudio del rendimiento y vida útil en fritura de alimentos precocinados y frescos en aceite de oliva, aceite de girasol y su mezcla. *Grasas Aceites* 54:32–40.

Behniwal, P.K., J. Gulati, G.L. Soni, and R. Singh. 1991. Toxicity of peroxidized oil and role of vitamins: Effect on osmotic fragility of erythrocyte membranes and membrane bound enzymes. *J. Food Sci. Technol.* 28:161–3.

Berglund, P. and K. Hutt. 2000. Biocatalytic synthesis of enantiopure compounds using lipases. In *Stereoselective Biocatalysis*, ed. R.N. Patel, 188–94. New York: Marcel Dekker.

Bernheim, F., K.M. Wilbur, C.B. Kenaston. 1952. The effect of oxidized fatty acids on the activity of certain oxidative enzymes. *Arch Biochem Biophys* 38:177–84.

Beynen, A.C. and M.B. Katan. 1985. Rapid sampling and long-term storage of subcutaneous adipose-tissue biopsies for determination of fatty acid composition. *Am. J. Clin. Nutr.* 42:317–22.

Blasi, F., S. Maurelli, L. Cossignani, G. D'Arco, M. Simonetti, and P. Damiani. 2009. Study of some experimental parameters in the synthesis of triacylglycerols with CLA isomers and structural analysis. *J. Am. Oil Chem. Soc.* 86:531–7.

Blatter, M.C., R.W. James, S. Messmer, F. Barja, and D. Pometta. 1993. Identification of a distinct human high-density lipoprotein subspecies defined by a lipoprotein-associated protein, K-85: Identity of K-85 with paraoxonase. *Eur. J. Biochem.* 211:871–9.

Blumenthal, M.M. 1991. A new look at the chemistry and physics of deep-fat frying. *Food Technol.* 45:68–71.

Bognár, A. 1998. Comparative study of frying to other cooking techniques. Influence on the nutritive value. *Grasas Aceites* 49:250–60.

Bornscheuer, U.T. 2000. *Enzymes in Lipid Modification*, ed. U.T. Bornscheuer, 394–414. Weinheim: Wiley-VCH.

Bornscheuer, U.T., M. Adamczak, and M.M. Soumanou. 2002. Lipase-catalyzed synthesis of modified lipids. In *Lipids as Constituents of Functional Foods*, ed. F.D. Gunstone 149–82. Bridgwater, England: Barnes & Associates.

Boskou, D. 1996. In *Olive Oil: Chemistry and Technology.* ed. D. Boskou, 161. Champaign, IL: AOCS Press.

Boskou, D. 2006. Olive oil. Monograph Series. In *Chemistry and Technology*. Champaign, IL: AOCS Press.

Bottino, N.R., G.A. Vanderburg, and R. Reiser. 1967. Resistance of certain long-chain polyunsaturated fatty acids of marine oils to pancreatic lipase hydrolysis. *Lipids* 2:489–93.

Bou, R., R.C. Tres, M.D. Baucells, and F. Guardiola. 2005. Increase of geometrical and positional fatty acid isomers in dark meat from broilers fed heated oils. *Poult. Sci.* 84:1942–54.

Breckenridge, W.C. 1978. Stereospecificity analysis of triglycerides. In *Fatty Acids and Glycerides. Handbook of Lipid Research, Vol. 1,* ed. A. Kuksis, 197–232. New York: Plenum Press.

Briante, R., M. Patumi, S. Terenziani, E. Bismuto, F. Febbraio, and R. Nucci. 2002. *Olea europaea* L. leaf extract and derivatives: Antioxidant properties. *J. Agric. Food Chem.* 50:4934–40.

Brockerhoff, H. 1965. Stereospecific analysis of triglycerides: An analysis of human depot fat. *Arch. Biochem. Biophys.* 110:586–92.

Brockman, H.L. 2000. Kinetic behaviour of the pancreatic lipase–colipase–lipid system. *Biochimie* 82:987–95.

Bruckner, G. 2008. Fatty acids lipids, and cellular signaling. In *Fatty Acids in Foods and Their Implications*, ed. C.K. Chow, 741–55. Boca Raton, FL: Taylor & Francis.

Bucolo, G. and H. David. 1973. Quantitative determination of serum triglycerides by the use of enzymes. *Clin. Chem.* 19:476–82.

Canales, A. and F.J. Sánchez-Muniz. 2003. La paraoxonasa, ¿algo más que un enzima? *Med. Clin. (Barc.)* 121:537–48.

Carey, M.C., D.M. Small, and C.M. Bliss. 1983. Lipid digestion and absorption. *Annu. Rev. Physiol.* 45:651–77.

Chao, P.M., C.Y. Chao, F.J. Lin, and C. Huang. 2001. Oxidized frying oil up-regulates hepatic acyl-CoA oxidase and cytochrome P450 4A1 genes in rats and activates PPAR alpha. *J. Nutr.* 131:3166–74.

Chapus, C., M. Rovery, L. Sarda, and R. Verger. 1988. Minireview on pancreatic lipase and colipase. *Biochemie* 70:1223–34.

Chen, J.C. and S.W. Tsai. 2000. Enantioselective synthesis of (S)-ibuprofen ester prodrug in cyclohexane by *Candida rugosa* lipase immobilized on accurel MP1000. *Biotechnol. Prog.* 16:986–92.

Chism, G.W. 1985. Soy lipoxygenase. In *Flavor Chemistry of Fats and Oil*, ed. D.B. Miu and T.H. Smouse, 175–87. Champaign, IL: AOCS Press.

Choe, E. and D.B. Min. 2007. Chemistry of deep-fat frying oils. *J. Food Sci.* 72:R77–88.

Chow, C.K. 1979. Nutritional influence on cellular antioxidant defense systems. *Am. J. Clin. Nutr.* 32:1066–81.

Chow, C.K. 1988a. Interrelationships of cellular antioxidant defense systems. In *Cellular Antioxidant Defense Mechanisms*, ed. C.K. Chow, 217–37. Boca Raton, FL: Taylor & Francis.

Chow, C.K. 1988b. Biological effects of oxidized fatty acids. In *Fatty Acids in Foods and Their Implications*, ed. C.K. Chow, 855–78. Boca Raton, FL: Taylor & Francis.

Chow, C.K. (Ed.). 2008. *Fatty Acids in Foods and Their Implications.* Boca Raton, FL: Taylor & Francis.

Chow, C.K. and A.L. Tappel. 1972. An enzymatic protective mechanism against lipid peroxidation damage to lungs of ozone-exposed rats. *Lipids* 7:518–24.

Chow, T.J., K.W. Bruland, K. Berline, A. Soutar, M. Koide, and E.D. Goldberg. 1973. Lead Pollution: Records in Southern California coastal sediments. *Science* 18:551–2.

Christie, W.W. 1986. The positional distribution of fatty acids in triglycerides. In *Analysis of Oils and Fats,* ed. R.J. Hamilton and J.B. Rossell, 313–39. London: Elsevier Applied Science.

Ciafardini, G. and B.A. Zullo. 2002. Microbiological activity in stored olive oil. *Int. J. Food Microbiol.* 75:111–8.

Clausen, I.G., S.A. Patkar, K. Borch, M. Barfoed, K. Clausen et al. 2000. Method for Reducing Phosphorus Content of Edible Oils, U.S. Patent 6,103,505.

Combe, N., M.J. Constantin, and B. Entressangles. 1981. Lymphatic absorption of nonvolatile oxidation products of heated oils in the rat. *Lipids* 16:8–14.

Cortesi, R. and O.S. Privett. 1972. Toxicity of fatty ozonides and peroxides. *Lipids* 7:715–21.

Crampton, E.W., R.H. Common, F.A. Farmer, A.F. Wells, and D. Crawford. 1953. Studies to determine the nature of the damage to the nutritive value of some vegetable oil from heat treatment. III. The segregation of toxic and nontoxic material from the esters of heat-polymerized linseed oil by destillation and by urea-adduct formation. *J. Nutr.* 49:333–46.

Cuesta, C. and F.J. Sánchez-Muniz. 2001. La fritura de los alimentos. Fritura en aceite de oliva y aceite de oliva virgen extra. In *Aceite de Oliva Virgen: Nuestro Patrimonio Alimentario,* Vol. 1, ed. J. Mataix, 173–209. Granada: Universidad de Granada and Puleva Food (Instituto Omega 3).

DGF (German Society for Fat Research). 2000. Proceedings of the 3rd international symposium of deep-fat frying. Final recommendations, *Eur. J. Lipid Sci. Technol.,* 102, 594.

De Lorgeril, M., P. Salen, J.L. Martin, I. Monjaud, J. Delaye, and N. Mamelle. 1999. Mediterranean diet, traditional risk factors, and the rate of cardiovascular complications after myocardial infarction—Final report of the Lyon Diet Heart Study. *Circulation* 99:779–85.

Deuel, Jr., H.J. 1955. The lipids, their chemistry and biochemistry. In *VII Biochemistry,* 227–40, New York: Interscience.

Dobarganes, M.C., M.C. Pérez-Camino, and G. Márquez-Ruiz. 1988. High-performance size-exclusion chromatography of polar compounds in heated and non-heated fat. *Fett. Wiss. Technol.* 90:308–11.

Dobarganes, M.C., J. Velasco, and A. Dieffenbacher. 2000. Determination of polar compounds polymerized and oxidized triacylglycerols, and diacylglycerols in oils and fats. Results of collaborative studies and the standardized method (Technical report). *Pure Appl. Chem.* 72:1563–75.

Dubouloz, P., J. Laurent, and J. Dumas. 1951. The metabolism of lipid peroxides. I. Characteristics of blood pigment which destroys peroxides. *Bull. Soc. Chim. Biol.* 33:1740–4.

Eckerson, H.W., J. Romson, C. Wyte, and B.N. La Du. 1983. The human serum paraoxonase/arylesterase polymorphism. *Am. J. Hum. Genet.* 35:214–27.

Elmadfa, I. and K.H. Wagner. 1999. Fat and nutrition. In *Frying of Food. Oxidation, Nutrient and Non-Nutrient Antioxidants, Biologically Active Compounds and High Temperatures,* ed. D. Boskou, and I. Elmadfa, 1–15. Lancaster, PA: Technomic Publishing.

Engeseth, N.J., B.P. Klein, and K. Warner. 1987. Lipoxygenase isoenzymes in soybeans: Effects on crude oil quality. *J. Food Sci.* 52:1015–19.

Esposito, S., M. Semeriva, and P. Desnuelle. 1973. Effect of surface pressure on the hydrolysis of ester monolayers by pancreatic lipase. *Biochim. Biophys. Acta* 302:293–304.

Esterbauer, H. 1993. Cytotoxicity and genotoxicity of lipid-oxidation products. *Am. J. Clin. Nutr.* 57(5 Suppl.):779S–785S.

Esterbauer, H., R.J. Schaur, and H. Zollner. 1991. Chemistry and biochemistry of 4-hydroxynonenal, malonaldehyde and related aldehydes. *Free Radic. Biol. Med.* 11:81–128.

Faber K. 2004. *Biotransformations in Organic Chemistry*, Berlin: Springer Verlag.

Findley, D.L., P.L. Walne, and R.W. Holton. 1970. The effects of light intensity on the ultrastructure of *Chlorogloea fritschii* Mitra grown at high temperature. *J. Phycol.* 6:182–8.

Flohe, L. and W.A. Gunzler. 1984. Assays of glutathione peroxidase. *Methods Enzymol.* 105:114–21.

Friedman, B. 1991. New control of frying process provides major reduction of oil in food. In *Fat and Cholesterol Reduced Foods. Technologies and Strategies*, ed. C. Haberstroch, and C.E. Morris, Advances in Applied Biotechnology series. Vol. 12. pp. 141–52. The Woodland, TX: Porfolio Publishing Company.

Funes, J. and M. Karel. 1981. Free radical polymerization and lipid binding of lysozyme reacted with peroxidizing linoleic acid. *Lipids* 16:347–50.

Funes, J.A., U. Weis, and M. Karel. 1982. Effects of reaction conditions and reactant concentrations of polymerization of lysozyme reacted with peroxidizing lipids. *J. Agric. Food Chem.* 30:1204–8.

Gardner, H.W. 1996. Lipoxygenase as a versatile biocatalyst. *J. Am. Oil Chem. Soc.* 73:1347–57.

Georgalaki, M.D., A. Bachmann, T.G. Sotiroudis, A. Xenakis, A. Porzel, and I. Feussner. 1998. Characterization of a 13-lipoxygenase from virgin olive oil and bodies of olive endosperm. *Lipid-Fett.* 100:554–60.

Georgalaki, M.D., T.G. Sotiroudis, and A. Xenakis. 1998. The presence of oxidizing activities in virgin olive oil. *J. Am. Oil Chem. Soc.* 75:155–9.

Gere, A. 1982. Studies on the changes in edible fats during heating and frying. *Nahrung* 26:923–32.

Gertz, C. 2000. Chemical and physical parameters as quality indicators of used fats. *Eur. J. Lipid Sci. Technol.* 102: 566–72.

Giani, E., J. Masi, and C. Galli. 1985. Heated fats, vitamin E and vascular eicosanoids. *Lipids* 20:439–48.

Gilham, D. and R. Lehner. 2005. Techniques to measure lipase and esterase activity in vitro. *Methods* 36:139–47.

González-Muñoz, M.J., C. Tulasne, R. Arroyo, and F.J. Sánchez-Muniz. 1996. Digestibility and absorption coefficients of palm olein—Relationships with thermal oxidation induced by potato frying. *Fett-Lipid* 98:104–8.

González-Muñoz, M.J., S. Bastida, and F.J. Sánchez-Muniz. 1998. Short-term in vivo digestibility of triglyceride polymers, dimers and monomers of thermoxidized palm olein used in deep-frying. *J. Agric. Food Chem.* 46:5188–93.

González-Muñoz, M.J., S. Bastida, and F.J. Sánchez-Muniz. 2003. Short term in vivo digestibility assessment of a highly oxidised and polimerised sunflower oil. *J. Sci. Food Agric.* 83:413–8.

Goupy, P., A. Fleriet, M.J. Amiot, and J.J. Macheix. 1991. Enzymatic browning, oleuropein content, and diphenol oxidase activity in olive cultivars (*Olea europea* L.). *J. Agric. Food Chem.* 39:92–5.

Griffith, O.W. 1980. Determination of glutathione and glutathione disulfide using glutathione reductase and 2-vinylpyridine. *Anal. Biochem.* 106:207–12.

Gunstone, F.D. and F.A. Norris, 1983. Deodorisation. In *Lipids in Foods Chemistry, Biochemistry and Technology*, 144–6. Oxford: Pergamon Press.

Gutiérrez González-Quijano, R. and M.C. Dobarganes. 1988. Analytical procedures for the evaluation of used frying fats. In *Frying of Food. Principles, Changes, New Approaches,* ed. G. Varela, A.E. Bender, I.D. Morton, 141–54. Chichester: Ellis Horwood Ltd.

Harel, M., A. Aharoni, L. Gaidukov, B. Brumshtein, O. Khersonsky, R. Meged et al. 2004. Structure and evolution of the serum paraoxonase family of detoxifying and anti-atherosclerotic enzymes. *Nat. Struct. Mol. Biol.* 11:412–9.

Hartley, D.P., J.A. Ruth, and D.R. Petersen. 1995. The hepatocellular metabolism of 4-hydroxynonenal by alcohol dehydrogenase, aldehyde dehydrogenase, and glutathione S-transferase. *Arch. Biochem. Biophys.* 316:197–205.

Hayam, I., U. Cogan, and S. Mokady. 1995. Dietary oxidized oil and the activity of antioxidant enzymes and lipoprotein peroxidation in rats. *Nutr. Res.* 15:1037–44.

Henderson, R.J., I.C. Burkow, and R.M. Millar. 1993. Hydrolysis of fish oils containing polymers and triacylglycerols by pancreatic lipase in vitro. *Lipids* 28:313–9.

Henry, C.J.K. 1998. Impact of fried foods on macronutrient intake, with special reference to fat and protein. *Grasas Aceites* 49:336–9.

Hermoso, J., D. Pignol, B. Kerfelec, I. Crenon, C. Chapus, and J.C. Fontecilla-Camps. 1996. Lipase activation by nonionic detergents. *J. Biol. Chem.* 271:18007–16.

Hernáiz, M.J., J.M. Sánchez-Montero, and J.V. Sinisterra. 1994a. Comparison of the enzymatic activity of commercial and semipurified lipase of *Candida cylindracea* in the hydrolysis of the esters of (R,S) 2-aryl propionic acids. *Tetrahedron* 50:10749–60.

Hernáiz, M.J., J.M. Sánchez-Montero, P. Medina, B. Celda, M. Rua, and J.V. Sinisterra. 1994b. Contribution to the study of alteration of lipase activity of *Candida rugosa* by ions and buffers. *Appl. Biochem. Biotechnol.* 44:213–29.

Hernáiz, M.J., J.M. Sánchez-Montero, and J.V. Sinisterra. 1999. Modification of purified lipases from *Candida rugosa* with polyethylene glycol: A systemic study. *Enzyme Microbiol. Technol.* 24:181–90.

Hochgraf, E., S. Mokady, and U. Cogan. 1997. Dietary oxidized linoleic acid modifies lipid composition of rat liver microsomes and increases their fluidity. *J. Nutr.* 127:681–6.

Huang, C.J., N.S. Cheung, and V.R. Lu. 1988. Effects of deteriorated frying oil and dietary protein levels on liver microsomal enzymes in rats. *J. Am. Oil Chem. Soc.* 65:1796–803.

Huang, K., H. Liu, Z. Chen, and H. Xu. 2002. Role of selenium in cytoprotection against cholesterol oxide-induced vascular damage in rats. *Atherosclerosis* 162:137–44.

Husain, S., G.S.R. Sastry, N. Prasada Raju, and R. Narasimha. 1988. High-performance size-exclusion chromatography of oils and fats. *J. Chromatogr.* 454:317–26.

Iritani, N., K. Inoguchi, M. Endo, E. Fukuda, and M. Morita. 1980. Identification of shell fish fatty acids and their effects on lipogenic enzymes. *Biochim. Biophys. Acta* 618:378–82.

Itabashi, Y., J.J. Myher, and A. Kuksis. 2000. High-performance liquid chromatographic resolution of reverse isomers of 1,2-diacyl-*rac*-glycerols as 3,5-dinitrophenylurethanes. *J. Chromatogr. A.* 893:261–79.

Jorge, N., G. Márquez-Ruiz, M. Martın, M.V. Ruiz-Méndez, and M.C. Dobarganes. 1996. Influence of dimethylpolysiloxane addition to frying oils: Performance of sunflower oil in discontinuous and continuous laboratory frying. *Grasas Aceites* 47:20–5.

Jump, D.B. 2004. Fatty acid regulation of gene transcription. *Crit. Rev. Clin. Lab. Sci.* 41:41–78.

Jump, D.B. and S.D. Clarke. 1999. Regulation of gene expression by dietary fat. *Annu. Rev. Nutr.* 19:63–90.

Jump, D.B., S.D. Clarke, A. Thelen, and M. Liimatta. 1994. Coordinate regulation of glycolytic and lipogenic gene expression by polyunsaturated fatty. *J. Lipid Res.* 35:1076–84.

Kajimoto, G. and K. Mukai. 1970. Toxicity of rancid oil. IX. Digestibility of polymerized fatty acid in thermally oxidized soybean oil. *J. Jpn. Oil Chem. Soc.* 19:66–70.

Kamata, T., K. Akasaka, H. Ohrui, and H. Meguro. 1993. A sensitive fluorometric assay of glutathione reductase activity with *N*-(9-acridinyl) maleimide. *Anal. Sci.* 9: 867–70.

Kelso, G.J., W.D. Stuart, R.J. Richter, C.E. Furlong, T.C. Jordan-Starck, and J.A.K. Harmony. 1994. Apolipoprotein J is associated with paraoxonase in human plasma. *Biochimie* 33:832–9.

Kersten, S. 2001. Mechanisms of nutritional and hormonal regulation of lipogénesis *EMBO Rep.* 2:282–6.

Keys, A., A. Menotti, M.J. Karvonen, C. Aravanis, H. Blackburn, R. Buzina et al. 1986. The diet and 15-year death rate in the seven countries study. *Am. J. Epidemiol.* 124:903–15.

Kim, H.Y. and L. Edsall. 1999. The release of polyunsaturated fatty acids and their lipoxygenation in the brain. *Adv. Exp. Med. Biol.* 447:75–85.

Kim, K.H., D.Y. Kwon, and J.S. Rhee. 1984. Effects of organic solvents on lipase for fat splitting. *Lipids* 19:975–7.

Klibanov, A.M. 1986. Enzymes that work in organic solvent. *Chem. Technol.* 16:354–9.

Kode, A., R. Rajagopalan, S. V. Penumathsa, and V. P. Menon. 2005. Effect of ethanol and thermally oxidized sunflower oil ingestion on phospholipid fatty acid composition of rat liver: Protective role of *Cuminum cyminum* L. *Ann. Nutr. Metab.* 49:300–3.

Koidis, A., E. Triantafilou, and D. Boskou. 2008. Endogenous microflora in turbid virgin olive oils and the physicochemical characteristics of these oils, *Eur. J. Lipid Sci. Tehnol.* 110:164–71.

Kris-Etherton, P.M. 1999. AHA Science Advisory monounsaturated fatty acids and risk of cardiovascular disease. American Heart Association. Nutrition Committee. *Circulation* 100:1253–8.

Lagocki, J., J. Law, and F. Kezdy. 1973. The kinetic study of enzyme action on substrate monolayers. Pancreatic lipase reactions. *J. Biol. Chem.* 248:580–7.

Lamboni, C. and E.G. Perkins. 1996. Effects of dietary heated fats on rat liver enzyme activity. *Lipids* 31:955–62.

Lang, K. 1978. Nutritional and physiological properties of frying fats. *Z. Ernahrungswiss* Suppl. 21 (suppl) 21:1–61.

Lanteaume, M., P. Ramel, A.M. Le Clerc, and J. Rannaud. 1966. Influence de la friture et du chauffage sur les effects physiologiques d'une huile três riche en acide linoléique. Huile de pepins de raisin (Influence of the frying and the heating on physiological effects of a high-linoleic acid oil. Raisin seed oil). *Rev. Fr. Corps Gras* 13:603–1.

Leake, L. and M. Karel. 1982. Polymerization and denaturation of lysozyme exposed to peroxidizing lipids. *J. Food Sci.* 47:737–9.

Lee, J.L. and D.H. Hwang. 2008. Dietary fatty acids and eicosanoids. In *Fatty Acids in Foods and Their Implications*, ed. C.K. Chow, 713–26. Boca Raton, FL: Taylor & Francis.

Le Floch, E., P. Acker, P. Ramel, M.T. Lanteaume, and A.M. Le Clerc. 1968. Les effects d'un chauffage de type culinaire sur les principaux corps gras alimentaires, ses incidences physiologiques et nutritionelles (The effects of heating in a culinary manner on the principal dietary fats. Physiological and nutritional effects). *Ann. Nutr. Alim.* 22:249–65.

Liese, K. 2001. Putting enzymes to work. *Industrial Biotransformations*. Biotechnol. Adv., ed. A. Liese, K. Seelbach, and C. Wandrey, 19:317–8. Weinheim: Wiley-VCH.

Linscheer, W.G. and A.J. Vergroesen. 1994. Lipids. In *Modern Nutrition in Health and Disease*, ed. M.E. Shils, J.A. Olson, M. Shike, 47–88. Philadelphia: Lea & Febiger.

Little, C. and P.J. O'Brien. 1968. The effectiveness of a lipid peroxide in oxidizing protein and non-protein thiols. *Biochem. J.* 106:419–23.

Lobo de Araujo, A. and F. Radvanyi. 1987 Determination of phospholipase A2 activity by a colorimetric assay using a pH indicator, *Toxicon* 25:1181–8.

López-Varela, S., F.J. Sánchez-Muniz, C. Garrido-Polonio, R. Arroyo, and C. Cuesta. 1995. Relationships between chemical and physical indexes and column and HPSE chromatography methods for evaluating frying oil. *Z. Ernährungswiss* 34:308–13.

Lorentz, K., W. Wirtz, and T. Weiss. 2001. Continuous monitoring arylesterase in human serum. *Clin. Chim. Acta* 30:69–78.

Lowe, M.E. 1997. Molecular mechanisms of rat and human pancreatic triglyceride lipases. *J. Nutr.* 127:549–57.

Luddy, F.E., R.A. Barford, S.F. Herb, P. Magidman, and R.W. Riemenschneider. 1964. Pancreatic lipase by hydrolysis of triglycerides by a semimicro technique. *J. Am. Oil Chem. Soc.* 41:693–6.

Mackness, M.I., B. Harty, D. Bhatnagar, P.H. Wincour, S. Arrol, M. Ishola, and P.N. Durrington. 1991. Serum paraoxonase activity in familial hypercholesterolaemia and insulin dependent diabetes mellitus. *Atherosclerosis* 86:193–9.

Maeda, H., H. Matsuno, M. Ushida, K. Katayama, K. Saeki, N. Itoh, and A. Angew. 2005. 2,4-Dinitrobenzenesulfonyl fluoresceins as fluorescent alternatives to Ellman's reagent in thiol-quantification enzyme assays. *Ang. Chem. Int. Ed.* 44: 2922–5.

Mahungu, S.M., W.E. Artz, and E.G. Perkins. 1999. Oxidation products and metabolic processes. In *Frying of Food. Oxidation, Nutrient and Non-Nutrient Antioxidants, Biologically Active Compounds and High Temperatures,* ed. D. Boskou and I. Elmadfa, 25–46. Lancaster, PA: Technomic Publishing.

Márquez-Ruiz, G. and M.C. Dobarganes. 1996. Nutritional and physiological effect of used frying fats. In *Deep frying. Chemistry, Nutrition and Practical Applications*, ed. E.G. Perkins, M.D. Erickson, 160–82. Champaign, IL: AOCS Press.

Márquez-Ruiz, G., M.C. Pérez-Camino, and M.C. Dobarganes. 1992a. In vitro action of lipase on complex glycerides from thermally oxidized oils. *Fat. Sci. Technol.* 94:307–12.

Márquez-Ruiz, G., M.C. Pérez-Camino, and M.C. Dobarganes. 1992b. Digestibility of fatty acid monomers, dimers and polymers in the rat. *J. Am. Oil Chem. Soc.* 69:930–34.

Márquez-Ruiz, G., G. Guevel, and M.C. Dobarganes. 1998. Applications of chromatographic techniques to evaluate enzymatic hydrolysis of oxidized and polymeric triglycerides by pancreatic lipase in vitro. *J. Am. Oil Chem. Soc.* 75:119–26.

Márquez-Ruíz, G., M.C. Pérez-Camino, and M.C. Dobarganes. 2006. In vitro action of pancreatic lipase on complex glycerides from thermally oxidized oils *Eur. J. Lipid Sci. Technol.* 94:307–14.

Martinelle, M., M. Holmquist, and K. Hult. 1995. On the interfacial activation of *Candida antarctica* lipase A and B as compared with *Humicola lanuginosa* lipase. *Biochim. Biophys. Acta* 1258:272–6.

Massaro, M., G. Basta, G. Lazzerini, M.A. Carluccio, F. Bosetti, G. Solaini et al. 2002. Quenching of intracellular ROS generation as a mechanism for oleate-induced reduction of endothelial activation and early atherogenesis. *Thromb. Haemost.* 88:335–44.

Massaro, M., E. Scoditti, M.A. Carluccio, and R. De Caterina. 2006. Epidemiology of olive oil and cardiovascular disease. In *Olive Oil and Health,* ed. J.L. Quiles, M.C. Ramírez-Tortosa, and P. Yaqoob, 152–63. Oxfordshire, U.K.: CAB International.

Mata, P., R. Alonso, and N. Mata. 2002. Los omega-3 y omega-9 en la enfermedad cardiovascular. In *Libro Blanco de los Omega-3. Los Acidos Grasos Poliinsaturados Omega 3 y Monoinsaturados Tipo Oleico y su Papel en la Salud*, ed. J. Mataix, and A. Gil, 50–58. Madrid: Instituto Omega 3, Puleva Food.

Matsuchita, S.J. 1975. Specific interactions of linoleic acid hydroperoxides and their secondary degraded products with enzyme proteins. *J. Agric. Food Chem.* 23:150–4.

Mensink, R.P. and M.B. Katan. 1992. Effect of dietary fatty acids on serum lipids and lipoproteins: A meta-analysis on 27 trials. *Arterioscler. Thromb.* 12:911–9.

Mitchell, D.Y. and D.R. Petersen. 1989. Oxidation of aldehydic products of lipid peroxidation by rat liver microsomal aldehyde dehydrogenase. *Arch. Biochem. Biophys.* 269:11–7.

Miyashita, K., T. Tagaki, and E.N. Frankel. 1990. Preferential hydrolysis of monohydroperoxides of linoleoyl and linolenoyl triacylglycerol by pancreatic lipase. *Biochim. Biophys. Acta* 1045:233–8.

Morales, M.T. and R. Aparicio. 1999. Effect of extraction conditions on sensory quality of virgin olive oil. *J. Am. Oil Chem. Soc.* 76:295–300.

Morales, M.T. and R. Przybylski. 2000. Olive oil oxidation. In *Handbook of Olive Oil. Analysis and Properties*, ed. J. Harwood, and R. Aparicio, 459–85. Gaythersburg, MD: Aspen Publ. Inc.

Mu, H.L., J.P. Kurvinen, H. Kallio, X.B. Xu, and C.E. Hoy. 2001. Quantitation of acyl migration during lipase-catalyzed acidolysis, and the regioisomers of structured triacylglycerols formed. *J. Am. Oil Chem. Soc.* 78:959–64.

Mukherjee, M. 2003. Human digestive and metabolic lipases—a brief review. *J. Mol. Catal. B-Enzym.* 22:369–76.

Narasimhamurthy, K. and P.L. Raina. 1999. Long-term feeding effect of thermally oxidised oils on antioxidant enzymes in rats. *Indian J. Exp. Biol.* 37:1042–5.

Navab, M., J.A. Berliner, A.D. Watson, S.Y. Hama, M.C. Territo, and A.J. Lusis. 1996. The Yin and Yang of oxidation in the development of the fatty streak. A review based on the 1994 George Lyman Duff Memorial Lecture. *Arteriosc1. Throm. Vas.* 16:831–42.

Negishi, H., K. Fujimoto, and T. Kaneda. 1980. Effect of autoxidized methyl linoleate on glutathione peroxidase. *J. Nutr. Sci. Vitaminol.* 26:309–17.

Nishio, E. and Y. Watanabe. 1997. Cigarette smoke extract inhibits plasma paraoxonase activity by modification of the enzyme's thiols. *Biophys. Res. Commun.* 236:289–93.

Nus, M., F.J. Sánchez-Muniz, and J.M. Sánchez-Montero. 2006a. A new method for the determination of arylesterase activity in human serum using simulated body fluid. *Atherosclerosis* 188:155–9.

Nus, M., F.J. Sánchez-Muniz, and J.M. Sánchez-Montero. 2006b. Methodological aspects and relevance of the study of vegetable oil, fat and lipoprotein oxidation using pancreatic lipase and arylesterase. *Food. Technol. Biotechnol.* 44:1–15.

Nus, M., F. Frances, J. Librelotto, A. Canales, D. Corella, J.M. Sánchez-Montero, and F.J. Sánchez-Muniz. 2007. Arylesterase activity and antioxidant status depend on PON1-Q192R and PON1-L55M polymorphisms in subjects with increased risk of cardiovascular disease consuming walnut-enriched meat. *J. Nutr.* 137:1783–8.

Nus, M., F.J. Sánchez-Muniz, J.V. Sinisterra Gago, E. López-Oliva, and J.M. Sánchez-Montero. 2008. Determination of rat and mice arylesterase activity using serum mimetics. *Enzyme Microb. Technol.* 43:252–6.

Ohfuji, T. and T. Kaneda. 1973. Characterization of toxic compounds in thermally oxidized oil. *Lipids* 8:353–9.

Ohfuji, T., K. Sukarai, and T. Kaneda. 1972. Relation between the nutritive value and the structure of polymerized oils. VII. Absorption and metabolism of the toxic substance separated from thermally oxidized oil in rats. *Yukagaku* 21:63–73.

Olcott, H.S. and A. Dolev. 1963. Toxicity of fatty acid ester hydroperoxides. *Proc. Soc. Exp. Biol. Med.* 114:820–2.

Olias, J.M., A.G. Pérez, J.J. Rios, and L.C. Sanz. 1993. Aroma of virgin olive oil: Biogenesis of the "green" odor notes. *J. Agric. Food Chem.* 41:2368–73.

Olivero-David, R., F.J. Sánchez-Muniz, S. Bastida, J. Benedí, M.J. González-Muñoz, 2010a. Gastric empyting and short-term digestibility of thermally oxidized sunflower oil used for frying in fasted and non-fasted rats. *J. Agric. Food Chem.* doi 10.102/jf101715g.

Ottolenghi, A. 1973. High phospholipase content of intestines of mice infected with *Hymenolepis nana*. *Lipids* 8:426–8.

Ottolenghi, A., J.P. Pickett, and W.B. Greene. 1966. Histochemical demonstration of phospholipase B (Lysolecithinase) activity in rat tissues. *J. Histochem. Cytochem.* 14: 907–15.

Ozimek, P., M. Veenhuis, and I. van der Klei. 2005. Alcohol oxidase: a complex peroxisomal, oligomeric flavoprotein, *FEMS Yeast Res.* 5:975–83.

Padley, F.B., F.D. Gunstone, and J.L. Harwood. 1986. Occurrence and characteristics of oils and fats. In *The Lipid Handbook*, ed. F.D. Gunstone, J.L Harwood, and F.B. Padley, 49–170. London: Chapman & Hall.

Paglia, D.E. and W.N. Valentine. 1967. Studies on the quantitative and qualitative characterization of erythrocyte glutathione peroxidase. *J. Lab. Clin. Med.* 70:158–69.

Parthasarathy, S., J. Barnett, and L.G. Fong. 1990. High-density lipoprotein inhibits the oxidative modification of low-density lipoprotein. *Biochim. Biophys. Acta* 1044:275–83.

Patterson, H.B.W. 1989. Quality standards for oils, fats, seeds and meals. *Handling and Storage of Oilseeds, Oils, Fats and Meal*, 105–7. London & New York: Elsevier Applied Science.

Paulose, M.M. and S.S. Chang. 1973. Chemical reactions involved in deep fat frying of foods. VI. Characterization of nonvolatile decomposition products of trilinolein. *J. Am. Oil Chem. Soc.* 50:119–26.

Potteau, B., M. Lhuissier, J. Le Clerc, F. Custot, R. Mezonnet, and R. Cluzan. 1970. Recherches sur la composition et les effets physiologiques de l'huile de soja chauffée et des differents fractions obtenues á partir de cette huile (Research study of the composition and the physiological effects of the heated soya oil and the different fractions obtained from this oil). *Rev. Fr. Corps. Gras.* 17:143–53.

Potteau, B., A. Grandgirard, M. Lhuissier, and J. Causeret. 1977. Recherches récents sur les effets physiopathologiques d'huiles végétales chauffées (Recent research studies of the physiopathological effects of heated vegetal oils). *Bibl. Nutr. Diet* 25:122–33.

Purushothama, S., H.D. Ramachandran, K. Narasimhamurthy, and P.L. Raina. 2003. Long-term feeding effects of heated and fried oils on hepatic antioxidant enzymes, absorption and excretion of fat in rats. *Mol. Cell. Biochem.* 247: 95–9.

Rahman, I., A. Kode, and S.K. Biswas. 2007. Assay for quantitative determination of glutathione and glutathione disulfide levels using enzymatic recycling method, *Nat. Protocols* 1:3159–65.

Ridolfi, M., S. Terenziani, M. Batumi, and G. Fontanazza. 2002. Characterization of the lipoxygenases in some olive cultivars and determination of their role in volatile compounds formation. *J. Agric. Food Chem.* 50:835–9.

Ringseis, R., A. Muschick, and K. Eder. 2007. Dietary oxidized fat prevents ethanol-induced triacylglycerol accumulation and increases expression of PPAR target genes in rat liver. *J. Nutr.* 137:77–83.

Romero, A., C. Cuesta, and F.J. Sánchez-Muniz, 2000a. Trans fatty acid production in deep fat frying of frozen foods with different oils and frying modalities, *Nutr. Res.* 20:599–608.

Romero, A., F.J. Sánchez-Muniz, and C. Cuesta, 2000b. Deep fat frying of frozen foods in sunflower oil. Fatty acid composition in fryer oil and frozen prefried potatoes. *J. Sci. Food. Agric.* 80:2135–41.

Romero, A., S. Bastida, and F.J. Sánchez-Muniz. 2007. Cyclic fatty acids in sunflower oils during frying of frozen foods with oil replenishment. *Eur. J. Lipid Sci. Technol.* 109:165–73.

Ros, E. 2001. Guía para una alimentación cardiosaludable. Aporte de grasa. In *Guías Alimentarias para la población Española. Recomendaciones para una dieta saludable*, ed. Sociedad Española de la Nutrición Comunitaria, 413–21. Madrid: IM&C.

Rossell, J.B., B. King, and M.J. Downes. 1983. Detection of adulteration. *J. Am. Oil Chem. Soc.* 60:333–9.

Saka, S., A. Bairi, and M. Guellati. 2003. Immuno-corticotropin interactions in a nociceptive environment in the Wistar rats. *J. Soc. Biol.* 197:67–71.

Salas, J.J., M. Williams, J.L. Harwood, and J. Sanchez. 1999. Lipoxygenase activity in olive (*Olea europaea*) fruit. *J. Am. Oil Chem. Soc.* 10:1163–8.

Sampath, H. and J.M. Ntambi. 2008. Polyunsarturated fatty acids and regulation of gene expression. In *Fatty Acids in Foods and Their Implications,* ed. C.K. Chow, 727–39. Boca Raton, FL: Taylor & Francis.

Sánchez-Montero, J.M., V. Hamon, D. Thomas, and M.D. Legoy. 1991. Modulation of lipase hydrolysis and synthesis reactions using carbohydrates. *Biochim. Biophys. Acta* 60:419–27.

Sánchez-Muniz, F.J. 2003. Los lípidos. In *Nutrición y Dietética.* ed. M.T. García-Arias, and M.C. García-Fernández. 119-33. León: Universidad de León. Secretariado de Publicaciones y Medios Audiovisuales.

Sánchez-Muniz, F.J. 2006. Oils and fats: Changes due to culinary and industrial processes. *Int. J. Vitam. Nutr. Res.* 76: 230–7.

Sánchez-Muniz, F.J. and S. Bastida. 2006. Effect of frying and thermal oxidation on olive oil and food quality. In *Olive Oil and Human Health,* ed. J.L. Quiles, M.C. Ramírez-Tortosa, and P. Yaqoob, 74–108. Oxfordshire: CAB International.

Sánchez-Muniz, F.J. and M. Nus. 2008. Importancia de la interacción dieta-genética en la prevención cardiovascular. In *Genética, Nutrición y Enfermedad.* Coordinator M.P. Vaquero, ed. Médicos SA, 125–44. Madrid: Instituto Tomás Pascual Sanz y Consejo Superior de Investigaciones Científicas.

Sánchez-Muniz, F.J. and J.M. Sánchez Montero. 1999. Enzymatic methods for the study of thermally oxidized oils and fats. In *Frying of Food. Oxidation, Nutrient and Non-Nutrient Antioxidants, Biologically Active Compounds and High Temperatures*, ed. D. Boskou and I. Elmadfa, 105–42. Lancaster, PA: Technomic Publishing Co.

Sánchez-Muniz, F.J., P. Oubiña, J. Benedi, S. Ródenas, and C. Cuesta. 1998. A preliminary study on platelet aggregation in postmenopausal women consuming extra-virgin olive oil and high-oleic acid sunflower oil. *J. Am. Oil Chem. Soc.* 75:217–23.

Sánchez-Muniz, F.J., S. Bastida, and M.J. González-Muñoz. 1999. Column and high-performance size exclusion chromatography applications to the in vivo digestibility study of a thermoxidized and polymerized olive oil. *Lipids* 34:1187–92.

Sánchez-Muniz, F.J., P. Oubiña, S. Ródenas, J. Benedi, and C. Cuesta. 2003. Platelet aggregation, thromboxane production and thrombogenic ratio in postmenopausal women consuming high oleic acid-sunflower oil or palmolein. *Eur. J. Nutr.* 42:299–306.

Sánchez-Muniz, F.J., S. Bastida, G. Márquez-Ruiz, and C. Dobarganes. 2008. Effects of heating and frying on oil and food fatty acids. In *Fatty Acids in Foods and Their Implications*, ed. C.K. Chow, 511–43. Boca Raton, FL: Taylor & Francis.

Sandra, S., E.A. Decker, and D.J. McClements. 2008. Effect of interfacial protein cross-linking on the in vitro digestibility of emulsified corn oil by pancreatic lipase. *J. Agric. Food. Chem.* 56:7488–94.

Sarandol, E., Z. Serdar, M. Dirican, and O. Savak. 2003. Effects of red wine consumption on serum paraoxonase/arylesterasa activities and on lipoprotein oxidizability in healthy men. *J. Nutr. Biochem.* 14: 507–12.

Schmid, U., U.T. Bornscheuer, M.M. Soumanou, G.P. McNeill, and R.D. Schmid. 1999. Highly-selective synthesis of 1,3-oleyl-2-palmitoyl-glycerol by lipase catalysis. *Biotechnol. Bioeng.* 64:678–84.

Schmitt, J., S. Brocca, R.D. Schmid, and J. Pleiss. 2002. Blocking the tunnel: Engineering of *Candida rugosa* lipase mutants with short chain length specificity. *Protein Eng.* 15:595–601.

Sciancalepore, V. and V. Longone. 1984. Polyphenol oxidase activity and browning in green olives. *J. Agric. Food Chem.* 32:320–1.

Sherwin, E.R. 1978. Oxidation and antioxidants in fat and oil processing. *J. Am. Oil Chem. Soc.* 55:809–14.

Shih, D.M., L. Gu, Y-R. Xia, M. Navab, W-F. Li, S. Hama, L.W. Castellani, C.E. Furlong, L.G. Costa, A.M. Fogelman, and A.J. Lusis. 1998. Mice lacking serum paraoxonase are susceptible to organophosphate toxicity and atherosclerosis. *Nature*. 394:284–7.

Soumanou, M.M. and U T. Bornscheuer. 2003. Lipase-catalyzed alcoholysis of vegetable oils. *Eur. J. Lipid Sci. Technol.* 105: 656–60.

Staprans, I., X.M. Pan, M. Miller, and J.H. Rapp. 1993. Effect of dietary lipid peroxides on metabolism of serum chylomicrons in rats. *Am. J. Physiol. Gastrointest. Liver Physiol.* 264: G561–8.

Sutherland, W., R. Walker, A.S. de Jong, A. van Rij, V. Phillips, and H.Walter. 1999. Reduced postprandial serum paraoxonase activity after a meal rich in used cooking fat. *Arterioscler. Thromb. Vasc. Biol.* 19:1340–7.

Sutherland, W.H.F., P.J. Manning, S.A. de Jong, A.R. Allum, S.D. Jones, and S.M. Williams. 2001. Hormone-replacement therapy increases serum paraoxonase/arylesterase activity in diabetic postmenopausal women. *Metabolism* 50:319–24.

Suzuki, A., O. Kozawa, Y. Oiso, and K. Kato. 1996. Protein kinase C activation inhibits stress-induced synthesis of heat shock protein 27 in osteoblast-like cells: Function of arachidonic acid. *J. Cell. Biochem.* 62:69–75.

Tietze, F. 1969. Enzymic method for quantitative determination of nanogram amounts of total and oxidized glutathione: Applications to mammalian blood and other tissues. *Anal. Biochem.* 27:502–22.

Timur, S., D. Odaci, D., A. Dincer, F. Zihnioglu, and A. Telefoncu. 2008. Biosensing approach for glutathione detection using glutathione reductase and sulfhydryl oxidase bienzymatic system. *Talanta* 74:1492–7.

Tosi, E., G. Ballerini, and E. Ré. 2007. Soybean phospholipase D activity determination: A comparison of two methods. *Grasas Aceites* 58:270–4.

Trichopoulou, A., T. Costacou, C. Bamia, and D.Trichopoulos. 2003. Adherence to a Mediterranean diet and survival in a Greek population. *N. Engl. J. Med.* 348:2599–608.

Trinder, P. 1969. Determination of glucose in blood using glucose oxidase with an alternative oxygen acceptor. *Ann. Clin. Biochem.* 6:24–7.

Turon, F., P. Bachain, Y. Caro, M. Pina, and J. Graille. 2002 A direct method for regiospecific analysis of TAG using alpha-MAG. *Lipids* 37:817–21.

Undurraga, D. A., S. Markovits, and E. Erazo. 2001. Cocoa butter equivalent through enzymatic interesterification of palm oil midfraction. *Process. Biochem.* 36:933–9.

Ursini, F., M. Maiorino, M. Valente, L. Ferri, and C. Gregolin. 1982. Purification from pig liver of a protein which protects liposomes and biomembranes from peroxidative degradation and exhibits glutathione peroxidase activity on phosphatidylcholine hydroperoxides, *Biochim. Biophys. Acta* 710:197–211.

Varela, G. 1988. Current facts about the frying of food. In *Frying of Food. Principles, Changes, New Approaches,* ed. G. Varela, A.E. Bender, and I.D. Morton, 9–25. Chichester, England: Ellis Horwood.

Varela, G. and B. Ruiz-Roso. 1998. Influence of the frying process on the real fat intake. *Grasas Aceites* 49:366–9.

Varela, G., O. Moreiras, B. Ruiz-Roso, and R. Conde. 1986. Influence of repeated frying on the digestive utilization of various fats. *J. Sci. Food Agric.* 37:487–90.

Vázquez-Velasco, M., R. Lucas, L.E. Diaz-Prieto, S. Gómez-Martinez, A. Marcos, and F.J. Sánchez-Muniz. 2009. El consumo de aceite de girasol enriquecido con hidroxitirosol mejora en voluntarios la actividad arilesterasa y disminuye la peroxidación de LDL. XI Congreso Nacional de la Sociedad Española de Nutrición. 11–13 de junio de 2009. Sitges (Barcelona).

Vemuri, M. and D.S. Kelley. 2008. The effects of dietary fatty acids on lipid metabolism. In *Fatty Acids in Foods and Their Implications,* ed. C.K. Chow, 591–630. Boca Raton, FL: Taylor & Francis.

Vilas, N.N., R.R. Bell, and H.H. Draper. 1976. Influence of dietary peroxides, selenium and vitamin E on glutathione peroxidase of the gastrointestinal tract. *J. Nutr.* 106:589–96.

Wallace, A.J., W.H.F. Sutherland, J.I. Mann, and S.M. Williams. 2001. The effect of meals rich in thermally stressed olive and safflower oils on postprandial serum paraoxonase activity in patients with diabetes. *Eur. J. Clin. Nutr.* 55:951–8.

Waltking, A.E. and H. Wessels. 1981. Chromatographic separation of polar and nonpolar components of frying fats. *J. AOAC Int.* 64:1329–30.

Watanabe, Y., Y. Shimada, A. Sugihara, and Y. Tominaga. 2001. Enzymatic conversion of waste edible oil to biodiesel fuel in a fixed-bed bioreactor. *J. Am. Oil Chem. Soc.* 78:703–7.

Welch, V.A. and J.T. Borlak. 2008. Absorption and transport of dietary lipids. In *Fatty Acids in Foods and Their Implications,* ed. C.K. Chow, 561–90. Boca Raton, FL: Taylor & Francis.

Williams, M., J.J. Salas, J. Sanchez, and J.L. Harwood. 2000. Lipoxygenase pathway in olive callus cultures (*Olea europaea*). *Phytochemistry* 53:13–9.

Winkler, K.K., A. D'Arcy, and W. Hunziker. 1990. Structure of human pancreatic lipase. *Nature* 343:771–4.

Xin, J.-Y., S.-B. Li, Y. Xu, J.-R. Chui, and C.-G. Xia. 2001. Dynamic enzymatic resolution of naproxen methyl ester in a membrane bioreactor. *J. Chem. Technol. Biotechnol.* 76:579–85.

Yang, L.Y., A. Kuksis, and J.J. Myher. 1990. Lipolysis of menhaden oil triacylglycerols and the corresponding fatty acid alkyl esters by pancreatic lipase in vitro: A re-examination. *J. Lipid Res*. 31:137–47.

Yoshida, H. and J.C. Alexander. 1983. Enzymatic hydrolysis of fractionated products from oils thermally oxidized in the laboratory. *Lipids* 18:402–7.

Yuan Y.V. and D.D. Kitts. 1997. Endogenous antioxidants: Role of antioxidant enzymes in biological systems. In *Endogenous Natural Antioxidants. Chemistry, Health Effects, and Applications*, ed. F. Shahidi, 258–70. Champaign, IL: AOCS Press.

WHO (2003). Diet, Nutrition and the Prevention of Chronic Diseases. WHO Technical Report Series 916. Geneva (Switzerland). http://www.fao.org/docrep/005/AC911E/AC911E00.HTM#Contents.

WHO (2009). Cardiovascular Disease: Prevention and Control. http://www.who.int/dietphysicalactivity/publications/facts/cvd/en/.

6 Determination of Oxidation Compounds and Oligomers by Chromatographic Techniques

M. Carmen Dobarganes, Gloria Márquez-Ruiz, Susana Marmesat, and Joaquín Velasco

CONTENTS

6.1 INTRODUCTION

Formation of new compounds during the frying process has been the subject of intensive investigations that have recently been reviewed (Dobarganes and Márquez-Ruiz, 2007). Degradation products include those associated with the autoxidation process; with polymerization, because of the high temperatures applied; and with hydrolysis, due to the presence of water in the foods subjected to frying. Although the main reactions involved are well known, it is difficult to predict the rate of oil degradation because of the great number of variables involved in the process. Some of these variables are linked to the process itself, such as temperature, length of heating, continuous or discontinuous heating, turnover period, etc., whereas others depend on whether the food is subjected to frying or on the composition and initial quality of the oil used (Choe and Min, 2007; Sánchez Muniz et al., 2007).

The quality of fried foods has to be guaranteed since frying has become one of the most popular methods for the preparation of foods in developed countries. Nowadays, popularity of frying continues to grow, and this is reflected in the remarkable increase of new fried and prefried foods suitable for storing. For the evaluation of the quality of fried foods, relatively simple analytical methods are used presently. These methods fulfill the minimum requirements for control and inspection, especially in places where modern laboratory facilities are not available. Limitations on polar compounds of around 25% have been established in all the countries where frying oils are fully regulated, although in some of them other criteria are also used, such as oxidized fatty acids insoluble in petroleum ether, smoke point, free fatty acids, and polymer content (Firestone, 2007). It is, however, necessary to obtain more precise information about the nature and quantity of the new compounds formed during frying that may impair the nutritional value of the food (Márquez-Ruiz and Dobarganes, 2007). This chapter deals with recent analytical techniques that enable quantitation of the main groups of compounds formed during frying, and focuses particularly on oxidation products and oligomers.

6.2 OXIDATION PRODUCTS AND OLIGOMERS

While hydrolytic reactions release compounds of well-known structures, that is, diglycerides, monoglycerides, and fatty acids, oxidative and thermal reactions give rise to a large number of new compounds that are difficult to analyze. Oxidized triglyceride monomers, characterized by the presence of extra oxygen in at least one of the fatty acyl groups of the molecule, are formed through autoxidation reactions and constitute final stable products that result from the breakdown or decomposition of primary oxidation compounds (hydroperoxides). Thus, this group includes triglycerides containing different oxygenated groups, hydroperoxy, hydroxy, keto, epoxy, etc., as well as short-chain fatty acyl and short-chain *n*-oxo fatty acyl groups as the main products (Frankel, 2005). Considering the number of oxygenated forms that may be present in one or more fatty acyl group of triglycerides, one can imagine the variety of compounds formed and the difficulties encountered during analysis, even if the simplest derivatives, fatty acid methyl esters (FAMEs), are used for this purpose.

With respect to the triglyceride dimers, trimers, and higher oligomers—referred to as oligomers throughout this chapter—the situation is much more complex because the number of combinations of molecules increases exponentially, from monomers to dimers, etc. Apart from the participation of oxidized monomers in the formation of oxidized oligomers, there are also interactions between molecules of nonpolar triglycerides, leading to the formation of thermal oligomers without extra oxygen (Dobarganes and Márquez-Ruiz, 2007). Because of the great number of new compounds formed, quantitation of each one of them is impossible in practice, either by analyzing the fat or oil or its fatty acid methyl esters. Thus, the first analytical goal is the global quantitation of total new products and then of the main groups of compounds.

6.3 QUANTITATION OF OXIDATION PRODUCTS AND OLIGOMERS BY LIQUID CHROMATOGRAPHY

Liquid chromatography is, in principle, the most appropriate chromatographic technique, taking into consideration the two main characteristics of the new compounds formed. First, excluding minor compounds originating through isomerization and cyclization reactions, new compounds possess higher polarity than their parent, intact triglycerides. Second, the three main groups of compounds formed, that is, hydrolytic products, oxidized monomeric triglycerides, and oligomers, have different molecular weight ranges.

Thus, adsorption chromatography has been widely used for the determination of the total polar compounds; the standard method uses a classical silica column for a gravimetric determination (IUPAC, 1987a). For the separation of different groups of compounds, methodologies combining adsorption and size exclusion chromatography are necessary, applied directly to the fat or oil or to their fatty acid methyl esters (Dobarganes et al., 1988; Márquez-Ruiz et al., 1990; Márquez-Ruiz et al., 1996a; Dobarganes et al, 2000).

6.3.1 Direct Analysis of Oils

6.3.1.1 Quantitation of Oligomers

Evaluation of oligomers by direct size-exclusion chromatography (HPSEC) in oil samples constitutes a simple and rapid analysis since it only demands a solution of the oil sample in a suitable solvent and a running time of 10–30 min. Preliminary studies on the use of HPSEC (Dobarganes and Márquez-Ruiz, 1993) indicated that a good separation of oligomers was possible in Sephadex LH-20 swelled in chloroform or ethanol, BioBeads in benzene or chloroform, or Styragel in tetrahydrofuran. Later, it was demonstrated that in used frying fats a good correlation existed between the amount of polymerized triglycerides and polar components separated by column chromatography (Schulte, 1982), which indicates that the technique provides a reasonably good measurement of alteration in frying fats. Further developments to optimize chromatographic matrices and organic solvents and efforts to find the best chromatographic conditions to obtain quantitative data have led to the column normally used now, which is approximately 30 × 0.8 cm i.d., with spherically shaped particles (5 or 10 μm) and a controlled pore size distribution. With this column, good resolution is achieved for sample concentrations between 30 and 50 mg/mL and loops of 10–20 μL (Márquez-Ruiz and Dobarganes, 2006).

Direct analysis of fats by HPSEC can be specially applied to quality evaluation of used frying fats in which oligomers are the most representative group of compounds formed from thermoxidative alteration. Hence, the IUPAC Commission on Oils, Fats and Derivatives adopted the method, after two interlaboratory tests, for samples containing not less than 3% oligomers (Wolff et al., 1991; IUPAC, 1992). The method proposes a single column of 30 × 0.77 cm i.d. packed with copolysterene divinyl benzene (5 μm particle size), tetrahydrofuran as the mobile phase, a refractive index detector, sample concentration of 50 mg/mL for an injection valve with a 10 μL loop, and a flow rate of 1 mL/min. Under these conditions, the analysis time is about 10 min. In Figure 6.1 (upper part), a typical chromatogram obtained with direct analysis of used frying oil by HPSEC is given. The so-called triglyceride oligomers are resolved into two peaks: dimers (TGD) and higher oligomers or polymers (TGP).

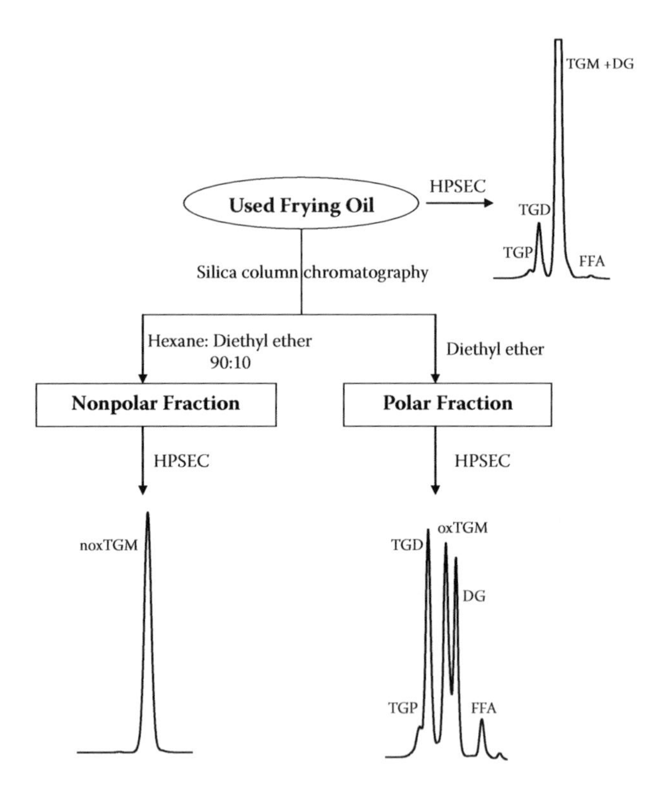

FIGURE 6.1 Analytical procedure for the determination of the total polar compounds and their distribution in used frying oils. High-performance size-exclusion chromatograms of total sample, nonpolar fraction, and polar fraction are included. Acronyms: TGP, triglyceride polymers; TGD, triglyceride dimers; TGM, total triglyceride monomers; noxTGM, nonoxidized triglyceride monomers; oxTGM, oxidized triglyceride monomers; DG, diglycerides; FFA, free fatty acids; HPSEC, high-performance size-exclusion chromatography.

Results obtained with a number of samples have clearly shown that oligomeric compounds constitute the major fraction among the different groups of alteration compounds formed during frying, normally accounting for more than 50%. The levels of oligomers have been found to correlate well with polar compounds (Perrin et al., 1985; Marmesat et al., 2007). Exceptions are used frying fats with low levels of polar compounds (Pérez-Camino et al., 1992) or oils with initial high content of diglycerides (Sébédio et al., 1991).

6.3.1.2 Analysis of Polar Compounds

As discussed earlier, quantitation of polar compounds by classical silica column constitutes the basis of legislation in some European countries for the control of quality of used frying fats. Further application of HPSEC to the isolated polar fractions offers substantial advantages for the quantitation of specific groups of altered compounds. The following methodology is a combination of adsorption and size-exclusion chromatography aimed not only at evaluating the quality of used frying fats, but also at checking the quality of the oils before use.

6.3.1.2.1 Methodology: Combined Silica Column Chromatography–HPSEC This analytical procedure combines two IUPAC methods, one for the determination of polar compounds and the other for determination of polymerized triglycerides (Dobarganes et al., 1988; Dobarganes et al., 2000). With this combination, quantitative data on groups of compounds characteristic of the different types of degradation in used frying fats are obtained. The procedure is briefly described below. Nonpolar and polar fractions are eluted with 150 mL of a mixture of hexane:diethyl ether (90:10, v/v) and 150 mL of diethyl ether, respectively. After gravimetric determination of polar compounds, the fractions are further analyzed by HPSEC, using 100 and 500 Å columns with polysterene divinylbenzene highly cross-linked macroporous packing (particle size: 5 µm) connected in series; mobile phase: tetrahydrofuran (flow rate: 1 mL/min); detector: refractive index. Figure 6.1 represents schematically the analytical procedure applied to used frying oils.

An alternative technique that aims at reducing the quantity of the sample and solvents and at shortening analysis time is based on the use of silica cartridges for the separation (solid-phase extraction) and monostearin as internal standard (Márquez-Ruiz et al., 1996a). Only 15 mL of elution solvents is required for each fraction. This modified procedure is useful for samples with a wide range of alteration products and especially for those of low levels of degradation. In this latter case, quantitation with an internal standard shows significantly lower errors compared to the gravimetric determination, as judged from relative standard deviations.

6.3.1.2.2 Applications

6.3.1.2.2.1 Analysis of Used Frying Fats and Oils The methodology based on a combination of silica column and HPSEC has proved to be an excellent alternative to the evaluation of used frying oil. Because a substantial improvement of the quantitation of polymers is achieved, the combination allows application to samples with less than 3% polymers, the minimum content established in the IUPAC method for analyses of total samples. Separate determination of oxidized triglyceride monomers from nonoxidized triglycerides eluting in the nonpolar fraction offers a measurement of oxidative deterioration. The simultaneous evaluation of diglycerides provides information on the hydrolytic alteration, since any overlap with the most abundant nonoxidized triglycerides peak is avoided. The last peak, of minor importance, corresponds to free fatty acids and also to the polar unsaponifiable material. Advantages offered by the combined technique are clearly reflected in Figure 6.1, which shows the profile obtained by simply analyzing the entire oil sample, and the improvement of conditions for the analysis of the fraction of polar compounds.

This analytical procedure has been widely applied to used frying fats and oils to better understand fat degradation under variable conditions. In fact, very different patterns of polar compound distribution have been obtained for samples with similar content of total polar compounds (Dobarganes et al., 1988). This is of particular nutritional significance since hydrolysis products, which are the same as those resulting from lipolysis in the gut (fatty acids and partial glycerides), are differentiated from oxidation and thermal degradation products.

The advantages offered by the combined technique have been discussed by Jorge et al. (1996a). The main variables involved in the frying process are length of heating, temperature, surface-to-oil volume ratio, and degree of unsaturation. From these laboratory experiments, carried out in the absence of food and under controlled conditions, the strong influence of the surface-to-oil volume ratio in discontinuous frying, the higher proportion of polymers as the degree of oil unsaturation increased, and the significant interaction between these two variables were deduced.

Additional information on hydrolytic products, provided by the combined technique, has been obtained in a number of studies related to the effect of the food during frying, using potatoes as food substrate. It was generally observed that hydrolysis was not substantial despite the high moisture content of potatoes, about 80% (Cuesta et al., 1993; Jorge et al., 1996b; Dobarganes, 1998; Romero et al., 1998; Rodrigues Machado et al., 2007).

6.3.1.2.2.2 Initial Quality of Frying Fats and Oils One of the most interesting applications of the combination of adsorption chromatography with HPSEC is the evaluation of frying oils before use (Dobarganes et al., 1988). At present, the main quality specifications for frying fats and oils are free fatty acids, peroxide value, flavor, color, stability (Active Oxygen Method or Oil Stability Index), smoke point, and fatty acid composition. However, similar values for such specifications do not guarantee similar performance during frying. One of the reasons is that such indices were introduced in the legislation mainly for checking if the oil is well refined. However, they provide no information about the oxidation state of the crude oil, which affects the performance of the refined oil during frying (Jawad et al., 1983).

The particular advantages of the combination of adsorption and size-exclusion chromatography when applied to determine initial quality of refined oils used in frying are illustrated in Figure 6.2. While only a single peak, which corresponds to the predominant nonoxidized triglycerides, can be observed in the HPSEC chromatogram

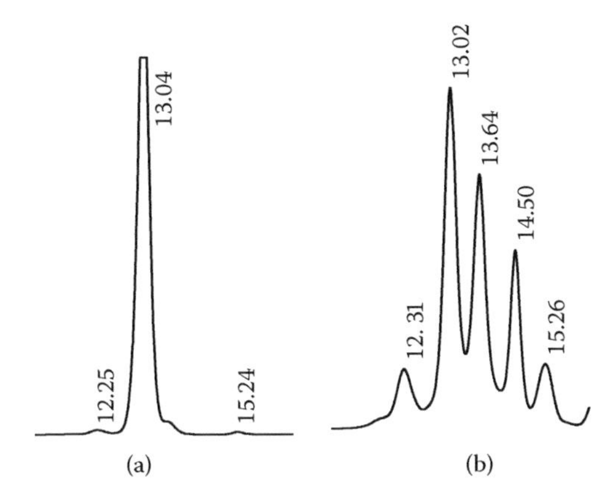

FIGURE 6.2 Significant parts of high-performance size-exclusion chromatograms of a refined oil: (a) total oil and (b) polar compounds; retention times (min): 12.3, triglyceride dimers; 13.0, total triglyceride monomers in (a) and oxidized triglyceride monomers in (b); 13.6, diglycerides; 14.5, monostearin (internal standard); 15.4, free fatty acids.

of the unfractionated oil, all groups of minor compounds are satisfactorily resolved in the chromatogram of the fraction isolated by solid-phase extraction.

Even if the fraction of polar compounds in refined fats and oils is low, as triglycerides normally constitute more than 95% of the sample, four peaks can be well resolved, as observed in Figure 6.2: triglyceride dimers, formed during the deodorization step; oxidized triglyceride monomers and diglycerides, whose levels remain close to those present in crude oils, as they are nonvolatile under refining conditions and are not eliminated in any of the other processing stages; and fatty acids, decreasing with respect to the crude oils due to the neutralization step (Dobarganes et al., 1989a).

From the analysis of initial frying oils, the following general deductions can be made:

1. The level of polar compounds in refined oils is an important first index of quality. The higher the polar compound percentage in the oils, the poorer their expected performance.
2. The three main peaks obtained after HPSEC analysis provide complementary information on the three main routes of oil degradation, initial polymerization, oxidation, and hydrolysis, which affect to different extents the subsequent performance of the oil during frying.

As an example, Table 6.1 shows the analysis of polar compounds and their distribution in three frying oils—palm olein (PO), high-oleic sunflower oil (HOSO), and conventional sunflower oil (SO)—before use; after industrial continuous frying of crisps at high turnover, for 20 h (Sébédio et al., 1996); and after discontinuous laboratory frying for 6 h (Jorge et al., 1996b). It is interesting to observe that the substantially lower increase in polar compounds for PO during frying, as compared to SO and HOSO, could have not been deduced without the analysis of initial oils.

TABLE 6.1
Polar Compounds and Polar Compound Distribution (wt% on Fat) in Initial Frying Oils, after Industrial Frying of Crisps for 20 h, and after Discontinuous Laboratory Frying of French Fries for 6 h

		Polar Compounds (%)				
Treatment	**Oil**	**Total**	**TGD**	**oxTGM**	**DG**	**FFA+UM**
Initial oil	Palm olein	7.7	0.5	0.6	6.5	0.2
	Sunflower oil	2.8	0.4	0.8	1.1	0.5
	High-oleic sunflower oil	3.1	0.3	0.7	1.6	0.5
Industrial frying	Palm olein	8.9	1.3	1.0	6.3	0.3
	Sunflower oil	5.2	1.8	1.5	1.3	0.6
	High-oleic sunflower oil	4.9	1.1	1.3	1.9	0.5
Laboratory frying	Palm olein	18.9	6.6[a]	5.4	6.6	0.3
	Sunflower oil	20.4	12.2[a]	6.6	1.1	0.5
	High-oleic sunflower oil	17.2	8.5[a]	6.8	1.5	0.5

Source: Sébédio et al., 1996. *Grasas Aceites* 47:5–13; Jorge et al., 1996b. *Grasas Aceites* 47:14–19.
Note: PO, palm olein; SO, sunflower oil; HOSO, high-oleic sunflower oil; TGD, triglyceride dimers; oxTGM, oxidized triglyceride monomers; DG, diglycerides; and FFA+UM, free fatty acids and polar unsaponifiable matter.
[a] Also including higher oligomers.

In general, PO is characterized by a high content of diacylglycerols which, in turn, contribute to a high level of starting polar compounds. From the detailed results here obtained, we know that amounts of polymerization and oxidation compounds were initially on the same order for the three oils, and that the main difference was in fact a high starting content of diglycerides in PO. When the level of polar compounds increased after frying, formation of new alteration compounds, such as polymerization and oxidation products, gave, as expected, lower levels in PO, although values of total polar compounds were equally high in the three oils. In summary, initial distribution of polar compounds in frying oils before use allows a better selection of a certain oil for frying among those of the same composition and a deeper knowledge of the changes occurring during the frying process. It is important to point out that applications of the foregoing methodology have been extended beyond its utilization for evaluating the quality of frying oils and fats before use.

The influence of the refining conditions on the quality of fats and oils (Dobarganes et al., 1989a; Hopia, 1993a; Ruiz-Méndez et al., 1997; Gomes and Caponio, 1997a); virgin and refined olive oil characterization (Dobarganes et al., 1989b; Gomes, 1992; Pérez-Camino et al., 1993; Gomes, 1995; Gomes and Caponio, 1997b; Gomes and Caponio, 1998; Caponio et al., 2005); and evolution of oxidation during storage (Pérez-Camino et al., 1990; Hopia, 1993b; Hopia et al., 1993; Márquez-Ruiz et al., 1996b; Martín-Polvillo et al., 1996; Martín Polvillo et al., 2004; Gómez-Alonso et al., 2004; Velasco et al., 2006; Márquez-Ruiz et al., 2008) are examples of successful application of the combination of adsorption and exclusion chromatography.

6.3.2 Analysis of Fatty Acid Methyl Esters

By applying a similar analytical methodology to fatty acid methyl esters (FAMEs) derived from used frying fats and oils, broader information on fat alteration can be acquired; this is obtained by the specific analysis of oxidized and oligomeric fatty acids included in the triglyceride molecules (Márquez-Ruiz et al., 1990).

6.3.2.1 Methodology

Briefly, 1 g of FAME is separated by silica column chromatography using 150 mL of hexane:diethyl ether (88:12, v/v) and 150 mL of diethyl ether to elute the nonpolar and polar fractions, respectively. Both fractions are gravimetrically determined and further analyzed by HPSEC. The combined chromatographic analysis permits quantitation of five types of fatty acyl groups: nonoxidized monomers and nonpolar dimers (representative of thermal, no oxygen-involving degradation) in the first fraction; and oxidation dimers, oxidized monomers, and polymers in the second fraction. This analytical procedure applied to FAME derived from used frying oil is shown in Figure 6.3.

6.3.2.2 Applications

Analysis of used frying fats after transesterification, when combined with the evaluation of the original oil before use, has been applied to a large number of samples from laboratory frying experiments or collected by Food Inspection Services in Spain (Márquez-Ruiz et al., 1995). While detection of hydrolytic alteration is attainable only by HPSEC analysis of polar fractions of original oil samples, differences due to thermoxidative degradation are better reflected in the values of altered fatty acid methyl esters, since the level of affected fatty acyl groups is measured. Some insight into the complexity of the triglyceride polymer structure can be also obtained by comparing triglyceride and methyl ester dimer and polymer values. The fatty acid polymer-to-triglyceride polymer ratio and the fatty acid dimer-to-triglyceride dimer ratio may give evidence of the degree of participation of dimeric linkages in the structures of trimeric and higher oligomeric triacylglycerols.

Results on real used frying samples have shown that fatty acid dimers and polymers, of low digestibility, were predominantly formed around the limit for fat rejection (25% polar compounds). Concomitantly, substantial amounts of oxidized fatty acid monomers, normally about 30 mg/g oil, were found. Considering that such oxidized monomeric fatty acids globally show high digestibility coefficients (Márquez-Ruiz et al., 1992), more research is needed to identify specific structures included in this group and to evaluate their nutritional significance.

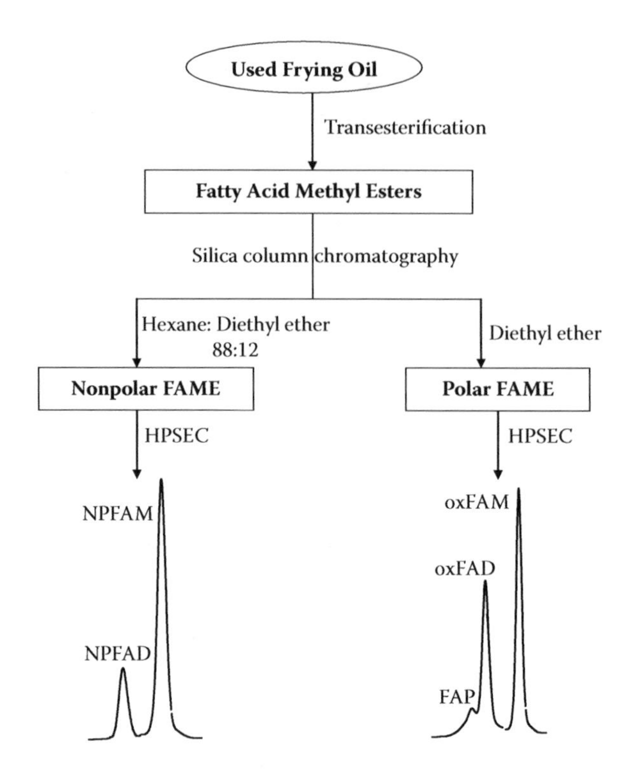

FIGURE 6.3 Analytical procedure for the quantitation of nonpolar and polar fatty acid methyl esters in used frying oils. Acronyms: NPFAD, nonpolar fatty acid dimers; NPFAM, nonpolar fatty acid monomers; FAP, fatty acid polymers; oxFAD, oxidized fatty acid dimers; oxFAM, oxidized fatty acid monomers; HPSEC, high-performance size-exclusion chromatography.

6.4 QUANTITATION OF OXIDIZED FATTY ACID METHYL ESTERS BY CAPILLARY GAS CHROMATOGRAPHY

Gas chromatography was used in the past for the indirect evaluation of polar FAME, that is, noneluted compounds under standard conditions (Waltking, 1975), and for direct quantitation of dimer FAME; the latter have sufficient volatility to elute from short columns giving one or two peaks (Paradis and Nawar, 1981; Dobarganes et al., 1984). The main objective of more recent applications is, however, the analysis of individual compounds, which may be based on the high resolution provided by capillary fused silica columns (Gardner et al., 1992).

As already mentioned, there is a growing interest in the group of oxidized monomers from a nutritional point of view because of their high absorbability and their presence in used frying fats at nonnegligible levels. Separation and identification of the main structures in used frying fats indicate that the fraction of oxidized monomers contains significant amounts of compounds with molecular weight lower than that of the original fatty acids. Among these compounds are the short-chain *n*-oxo FAME, originally triglyceride-bound aldehydes, resulting from hydroperoxide breakdown

(Márquez-Ruiz and Dobarganes, 1996; Kamal-Eldin et al., 1997). Knowledge of the levels of the nonvolatile aldehyde derivatives in used frying fats and fried products is of great importance from the nutritional point of view since they remain attached to the triglyceride molecule and are thus retained in the fried food ingested by the consumer (Kamal-Eldin and Appelqvist, 1996). A number of studies have been carried out on the physiological effects of volatile aldehydes, particularly on 4-hydroxy-2-nonenal (Esterbauer et al., 1991; Uchida, 2005) and on their analysis in foods (Guillén y Goicoechea, 2008), but the group of esterified aldehydes, which accumulate as degradation increases, has not been thoroughly investigated. There are, however, some reports on 9-oxononanoic acid, the major esterified aldehyde in oxidized lipids, the results of which indicate that such structures could induce lipid peroxidation and affect hepatic metabolism (Minamoto et al., 1988; Kanazawa and Ashida, 1991).

Concerning oxidized compounds of molecular weight similar to that of the starting fatty acids, the main groups present corresponded to epoxy acids, keto acids, and hydroxy acids (Capella, 1989; Velasco et al., 2008). In recent years, methods for accurate quantitation of these groups of oxidized compounds have been proposed and results in model systems, thermoxidized oils, and used frying fats and oils have been reported (Berdeaux et al., 1999a; Berdeaux et al., 1999b; Berdeaux et al., 2002; Velasco et al., 2002; Velasco et al., 2004a; Velasco et al., 2005; Marmesat et al., 2008; Berdeaux et al., 2009). Base-catalyzed transesterification, using either sodium methoxide in methanol (Cecchi et al., 1985) or potassium hydroxide in methanol (IUPAC, 1987b) at room temperature, is the only derivatization method that prevents formation of artifacts and losses of some compounds of interest (Berdeaux et al., 1999a).

6.4.1 Short-Chain Oxidation Compounds

For the chromatographic analysis of short-chain compounds, a second methylation step with diazomethane is necessary, as significant amounts of short chains bearing a free carboxylic group are present as a result of further oxidation of aldehydes to corresponding acids. An Innowax capillary column of 30 m × 0.25 mm i.d. and 0.25 μm film thickness was used under the following temperature program: 90°C (2 min); 4°C/min to 240° (25 min), and methyl tridecanoate and pentadecanoate were used as internal standards. Methyl octanoate (C8:0), methyl 9-oxo-nonanoate (9-oxo-C9:0), and dimethyl nonanodiate were the major short-chain compounds identified in model systems of methyl linoleate and oleate, while methyl heptanoate, methyl 8-oxo-octanoate, and dimethyl octanodiate, derived from 13-hydroperoxide breakdown of linoleic acid and further oxidative reactions and from 8-hydroperoxide of oleic acid, were detected in lower amounts (Berdeaux et al., 1999a; Berdeaux et al., 2002).

Detailed quantitation of the six compounds in thermoxidized olive and sunflower oils as well as in used frying fats and oils of different origins has been reported. Among them, the most abundant compounds in thermoxidized olive and sunflower oils were C8:0 and 9-oxo-C9:0, which derive from the 9-hydroperoxide of unsaturated fatty acids. A more significant participation of oleic acid as the level of alteration increased led to higher formation of compounds derived from the 8-hydroperoxide. In samples of used frying fats with levels of around 25% on oil, the total content of short-chain compounds was close to 3 mg/g oil (Velasco et al., 2005). Figure 6.4

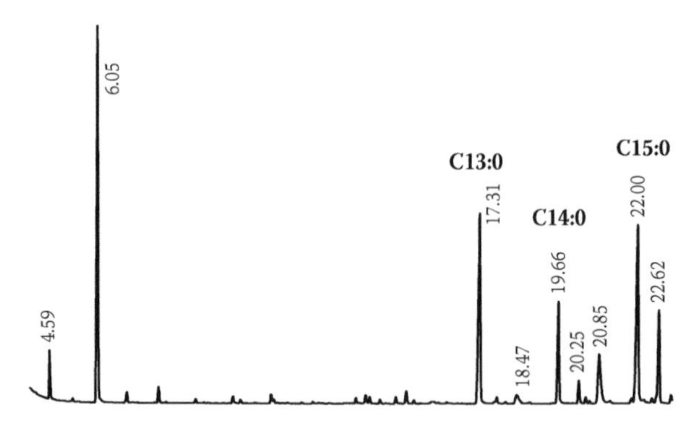

FIGURE 6.4 Significant part of gas chromatogram corresponding to the zone of short-chain compounds of used frying oil at the limit of rejection (25% polar compounds); retention times (min): 4.59, methyl heptanoate; 6.05, methyl octanoate; 17.31, methyl tridecanoate (IS); 18.47, methyl 8-oxo-octanoate; 19.96, methyl myristate; 20.25 dimethyl octanodiate; 20.85, methyl 9-oxononanoate; 22.00 methyl pentadecanoate (IS); 22.62, dimethyl nonanodiate.

illustrates the significant part of the chromatogram obtained in a used frying oil with 24.7% polar compounds. All the compounds elute at retention times shorter than the main oil FAME.

6.4.2 Epoxy, Keto, and Hydroxy Acids

The quantitative analysis of monoepoxy compounds in model systems, in thermoxidized oils, and in used frying fats has also been described (Berdeaux et al., 1999b; Velasco et al., 2002; Velasco et al., 2005; Kalogeropoulos et al., 2007; Marmesat et al., 2008). Two distinct mechanisms have been proposed for the epoxide formation at 50°C, either at the site of the double bond or near the double bond. In the latter case, the original double bond remains intact (Neff and Byrdwell, 1998). However, only the compounds formed from the addition of oxygen to an existing double bond have been detected in samples heated at 180°C by a combination of gas chromatography and mass spectrometry (Berdeaux et al., 1999b; Lercker et al., 2003; Giuffrida et al., 2004).

Two saturated epoxides, *trans*-9,10- and *cis*-9,10-epoxystearate, have been detected in methyl oleate and triolein samples heated at 180°C and four monounsaturated epoxides, *trans*-12,13-, *trans*-9,10-, *cis*-12,13-, and *cis*-9,10-epoxyoleate, have been found in methyl linoleate and trilinolein samples under the same experimental conditions (Berdeaux et al., 1999b). With regard to heated fats and oils, accurate quantitation of these compounds requires removal by solid-phase extraction of other components that interfere in the analysis by co-eluting with the analytes of interest, such as the FAME of 22 carbon atoms (Velasco et al., 2002).

Quantitation of epoxides in thermoxidized olive and sunflower oils, and in used frying oils, has demonstrated that, for similar levels of polar compounds, monounsaturated oils contain higher contents of monoepoxides than polyunsaturated oils. This fact is attributable to two reasons: first, a lower tendency for polymerization of monounsaturated oils, and, second, a greater stability and hence accumulation of the major monoepoxides formed in monounsaturated oils, that is, monoepoxystearates, in contrast to the susceptibility to further reactions of the most abundant monoepoxides found in polyunsaturated oils, that is, monoepoxyoleates. It has also been found that monoepoxides are major oxidized compounds, as they constitute about 25% of the total oxidized monomers in real used frying oils at a the limit of rejection (Velasco et al., 2004b).

In contrast to epoxy FAME, keto and hydroxy FAME are found as numerous minor peaks that are not well resolved. They correspond to diene keto and hydroxy FAME from linoleyl groups, and monoene keto and hydroxy FAME from oleyl groups. Although saturated keto FAME eluted at a lower retention time than saturated hydroxy FAME, the elution order of the unsaturated molecules is not that expected from the saturated compounds. Thus, ketomonoene FAME elute at longer retention times than their parent hydroxymonoenes, and this applies to keto- and hydroxy-dienes. This fact is attributed to an increased polarity of keto FAME due to the conjugation of the carbonyl group with the double bond or with the diene structure, respectively (Marmesat et al., 2008).

Reduction of double bonds by hydrogenation has been proposed to simplify the structures of the compounds and, thus, to reduce the number of analytes (Marmesat et al., 2008). Figure 6.5 illustrates the steps of the complete analysis of oxidized monomers. In Figure 6.5a, the initial GC chromatogram obtained for the sample of heated sunflower oil after transesterification is observed. Figures 6.5b and 6.5c show the excellent separation of nonpolar and polar FAME by solid-phase extraction, the resulting concentration effect of the oxidized compounds in the polar fraction, and the elimination of the interference of C22 and C24 FAME. Finally, Figure 6.5d shows the separation of the three main groups of compounds after polar FAME hydrogenation. As can be observed, this approach made it possible to separate keto FAME and hydroxy FAME from each other. The former eluted as two peaks that were identified as methyl 9-oxoestearate and methyl 13-oxoestearate. Hydroxy FAME eluted as a main peak formed by different isomers bearing the hydroxyl group at different positions. In addition, it is evident that hydrogenation also gives rise to an increase in sensitivity due to a pseudo concentration effect. An Innowax capillary column of 30 m × 0.25 mm i.d. and 0.25 μm film thickness was used for the analyses under isothermal conditions using 240°C for 25 min. Methyl heneicosanoate (C21:0) was selected as the internal standard, and methyl *t*-9,10-epoxyestearate, methyl 12-oxoestearate, and methyl 12-hydroxyestearate were used to obtain the response factors that were applied to quantitate epoxy FAME, keto FAME, and hydroxyl FAME (Marmesat et al., 2008).

6.5 QUANTITATION OF OXIDIZED MONOMERS AND DIMERS BY HIGH-TEMPERATURE GAS CHROMATOGRAPHY

Application of high-temperature gas chromatography (HTGC) in the analysis of biodiesels for controlling the unreacted material (triacylglycerols, partial glycerides, fatty acids) as well as the reaction products, that is, FAME and glycerol (European Standard, 2003), offers new possibilities in the analysis of frying oils.

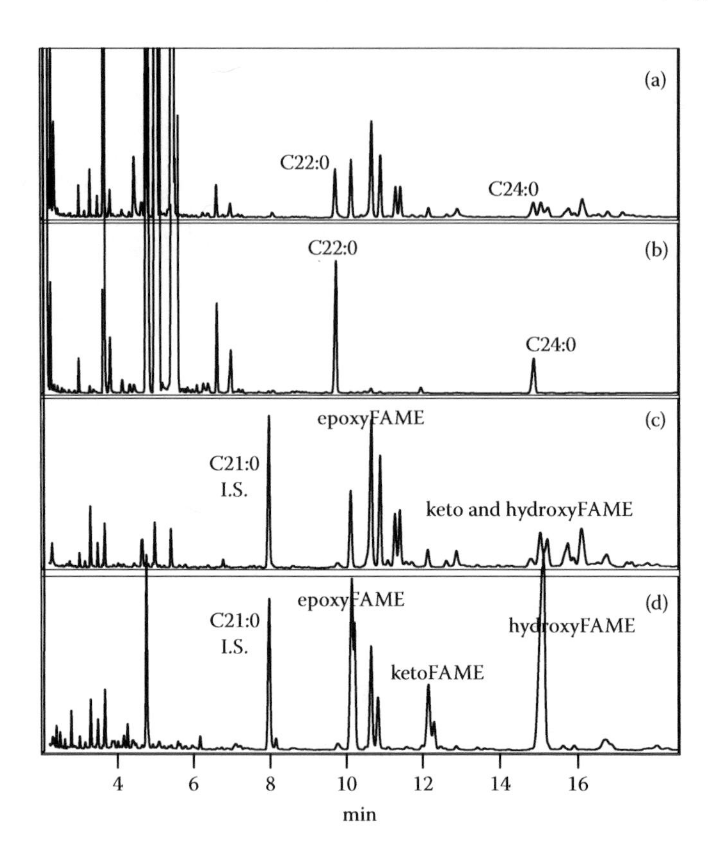

FIGURE 6.5 Gas chromatograms of (a) total FAME, (b) nonpolar FAME, (c) polar FAME, and (d) hydrogenated polar FAME of conventional sunflower oil after heating at 180°C for 10 h. (Reprinted from Marmesat et al., 2008, *J. Chromatogr. A*, 1211, p. 132. With permission from Elsevier Science.)

In a recent study, the presence of dimeric FAME in significant amounts after transesterification of used frying oils was shown to be the most characteristic group of compounds to detect used frying oil as raw material in biodiesel production (Ruiz Mendez et al., 2008). Dimeric FAME elute as a group of unresolved peaks whose total area can be easily quantified.

Figure 6.6 shows a typical chromatogram corresponding to the polar FAME of olive oil used in frying. The two groups of modified FAME, oxidized monomers and dimers, can be easily observed in the chromatogram along with a part of the polar unsaponifiable matter, such as sterols. Separations were performed with split injection on a VF-5ht ultimetal (15 m × 0.25 mm, 0.1 μm) with the following oven temperature program: 150°C (held for 5 min) rising at 5°C/min to 350°C (held for 5 min).

At present, the analysis is being applied to the polar compound fractions obtained by adsorption chromatography, after base-catalyzed transesterification, using methyl tricosanoate as the internal standard. This is considered the best approach to obtaining rapid complementary information on used frying oil degradation due to two

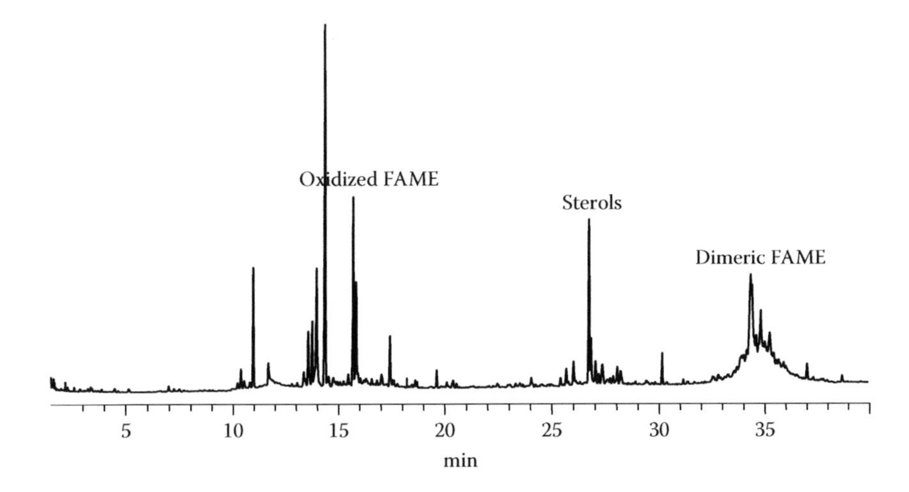

FIGURE 6.6 High-temperature gas chromatogram of polar FAME of olive oil used in frying (18.5% polar compounds). Chromatographic conditions: VF-5ht Ultimetal (15 m × 0.25 mm, 0.1 μm) capillary column; temperature program, 150°C (5 min); 5°C/min to 350°C (5 min).

reasons. On the one hand, polar compound determination is normally applied in laboratories to used frying oils and, on the other hand, prior elimination of the nonpolar triglycerides produces a concentration of the compounds of interest that is necessary for a more accurate quantitation.

6.6 CONCLUSIONS

1. Determination of all changes occurring in frying fats and oils is very difficult at present, but efforts made during the last few years to develop chromatographic techniques resulted in useful information on the types of degradation in relation to the conditions of frying.
2. Determination of the main groups of compounds, namely, oxidized, oligomeric, and hydrolytic products, constitutes an excellent analytical tool not only for quality evaluation of used frying oils, but also as a complementary method to better assess the initial quality of the frying oils before use.
3. Determination of oxidized fatty acids of high digestibility is useful for future evaluation of the nutritional implications of used frying fats.
4. Much work remains to be done to clarify the specific structures of the dimers and oligomers formed during frying due to a large number of individual components with similar polarity and molecular weight.

6.7 SUMMARY

In this chapter, the most recent techniques applied for the quantitation of oxidized compounds and oligomers in frying fats and oils are presented. Given the complex mixture of alteration compounds formed during heating of oils at frying temperatures, quantitation of groups of compounds has been shown to be more practical than specific analysis of individual structures. High-performance liquid-chromatographic techniques, mainly based on the molecular size of the new compounds formed, enable rapid determination of total oligomeric compounds, while a previous separation by adsorption chromatography of the less polar compounds, that is, intact triglycerides, allows the determination of oxidized, polymerized, and hydrolytic compounds. Methodologies applied directly to fat or to its simpler derivatives—fatty acid methyl esters—are described and applications discussed. At present, a more detailed quantitation of the main oxidized fatty acyl groups included in glyceridic molecules is feasible. In this context, quantitation of short-chain compounds, originally attached to the glyceridic backbone, resulting from hydroperoxide breakdown, as well as epoxy, keto, and hydroxy fatty acyl groups by capillary gas-liquid chromatography, is detailed. Finally, the current possibilities of high-temperature gas chromatography for determination of oxidized fatty acids and dimers are described.

ACKNOWLEDGMENTS

This work was funded in part by Ministerio de Ciencia e Innovación (projects AGL 2007-62922 and AGL 2007-63647).

REFERENCES

Berdeaux, O., G. Márquez-Ruiz, and M.C. Dobarganes. 1999a. Selection of methylation procedures for quantitation of short-chain glycerol-bound compounds formed during thermoxidation. *J. Chromatogr. A* 863:171–181.

Berdeaux, O., G. Márquez-Ruiz, and M.C. Dobarganes. 1999b. Characterization, quantitation and evolution of monoepoxy compounds formed in model systems of fatty acids methyl esters and monoacid triglycerides heated at high temperature. *Grasas Aceites* 50:53–59.

Berdeaux, O., J. Velasco, G. Márquez-Ruiz, and M.C. Dobarganes. 2002. Evolution of short-chain glycerol-bound compounds in fatty acid methyl esters and monoacid triacylglycerols during thermoxidation. *J. Am. Oil Chem. Soc.* 79:279–285.

Berdeaux, O., P.C. Dutta, M.C. Dobarganes, and J.L. Sébédio. 2009. Analytical methods for quantification of modified fatty acids and sterols formed as a result of processing. *Food Anal. Methods* 2:30–40.

Capella, P. 1989. Les produits de l'evolution des hydroperoxydes. *Rev. Franc. Corps Gras* 36:313–323.

Caponio, F., M.T. Bilancia, A. Pasqualone, E. Sikorska, and T. Gomes. 2005. Influence of the exposure to light on extra virgin olive oil quality during storage. *Eur. Food Res. Technol.* 221:92–98.

Cecchi, G., S. Biasini, and J. Castano. 1985. Méthanolyse rapid des huiles en solvant. *Rev. Franc. Corps Gras* 32:163–164.

Choe, E. and D.B. Min. 2007. Chemistry of deep-fat frying oils. *J. Food Sci.* 72:R77–R86.

Cuesta, C., F.J. Sánchez-Muniz, C. Garrido-Polonio, S. López-Varela, and R. Arroyo. 1993. Thermoxidative and hydrolytic changes in sunflower oil used in fryings with a fast turnover of fresh oil. *J. Am. Oil Chem. Soc.* 70:1069–1073.

Dobarganes, M.C., M.C. Pérez-Camino, and J.J. Ríos. 1984. Métodos analíticos de aplicación en grasas calentadas. II. Determinación de ésteres metílicos de dímeros no polares. *Grasas Aceites* 35:351–357.

Dobarganes, M.C., M.C. Pérez-Camino, and G. Márquez-Ruiz. 1988. High performance size exclusion chromatography of polar compounds in heated and non-heated fats. *Fat Sci. Technol.* 90:308–311.

Dobarganes, M.C., M.C. Pérez-Camino, G. Márquez-Ruiz, and M.V. Ruiz-Méndez. 1989a. New analytical possibilities in quality evaluation of refined oils. In *Edible Fats and Oils Processing: Basic Principles and Modern Practices,* ed. D.R. Erikson, 427–429. Champaign, IL: AOCS Press.

Dobarganes, M.C., M.C. Pérez-Camino, and G. Márquez-Ruiz. 1989b. Application of minor glyceridic component determination to the evaluation of olive oil. In *Actes du Congres International Chevreul pour l'Etude des Corps Gras.* Premier congrès Eurolipid. Tome 2, 578–84. Angers, France.

Dobarganes, M.C. and G. Márquez-Ruiz. 1993. Size exclusion chromatography in the analysis of lipids. In *Advances in Lipid Methodology—Two*, ed. W.W. Christie, 113–137. Dundee, Scotland: The Oily Press.

Dobarganes, M.C. 1998. Formation and analysis of high-molecular weight compounds in frying fats and oils. *OCL* 5:41–47.

Dobarganes, M.C., J. Velasco, and A. Dieffenbacher. 2000. Determination of polar compounds, polymerized and oxidized triacylglycerols, and diacylglycerols in oils and fats. *Pure Appl. Chem.* 72:1563–1575.

Dobarganes, M.C. and G. Márquez-Ruiz. 2007. Formation and analysis of oxidized monomeric, dimeric and higher oligomeric triglycerides. In *Deep Frying: Chemistry, Nutrition and Practical Applications* (2nd edition), ed. M.D. Erickson, 87–110. Champaign, IL: AOCS Press.

Esterbauer, H., R.J. Schaur, and H. Zollner. 1991. Chemistry and biochemistry of 4-hydroxynonenal, malonaldehyde and related aldehydes. *Free Radic. Biol. Med.* 11:81–128.

European Standard 2003. Standard method EN14105. Fat and oil derivatives—fatty acid methyl esters. determination of free and total glycerol and mono-, di-, and triglyceride contents. European Committee for Standardization, Brussels, Belgium.

Firestone, D. 2007. Regulation of frying fats and oils. In *Deep Frying: Chemistry, Nutrition and Practical Applications* (2nd edition), ed. M.D. Erickson, 373–385. Champaign, IL: AOCS Press.

Frankel, E.N. 2005. Hydroperoxide decomposition In *Lipid Oxidation* (2nd edition), 67–98. Bridgwater, England: The Oily Press.

Gardner, D.R., R.A. Sanders, D.E. Henry, D.H. Tallmadge, and H.W. Wharton. 1992. Characterization of used frying oils. Part 1: Isolation and identification of compound classes. *J. Am. Oil Chem. Soc.* 69:499–508.

Giuffrida, F., F. Destaillats, F. Robert, L.H. Skibsted, and F. Dionisi. 2004. Formation and hydrolysis of triacylglycerol and sterol epoxides: Role of unsaturated triacylglycerol peroxyl radicals. *Free Radic. Biol. Med.* 37:104–114.

Gomes, D. 1992. Oligopolymer, diglyceride and oxidized triglyceride contents as measures of olive oil quality. *J. Am. Oil Chem. Soc.* 69:1219–1223.

Gomes, D. 1995. A survey of the amounts of oxidized triglycerides and triglyceride dimers in virgin and "lampante" olive oils. *Fat Sci. Technol.* 97:368–372.

Gomes, T. and F. Caponio. 1997a. A study of oxidation and polymerization compounds during vegetable oil refining. *Riv. Ital. Sost. Grasse* 75:97–100.

Gomes, T. and F. Caponio. 1997b. Investigation on the degree of oxidation and hydrolysis of refined olive oils. An approach for better product characterisation. *Ital. J. Food Sci.* 9:277–285.

Gomes, T. and F. Caponio. 1998. Evaluation of the state of oxidation of olive-pome oils. Influence of the refining process. *J. Agric. Food Chem.* 46:1137–1142.

Gómez-Alonso, S., M.D. Salvador, and G. Fregapane. 2004. Evolution of the oxidation process in olive oil triacylglycerol under accelerated storage conditions (40–60°C). *J. Am. Oil Chem. Soc.* 81:177–184.

Guillén, M.D. and E. Goicoechea. 2008. Toxic oxygenated α,β-unsaturated aldehydes and their study in foods: A Review. *Crit. Rev. Food Sci. Nutr.* 48:1549–7852.

Hopia, A. 1993a. Analysis of high molecular weight autoxidation products using high performance size exclusion chromatography. I. Changes during processing. *Food. Sci. Technol.* 26:568–571.

Hopia, A. 1993b. Analysis of high molecular weight autoxidation products using high performance size exclusion chromatography. I. Changes during autoxidation. *Food. Sci. Technol.* 26:563–567.

Hopia, A., A. Lampi, V. Piironen, L. Hyvönen, and P. Koivistoinen. 1993. Application of high-performance size-exclusion chromatography to study the autoxidation of unsaturated triacylglycerols. *J. Am. Oil Chem. Soc.* 70:779–784.

IUPAC. 1987a. Standard method 2.507: Determination of polar compounds in frying fats. In *Standard Methods for the Analysis of Oils, Fats and Derivatives,* 7th edition, ed. International Union of Pure and Applied Chemistry. Oxford, England: Blackwell.

IUPAC. 1987b. Standard method 2.301: Preparation of the fatty acid methyl esters. In *Standard Methods for the Analysis of Oils, Fats and Derivatives,* 7th edition, ed. International Union of Pure and Applied Chemistry. Oxford, England: Blackwell.

IUPAC. 1992. Standard method 2.508: Determination of polymerized triglycerides in oils and fats by high performance liquid chromatography. In *Standard Methods for the Analysis of Oils, Fats and Derivatives,* 1st Supplement to the 7th edition, ed. International Union of Pure and Applied Chemistry. Oxford, England: Blackwell.

Jawad, I.M., S.P. Kochhar, and B.J.F. Hudson. 1983. Quality characterization of physically refined soybean oils: Effect of pretreatment and processing time and temperature. *J. Food Technol.* 18:353–360.

Jorge, N., G. Márquez-Ruiz, M. Martín-Polvillo, M.V. Ruiz-Méndez, and M.C. Dobarganes. 1996a. Influence of dimethylpolysiloxane addition to edible oils: Performance of sunflower oil in discontinuous and continuous laboratory frying. *Grasas Aceites* 47:20–25.

Jorge, N., G. Márquez-Ruiz, M. Martín-Polvillo, M.V. Ruiz-Méndez, and M.C. Dobarganes. 1996b. Influence of dimethylpolysiloxane addition to edible oils: Dependence on the main variables of the frying process. *Grasas Aceites* 47:14–19.

Kalogeropoulos, N., F.N. Salta, A. Chiou, and N.K. Andrikopoulos. 2007. Formation and distribution of oxidized fatty acids during deep- and pan-frying of potatoes. *Eur. J. Lipid Sci. Technol.* 109:1111–1123.

Kamal-Eldin, A. and L.A. Appelqvist. 1996. Aldehydic acids in frying oils: formation, toxicological significance and analysis. *Grasas Aceites* 47:342–348.

Kamal-Eldin, A., G. Márquez-Ruiz, M.C. Dobarganes, and L.A. Appelqvist. 1997. Characterisation of aldehydic acids in used and unused frying oils. *J. Chromatogr. A* 776:245–254.

Kanazawa, K. and H. Ashida. 1991. Target enzymes on hepatic dysfunction caused by dietary products of lipid peroxidation. *Arch. Biochem. Biophys.* 288:71–78.

Lercker, G., M.T. Rodriguez-Estrada, and M. Bonoli. 2003. Analysis of the oxidation products of *cis*- and *trans*-octadecenoate methyl esters by capillary gas chromatography–ion-trap mass spectrometry. I. Epoxide and dimeric compounds. *J. Chromatogr. A* 985:333–342.

Marmesat, S., E. Rodrigues, J. Velasco, and M.C. Dobarganes. 2007. Used frying fats and oils: Comparison of rapid tests based on chemical and physical oil properties. *Int. J. Food Sci. Technol.* 42:601–608.

Marmesat, S., J. Velasco, and M.C. Dobarganes. 2008. Quantitative determination of epoxyacids, ketoacids and hydroxyacids formed in fats and oils at frying temperatures. *J. Chromatogr. A* 1211:129–134.

Márquez-Ruiz, G., M.C. Pérez-Camino, and M.C. Dobarganes. 1990. Combination of adsorption and size-exclusion chromatography for the determination of fatty acid monomers, dimers and polymers. *J. Chromatogr. A* 514:37–44.

Márquez-Ruiz, G., M.C. Pérez-Camino, and M.C. Dobarganes. 1992. Digestibility of fatty acid monomers, dimers and polymers in the rat. *J. Am. Oil Chem. Soc.* 69:930–934.

Márquez-Ruiz, G., M. Tasioula-Margari, and M.C. Dobarganes. 1995. Quantitation and distribution of altered fatty acids in frying fats. *J. Am. Oil Chem. Soc.* 72:1171–1176.

Márquez-Ruiz, G. and M.C. Dobarganes. 1996. Short-chain fatty acid formation during thermoxidation and frying. *J. Sci. Food Agric.*, 70:120–126.

Márquez-Ruiz, G., N. Jorge, M. Martín-Polvillo, and M.C. Dobarganes. 1996a. Rapid, quantitative determination of polar compounds in fats and oils by solid-phase extraction and exclusion chromatography using monostearin as internal standard. *J. Chromatogr. A* 749:55–60.

Márquez-Ruiz, G., M. Martín-Polvillo, and M.C. Dobarganes, 1996b. Quantitation of oxidized triglyceride monomers and dimers as a useful measurement of early and advanced stages of oxidation. *Grasas Aceites* 47:48–53.

Márquez-Ruiz, G. and M.C. Dobarganes. 2006. HPSEC for lipid analysis in organic media. In *New Techniques and Applications in Lipid Analysis and Lipidomics*, ed. M.M. Mossoba, 205–238. Champaign: IL: AOCS Press.

Márquez-Ruiz, G. and M.C. Dobarganes. 2007. Nutritional and physiological effects of used frying fats. In *Deep Frying: Chemistry, Nutrition and Practical Applications* (2nd edition), ed., M.D. Erickson, 173–203. Champaign, IL: AOCS Press.

Márquez-Ruiz, G., M. Martín-Polvillo, J. Velasco, and M.C. Dobarganes. 2008. Formation of oxidation compounds in olive and sunflower oils under oxidative stability index (OSI) conditions. *Eur. J. Lipid Sci. Technol.* 110:465–471.

Martín-Polvillo, M., G. Márquez Ruiz, N. Jorge, M.V. Ruiz-Méndez, and M.C. Dobarganes. 1996. Evolution of oxidation during storage of crisps and french fries prepared with sunflower oil and high oleic sunflower oil. *Grasas Aceites* 47:54–58.

Martín Polvillo, M., G. Márquez-Ruiz, and M.C. Dobarganes. 2004. Oxidative stability of sunflower oils differing in unsaturation degree during long-term storage at room temperature. *J. Am. Oil Chem. Soc.* 81:577–583.

Minamoto, S., K. Kanazawa, H. Ashida, and M. Natake. 1988. Effect of orally administered 9-oxononanoic acid on lipogenesis in rat liver. *Bioch. Biophys. Acta* 958:199–204.

Neff, W.E. and W.C. Byrdwell. 1998. Characterization of model triacylglycerol (triolein, trilinolein and trilinolenin) autoxidation products via high-performance liquid chromatography coupled with atmospheric pressure chemical ionization mass spectrometry. *J. Chromatogr. A* 818:169–186.

Paradis, A.J. and W.W. Nawar. 1981. Evaluation of methods for the assessment of used frying oils. *J. Am. Oil Chem. Soc.* 58:635–638.

Pérez-Camino, M.C., G. Márquez-Ruiz, M.V. Ruiz-Méndez, and M.C. Dobarganes. 1990. Quantitation of oxidized triglycerides for the evaluation of the total oxidation level in edible fats and oils. *Grasas Aceites* 41:366–370.

Pérez-Camino, M.C., G. Márquez-Ruiz, M.V. Ruiz-Méndez, and M.C. Dobarganes. 1992. Lipid changes during frying of frozen prefried foods. *J. Food Sci.* 56:1644–1648.

Pérez-Camino, M.C., M.V. Ruiz-Méndez, G. Márquez Ruiz, and M.C. Dobarganes. 1993. Aceites de oliva vírgenes y refinados: diferencias en componentes menores glicerídicos. *Grasas Aceites* 44:91–96.

Perrin, J.L., P. Perfetti, C. Dimitriades, and M. Naudet. 1985. Etude analytique approfondie d'huiles chauffées I. techniques analytiques et essais préliminaires. *Rev. Franc. Corps Gras* 32:151–158.

Rodrigues Machado, E., S. Marmesat, S. Abrantes, and M.C. Dobarganes. 2007. Uncontrolled variables in frying studies: differences in repeatability between thermoxidation and frying experiments. *Grasas Aceites* 58:283–288.

Romero, A., C. Cuesta, and F.J. Sánchez-Muniz. 1998. Effect of oil replenishment during deep-fat frying of frozen foods in sunflower oil and high-oleic acid sunflower oil. *J. Am. Oil Chem. Soc.* 75:161–167.

Ruiz-Méndez, M.V., G. Márquez-Ruiz, and M.C. Dobarganes. 1997. Relationships between quality of crude and refined edible oils based on quantitation of minor glyceridic compounds. *Food Chem.* 60:549–554.

Ruiz-Méndez, M.V., A. Liotta, S. Marmesat, and M.C. Dobarganes. 2008. Characterization of the presence of used frying oil as raw material in biodiesel production. *Grasas Aceites* 59:383–388.

Sánchez-Muniz, F.J., S. Bastida, G. Márquez-Ruiz, and M.C. Dobarganes. 2007. Effect of heating and frying on oil and food fatty acids. In *Fatty Acids in Foods and Their Health Implications* (3rd edition), ed. C.K. Chow, 511–542. Philadelphia, PA: Taylor & Francis Group.

Schulte, V.E. 1982. Gelchromatographische bestimmung polymerisierter triglyceride. *Fette Seifen Anstrichm.* 84:178–180.

Sébédio, J.L., J. Kaitaranta, A. Grandgirard, and Y. Malkki. 1991. Quality assessment of industrial prefried french fries. *J. Am. Oil Chem. Soc.* 68:299–302.

Sébédio, J.L., M.C. Dobarganes, G. Márquez, I. Wester, W.W. Christie, G. Dobson et al. 1996. Industrial production of crisps and prefried french fries using sunflower oils. *Grasas Aceites* 47:5–13.

Uchida, K. 2005. Protein-bound 4-hydroxy-2-nonenal as a marker of oxidative stress. *J. Clin. Biochem. Nutr.* 36:1–10.

Velasco, J., O. Berdeaux, G. Márquez-Ruiz, and M.C. Dobarganes. 2002. Sensitive and accurate quantitation of monoepoxy fatty acids in thermoxidized oils by gas-liquid chromatography. *J. Chromatogr. A* 982:145–152.

Velasco, J., S. Marmesat, O. Bordeaux, G. Márquez-Ruiz, and M.C. Dobarganes. 2004a. Formation and evolution of monoepoxy fatty acids in thermoxidized olive and sunflower oils and quantitation in used frying oils from restaurants and fried food outlets. *J. Agric. Food Chem.* 52:4438–4443.

Velasco, J., S. Marmesat, O. Bordeaux, G. Márquez-Ruiz, and M.C. Dobarganes. 2004b. Formation of short-chain glycerol-bound oxidation products and oxidised monomeric triacylglycerols during deep-frying and occurrence in used frying fats. *Eur. J. Lipid Sci. Technol.* 106:728–735.

Velasco, J., S. Marmesat, O. Bordeaux, G. Márquez-Ruiz, and M.C. Dobarganes. 2005. Quantitation of short-chain glycerol-bound compounds in thermoxidized and used frying oils. A monitoring study during thermoxidation of olive and sunflower oils. *J. Agric. Food Chem.* 53:4006–4011.

Velasco, J., S. Marmesat, M.C. Dobarganes, and G. Márquez-Ruiz. 2006. Heterogeneous aspects of lipid oxidation in dried microencapsulated oils. *J. Agric. Food Chem.* 54:1722–1729.

Velasco, J., S. Marmesat, and M.C. Dobarganes. 2008. Chemistry of frying. In *Deep Fat Frying of Foods,* ed. S. Sahin and G. Sumnu, 33–56. Philadelphia, PA: Taylor & Francis.

Waltking, A.E. 1975. Evaluation of methods for the determination of polymers and oxidation products of heated vegetable oils: Collaborative study of the gas-liquid chromatographic method for non-elution materials. *J. Assoc. Off. Anal. Chem.* 58:898–901.

Wolff, I.P., F.X. Mordret, and A. Dieffenbacher. 1991. Determination of polymerized triglycerides in oils and fats by high-performance liquid chromatography. *Pure Appl. Chem.* 63:1163–1171.

7 Nutrient Antioxidants and Stability of Frying Oils (Tocochromanols, β-Carotene, Phylloquinone, Ubiquinone 50, and Ascorbyl Palmitate)

Karl-Heinz Wagner and Ibrahim Elmadfa

CONTENTS

7.1 INTRODUCTION

The delay in fat oxidation is of major interest to oil processors and for cooking and frying operators in order to guarantee good oil performance and stability at room and at elevated temperatures. Further, the use of fats and oils under nonstandard household conditions requires optimization and a long shelf life at any storage condition or cooking procedure in order to obtain food with low oxidation products, which are known to be involved in the development of various human diseases such as cardiovascular disease, cancer, or type 2 diabetes.

All different crude and refined plant oils and fats, as well as foodstuffs containing them, undergo autoxidation and other transformations during processing and heating. Lipid oxidation of oils and fats is the most important one, causing rapid deterioration of their organoleptic and functional properties, and an impaired nutritional quality of the oils as well as the food containing them.

One efficient way to prolong the shelf life of oil is the addition of natural and synthetic substances. Synthetic compounds such as BHA or BHT are known as active antioxidants but, at the same time, they may contribute to carcinogenicity or tumorogenicity. Therefore, the focus is drawn to the natural substances tocochromanols (tocopherols and tocotrienols), β-carotene, phylloquinone, and ubiquinone 50 (coenzyme Q10) as important primary (and ascorbate or its palmitate as secondary) oxidation inhibitors. Many of these antioxidants, apart from their effects on the performance of oils, act also as vitamins and biologically active metabolites, and increase concurrently the nutritional value of foods.

This chapter discusses briefly the basics of lipid oxidation mechanisms and focuses on the potential of nutrient antioxidants to stabilize fats and oils at heating and frying conditions.

7.1.1 Lipid Peroxidation

Exogenic and endogenic factors affecting the stability of oils may result in an increase in the formation of radicals and subsequent organoleptic impairment. Rancidity can be caused by atmospheric oxygen or by hydrolytic reactions that are catalyzed by enzymes.

Oxidative rancidity is one of the most critical factors affecting the quality of vegetable oils and fats. While hydrolytic reactions can be minimized by reduced temperatures, dark storage, careful packaging and transportation and, in general, by technological improvements (e.g., oxygen absorbers in foods, inert gas atmosphere in the headspace), it is not always easy to inhibit oxidative rancidity, which may take place even at low temperatures.

The basic mechanism of autoxidation has been discussed in several books and reviews (e.g., Jadhav et al., 1996; St Angelo, 1996; Frankel, 2007; Min and Kamal-Eldin, 2008) and will therefore only be summarized. Free radicals ($R^{\bullet}$) are formed from lipid molecules RH by various mechanisms involving energy sources such as light and heat or the presence of metal ions or metalloproteins. The initiation of radical formation occurs at the bond that requires the least energy for hydrogen removal. On average, it requires 80 kcal/mol but, for unsaturated fatty acids, it can be much

$$RH \longrightarrow R^{\bullet} + H$$
$$R^{\bullet} + O_2 \longrightarrow ROO^{\bullet}$$
$$ROO^{\bullet} + RH \longrightarrow ROOH + R^{\bullet}$$
$$RO^{\bullet} + RH \longrightarrow ROH + R^{\bullet}$$
$$ROOH \longrightarrow RO^{\bullet} + HO^{\bullet}$$
$$2\,ROOH \longrightarrow ROO^{\bullet} + RO^{\bullet} + H_2O$$

$ROO^{\bullet}$...... peroxy- $RO^{\bullet}$ ......alkoxy- $R^{\bullet}$....... alkyl-radical

FIGURE 7.1 Main lipid oxidation reactions.

less. Once a free radical is generated, the chain reaction of oxidation is initiated, new free radicals are formed, and the process is easily propagated (a self-propagating chain reaction process). The free radicals produced react with oxygen to form peroxy radicals ($ROO^{\bullet}$) that can further react with lipids to form a hydroperoxide (ROOH) and a new unstable lipid radical (Figure 7.1).

This lipid radical will then react with oxygen to produce another peroxy radical, resulting in a cyclical, self-catalyzing oxidative mechanism unless polyunsaturated fatty acids are available in the system. Hydroperoxides, which are formed by various pathways, including the reaction with singlet oxygen, are unstable and able to degrade and produce radicals such as alkoxy or hydroxyl radicals that further accelerate propagation reactions. These radicals undergo carbon–carbon cleavage to form breakdown products including aldehydes, ketones, alcohols, hydrocarbons, esters, furans, and lactones. Alternatively, the hydroperoxides can condense into dimers and polymers.

The self-propagating process is stopped by termination reactions, where two radicals are combined to form products that do not tend to form new free radicals.

In model systems under defined conditions, it was found that autoxidation reactions in foods lead to overall oxidative breakdown of poly- and mono-unsaturated fatty acids. These models also showed that the speed of the oxidation is mainly dependent on the fatty acid pattern, the amount and effectiveness of pro- and antioxidants, and the storage conditions such as temperature and exposure to oxygen or light.

In particular, during frying the oil and food are continuously and repeatedly exposed to elevated temperatures in air and moisture. A number of chemical reactions occur during this time, including oxidation and hydrolysis, as well as changes due to thermal decomposition. These decomposition products formed during frying are mainly volatiles such as aldehydes, ketones, or alcohols and nonvolatiles due to chemical changes (breakdown of fatty acids), followed by an increase in the free fatty acids content, saponification value, and by physical oil changes (e.g., color changes, increase in viscosity).

7.1.2 Evaluation of the Stability of Oils and Fats

There are many methods described in the literature to determine the resistance of oils against autoxidation, for example, measurement of the amount of oxygen absorbed;

following the development of primary and secondary oxidation products such as peroxides, conjugated dienes, hexanal; and many more. Further, fully automatic methods are available such as the change of conductivity of solutions containing secondary oxidation products, also known as the *Rancimat method.*

Generally, the changes in chemical, physical, or sensory properties during oxidation are monitored as primary and secondary ones.

Oxidation of fats and oils is slow until their level of oxygen is increasing and their natural antioxidant potential overpasses, a point at which oxidation accelerates and becomes very rapid. The time period that elapses until the oxidation process accelerates is the measure to determine the resistance against oxidation and is very often referred to as the *induction period* or *induction time.* The course of oxidation has two different phases. The first one is characterized by slow oxidation, protective mechanisms being active in the oil. When there is less or no more protection present, the reaction enters a second phase and the rate of oxidation increases rapidly.

In order to evaluate the effect of various substances on their antioxidative potential, it is important to consider various oxidation temperatures since the mode of action of a substance might change in terms of the temperature used and cannot be generalized (Frankel, 2005).

7.1.3 Impact of the Fatty Acid Pattern on Fat Stability

The fatty acid pattern of a fat is one of the most important factors influencing lipid oxidation. The P/S ratio (polyunsaturated (PUFA):saturated fatty acids (SFA)), and S/M/P ratio (SFA:monounsaturated (MUFA):PUFA) of a fat are calculated from the fatty acid composition. The fatty acid pattern is typical for the fat or oil (Table 7.1) and is generally the limiting factor for shelf life.

A low P/S ratio due to a low content of PUFAs decreases the susceptibility of the system to oxidation and results in much longer shelf life. Therefore, frying fats are not recommended to have a high PUFA content, especially long-chain PUFAs, since their speed of oxidation is much higher compared to MUFAs ad SFAs. The shelf life of olive oil at 120°C determined with a Rancimat was found to be about 6 h, whereas that of linseed oil at the same condition was only 1.1 h. Linking it to their fatty acid composition, the P/S ratio of olive oil is 0.5, that of linseed oil is 7.7, due to its higher concentration of linoleic and, in particular, alpha-linolenic acid (Wagner and Elmadfa, 2000). The IP of sunflower oil under the same conditions was found to be 2.3 h (Figure 7.2), a fact that is attributed to the content of linoleic acid (64.6% of total fatty acids).

An increasing number of allylic groups in the fatty acid molecule increases the speed of lipid oxidation; at room temperature, the relative speed of oxidation for oleic:linoleic:linolenic acids is approximately 1:100:2500. The different speed in lipid oxidation among the fatty acids can be explained by various energies required for the abstraction of one H atom from a fatty acid molecule as the initial step in lipid oxidation. Dissociation energies of stearic, oleic, linoleic, and linolenic acid are equivalent to 410, 322, 272, and 167 kJ/mol, respectively (Belitz and Grosch, 1992),

TABLE 7.1
Fatty Acids Ratio in Fats and Oils

Fat Source	S/M/P[a]	Fat Source	S/M/P[a]
Animal Fats		**Plant Fats and Oils**	
Butter	62:31:2	Coconut fat	90:7:2
Lard	41:58.11	Palm oil	47:43:8
Tallow	27.5:27-59:3-7.5	Olive oil	14:72:12
Cod-liver oil	21:57.5:16	Peanut oil	15:49:30
		Rapeseed oil	5:71:24
		Lupine oil	19:44:37
		Sesame seed oil	15:38:46
		Corn oil	15.5:30.5:53
		Soybean oil	14:25:59
		Rice seed oil	18:42:38
		Wheat germ oil	14:20:66

Source: Elmadfa, I. and Leitzmann, C. 2004. *Ernaehrung des Menschen*. 4th ed., Eugen Ulmer, Stuttgart.

[a] S = saturated, M = monounsaturated, P = polyunsaturated.

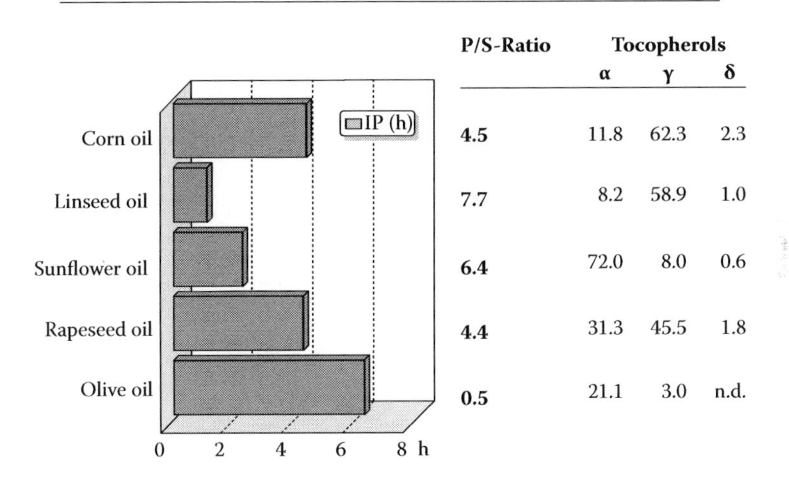

FIGURE 7.2 Shelf life (induction period; IP) of different plant oils in relation to their P/S ratio and their tocopherol content (mg/100 g) at 120°C.

which explains the higher susceptibility of unsaturated fatty acids to an attack by peroxy radicals than of saturated ones.

One strategy to extend shelf life is the increase in the MUFA content such as high-oleic sunflower oil (HOSuO) or high-oleic safflower oil (HOSaO) either by breeding techniques (also including gene technology such as gene-silencing techniques) or simply by blending PUFA-rich oils with MUFA-rich ones to end up with a high-oleic acid oil (Matthäus, 2006; Normand et al., 2006; Farhoosh et al., 2009).

7.1.4 Methods Used for the Determination of Shelf Life

For shelf-life assessments, various methods are available with many different end points; however, this wide range requires knowledge to really choose the best and, for the individual need, most appropriate methodology.

Direct shelf-life measurements use the active oxygen method (AOM; AOCS Official Method Cd 12-57) or the Rancimat method (measuring the oils stability index, OSI; AOCS Official Method Cd 12b-92). Most popular primary or secondary oxidation products investigated are hydroperoxides (Peroxide value, POV; AOCS Official Method Cd 8-53), conjugated dienes (CD, AOCS Official Method Ti 1a-64), malondialdehyde (Frankel, 2005), *p*-anisidine value (Cd 18-90), the TOTOX value that considers PV and *p*-anisidine (Rossell, 1983), volatile carbonyls such as hexanal formed from n-6 fatty acids or propanal from n-3 fatty acids (Frankel, 2005), or short-chain hydrocarbons such as ethane, propane, or pentane.

The first accepted accelerated method was the Schaal Oven test, where daily samples are drawn at 65°C, and the Swift Stability Test, where samples are maintained at 98°C. Peroxide value is determined in intervals until a defined value is obtained. In a similar way, but under more defined conditions, the Active Oxygen Method (AOM) determines the time (expressed in hours) required for a fat or oil sample to attain a predetermined peroxide value, thereby using specific conditions. This time is assumed to be an index of resistance to rancidity. Induction periods are also measured by electrochemical instruments. Oxidation is monitored by the change in conductivity of distilled water due to the formation of volatile oxidation products. Purified air passes through a heated oil sample. The effluent air contains volatile organic acids that increase the conductivity. The Oil Stability Index (OSI) value is defined as the point of rapid change of the rate of oxidation. Two electrochemical instruments, the Rancimat and the OSI instrument, are available. Investigations showed a linear relationship ($r = 0.997$ (Hill, 1994); $r = 0.987$, 0.976, and 0.905 at 100°C, 110°C, and 120°C, respectively (Hasenhuettl and Wan, 1992)) between OSI values, determined with the Rancimat, and AOM values using different fats and oils. Good correlation ($r = 0.966$) between the OSI method measured with the Rancimat and the peroxide development for six edible oils was found by Gordon and Mursi (1994). These direct methods are very often used as "frying models" in order to determine the potential of added substances in a defined system; the major disadvantage is that there is no food in the oil and, therefore, the oil–food interactions are neglected.

During deep frying, a complex series of chemical reactions takes place, which is characterized by the formation of a series of products. The products formed and changes of the properties (such as increase in free fatty acids, cyclic fatty acids, color changes, increase in viscosity, polar matter, and polymeric compounds) are important markers to evaluate the extent of lipid oxidation. At the same time, sensory changes appear and the smoke point of the oil increases.

7.2 MINOR CONSTITUENTS AFFECTING THE SHELF LIFE OF FATS AND OILS

A plethora of research projects in the past focused on the addition of substances to fats and oils, either of natural origin or synthesized, in order to delay the onset of oxidation and extend shelf life. According to their mode of mechanisms, they can be broadly classified into primary, mainly chain breaking, and secondary antioxidants. Primary antioxidants mostly react with lipid and peroxy radicals, mainly by donating hydrogen to them and converting them into stable, nonreactive products. Among them are phenolic compounds or quinones, such as tocochromanols, phylloquinone, or ubiquinone 50. β-Carotene and related carotenoids, also primary antioxidants, are effective quenchers of singlet oxygen and also prevent the formation and decomposition of hydroperoxides (Huang et al., 1995). Secondary antioxidants have different mechanisms such as chelation with metals, regeneration of oxidized primary antioxidants by donating hydrogen to them, and deactivation of single oxygen; they also act as direct scavengers of oxygen or decompose hydroperoxides.

7.2.1 Tocochromanols

Tocochromanols, or more generally vitamin E, represent a family of natural, structurally related compounds known as *tocopherols* and *tocotrienols*. They are mainly found in plant oils, margarines, seeds, and nuts and in smaller quantities in meat, fruits, and vegetables. The compounds include α-, β-, γ-, and δ- homologues and exhibit, based on the structure, different antioxidative and vitamin E activities. For all homologues, the basic structural unit is a chromanol ring system (2-methyl-6-

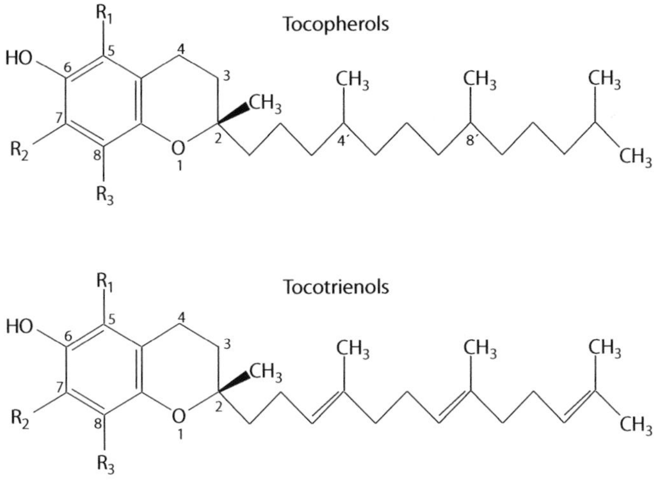

Tocopherols

Tocotrienols

Name	R1	R2	R3
RRR-α-Tocopherol	CH_3	CH_3	CH_3
RRR-β-Tocopherol	CH_3	H	CH_3
RRR-γ-Tocopherol	H	CH_3	CH_3
RRR-δ-Tocopherol	H	H	CH_3

FIGURE 7.3 Structure of tocopherols and tocotrienols.

hydroxy-chroman) with a phytyl side chain (Figure 7.3). The homologues differ in the number of methyl groups bound to the aromatic ring, which are responsible for their vitamin E activity. The side chain is saturated for tocopherols and unsaturated for tocotrienols. In addition to the natural forms, synthetic tocopherols or their esters are used in food industry to increase food stability (Ebermann und Elmadfa, 2008). Tocopherols are antioxidants possessing a "carry-through" property, which is defined as the ability of an antioxidant to survive the technological process, such as heat treatment or the refining process of oils, and transfer the stabilizing activity to the final product (Min and Kamal-Eldin, 2008).

Acting as chain-breaking antioxidants, tocopherols react with lipid radicals to convert them into more stable products. The major lipid radical formed at normal oxygen pressure is the peroxyl radical $ROO^{\bullet}$, which is stabilized to a hydroxyperoxide by a hydrogen donator.

The ability to donate the phenolic hydrogen is very important for the antioxidant activity of tocopherols as they scavenge the peroxyl radicals. The lack of the C-5 methyl group decreases the electron density in the phenolic ring, making α-tocopherol a less potent hydrogen donor than γ-tocopherol. The bond dissociation energies for the phenolic hydrogens are 75.8 and 79.6 kcal/mol in the case of α- and γ-tocopherols, respectively. This makes α-tocopherol a more efficient hydrogen donor and radical scavenger than γ-tocopherol (Wagner et al., 2004a). The resulting tocopheryl semiquinone radical molecule (Figure 7.4), which has lost the antioxidative properties, itself shows different possibilities for further reactions.

Two tocopheryl semiquinone radical molecules can form one molecule of the stable tocopherylquinone and one regenerated molecule of tocopherol. The reaction between the two radicals may also form tocopherol dimers, especially the γ-tocopherol-biphenyl-dimer and the γ-tocopherol-ether-dimer, which have antioxidative properties (Schuler, 1990). Based on a model study, Gottstein and Grosch (1990) concluded that γ-tocopherol was superior to α-tocopherol as antioxidant because it oxidizes to more stable compounds that are still effective as antioxidants. Similar results, which highlighted the antioxidative potential γ-tocopherol, were published later (e.g., Wagner et al., 2000, 2001, 2003, 2004a; Neff et al., 2003; Isnardy et al., 2004).

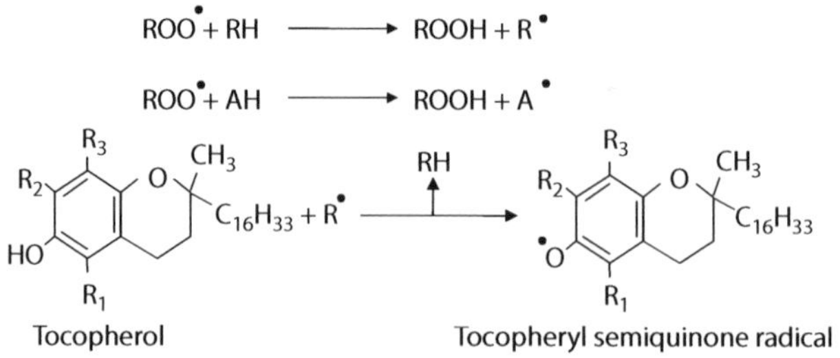

$$ROO^{\bullet} + RH \longrightarrow ROOH + R^{\bullet}$$

$$ROO^{\bullet} + AH \longrightarrow ROOH + A^{\bullet}$$

FIGURE 7.4 Tocopherol as a chain-breaking antioxidant.

7.2.2 Ubiquinone 50 (Coenzyme Q10)

Ubiquinone (2,3-dimethoxy-5-methyl-6-polyprenyl-1,4-benzoquinone), also called coenzyme Q, is a lipid-soluble compound that comprises a redox-active quinoid nucleus and a hydrophobic side chain containing a number of monounsaturated *trans*-isoprenoid units. The predominant form of ubiquinone is ubiquinone 50, which is also called coenzyme Q10 from the number of isoprenoid units in the side chain (Figure 7.5).

The biochemical and physiological functions of ubiquinone *in vivo* are well known. Ubiquinone and the reduced form ubiquinol have been implicated as biological antioxidants (Crane, 2001). A protective effect of ubiquinol against lipid peroxidation was found in fatty acid emulsions, mitochondria, and submitochondrial particles and membranes (Forsmark-Andree et al., 1995; Nohl et al., 1998). Antioxidant properties of coenzyme Q10 in food systems are not well established; Lambelet et al. (1992) demonstrated that neither ubiquinone nor ubiquinol is a good food antioxidant.

The mechanism by which the coenzyme Q10 and the reduced forms act as antioxidants has been the subject of considerable debate. One possibility is a direct reaction of ubiquinone with lipid radicals, serving as hydrogen donator:

$$QH + ROO^{\bullet} \rightarrow Q^{\bullet} + ROOH$$

Coenzyme Q10 as a quinone does not possess any labile hydrogen atom; only its reduced form ubiquinol ($Q10H_2$) with two labile hydrogen atoms is able to act as free radical scavenger. Results obtained so far actually show a reaction between lipid peroxy radicals and $Q10H_2$ to form semiquinone radicals (Lambelet et al., 1992; Kagan et al., 1996) by a hydrogen atom exchange reaction. Experimentally, *in vitro* studies showed a poor antioxidative activity for Q10 and $Q10H_2$, explained by a spontaneously oxidation to its corresponding inactive quinone (Lambelet et al., 1992; Elmadfa and Wagner, data not published).

The alternative mechanism that is more likely is the reaction of ubiquinone with the oxidized form of vitamin E, the tocopherol phenoxy radical. This action results in a vitamin E recycling:

$$QH + T\text{-}O^{\bullet} \rightarrow Q^{\bullet} + T\text{-}OH$$

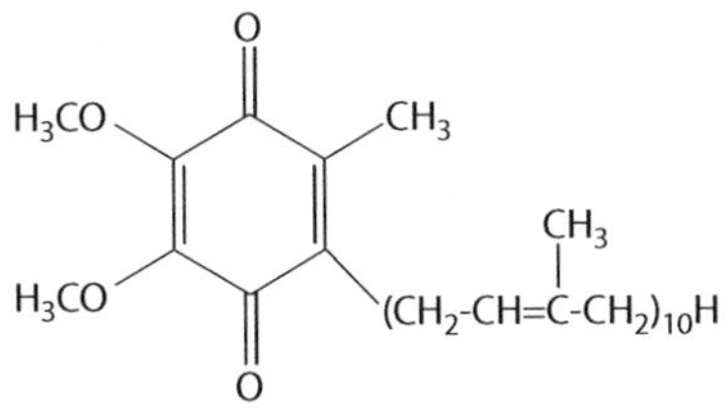

FIGURE 7.5 Structure of ubiquinone 50 (coenzyme Q10).

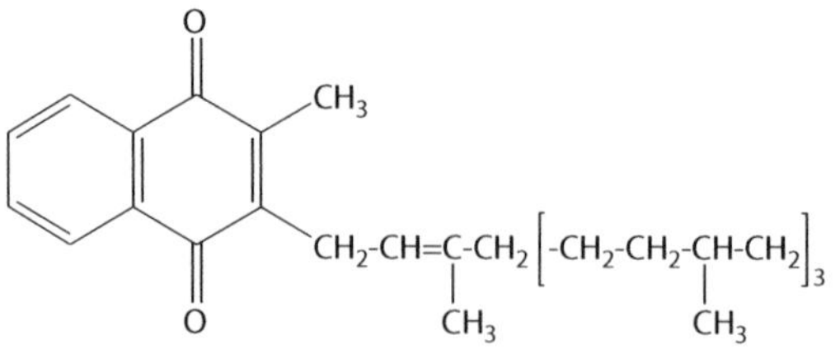

FIGURE 7.6 Structure of phylloquinone.

Again, $Q10H_2$ with its phenolic structure is more important than the Q10. The latter mechanism suggests that a reduction of ubiquinone by electron transport would force a ubiquinol-dependent regeneration of vitamin E, while the direct chain-breaking effect of ubiquinol may be less significant (Kagan et al., 1996).

7.2.3 Phylloquinone

Vitamin K_1, also called phylloquinone (Figure 7.6), has been well described as an essential nutrient since the mid-1930s; its role as a substrate for the enzyme catalyzing the posttranslational conversion of specific glutamyl residues in a limited number of proteins to γ-carboxyglutamyl residues *in vivo* was not established until the 1970s.

The primary dietary source of vitamin K activity is phylloquinone or vitamin K_1. Less described are the antioxidant properties *in vitro*. In biological tissues, the action of phylloquinone has been ascribed to their electron acceptor properties and their function as primary antioxidants. Canfield et al. (1985) assessed an 80% effectiveness of vitamin K compared to vitamin E in the prevention of linoleic acid oxidation processes.

The antioxidative action of phylloquinone and of its reduced form VKH_2 may be explained by examining two pathways. First, it acts as a hydrogen donator to stabilize lipid radicals:

$$KH + ROO^{\bullet} \rightarrow K^{\bullet} + ROOH$$

The second mode of action is the regeneration of one vitamin E phenoxy radical by the reduced VKH_2:

$$KH + T\text{-}O^{\bullet} \rightarrow K^{\bullet} + T\text{-}OH$$

The biologically important hydroquinone VKH_2, according to the phenolic structure, showed a higher reactivity than the not-reduced quinone compound (Mukai et al., 1993).

7.2.4 Carotenoids

Carotenoids are a group of fat-soluble pigments that contribute to yellow, orange, or green shades in fruits and vegetables, as well as in plant oils. More than 600 different molecules have been identified so far; about 50 possess the biological activity of vitamin A.

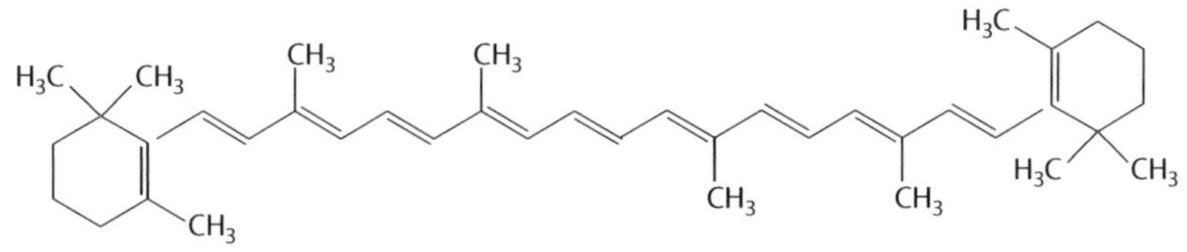

FIGURE 7.7 Structure of β-carotene.

Carrots, green leafy vegetables, and tropical fruits are rich sources and so are some oils; red palm oil contains α- and β-carotene, and is one of the best sources of carotenoids. Two main structures in the group of carotenoids are the oxygen-free carotenes such as β-carotene, α-carotene, and lycopene, and the oxygenated carotenoids such as lutein, zeaxanthine, and β-cryptoxanthine.

Structurally, β-carotene consists of two β-ionone rings and an 18-carbon polyene side chain (Figure 7.7). It is insoluble in water and soluble in fats and oils. β-Carotene is very sensitive to oxidative decomposition when exposed to air. The antioxidant action is limited to low oxygen partial pressure, up to 150 mmHg; higher pressure may change the antioxidant activity into a prooxidant one. β-Carotene and related oxygen carotenoids act contrary to hydrogen donators as singlet oxygen quenchers and free radicals, trapping agents (Jorgensen and Skibsted, 1993; Handelman, 1996).

Carotenoid radicals are produced in different ways (Britton, 1995; Rice-Evans et al., 1997):

Oxidation: $CAR - e^{+\bullet}$
Reduction: $CAR + e^{-}$
Hydrogen donating: $CAR = X - H;\ X - H + R^{\bullet} \rightarrow X^{\bullet} + RH$
Addition: $CAR + ROO^{\bullet} \rightarrow {}^{\bullet}CAR - OOR$

β-Carotene inhibits lipid peroxidation by forming a transit free radical complex with a peroxy or hydroxy radical. The addition of another peroxy radical results in the formation of a polar product:

$$^{\bullet}CAR - OOR + ROO^{\bullet} \rightarrow \text{polar product}$$

This latter activity is probably due to the presence of a conjugated double bond system. It has been postulated that β-carotene, as well as other carotenoids, acts as an effective antioxidant at low oxygen pressure under conditions where singlet oxygen is not formed, because it reacts rapidly with peroxy radicals to form a carbon-centered radical, which is resonance stabilized (Burton and Ingold, 1984).

7.2.5 Ascorbyl Palmitate

Ascorbyl palmitate (Figure 7.8), a fat-soluble form of vitamin C, is an ester formed from ascorbic acid and palmitic acid.

It has better fat solubility than vitamin C and functions as a primary, but more often as secondary antioxidant. It is a hydrogen donator, thereby regenerating

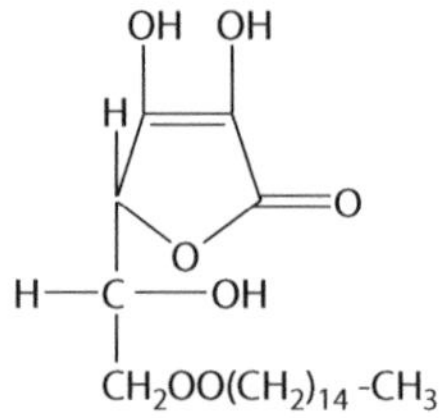

OH OH
H
O
O
H—C—OH
$CH_2OO(CH_2)_{14}$-CH_3

FIGURE 7.8 Structure of ascorbyl palmitate.

primary antioxidants such as an tocopherols, but it is also highly effective as an oxygen scavenger. Due to its structure, it acts as a hydrogen donator to regenerate tocopherol peroxy radicals. In fats and oils, technology levels of 0.01%–0.02% of ascorbyl palmitate are used. Higher concentrations are somewhat problematic as the substance is poorly soluble in edible fats and oils. It can be poured directly into the oils or dissolved in a solvent such as ethanol and then added to the oils.

7.3 MINOR CONSTITUENTS AND OIL STABILITY AT COOKING AND FRYING TEMPERATURES

7.3.1 Tocopherols, Ascorbyl Palmitate, and Their Synergism

Many studies have been reported on the antioxidant activity of tocopherols using different lipid systems and different temperatures (e.g., Huang et al., 1995; Elmadfa and Wagner, 1997; Wagner et al., 2000, 2001, 2003, 2004b; Isnardy et al., 2004; Yanishlieva et al., 2002; Nogala-Kalucka et al., 2005). The optimum concentration to inhibit hydroperoxide formation varied for specific systems, for example, using full oil, stripped oils, lard, or emulsions. In general, it is important to evaluate the radical scavenging effects of different tocopherols and their mixtures, and understand the specific mode of action that depends, among others, on the existing conditions.

Plant oils are different in their initial content of tocopherols and their fatty acid pattern. In spite of the fact that almost all plant oils contain tocopherols, their shelf life varies, mainly depending on their P/S ratio and the content of antioxidants (Figure 7.2).

The single addition of α-, γ-, and δ-tocopherol in a concentration of 0.1% prolonged the shelf life differently at a temperature of 120°C (Table 7.2). Although olive, linseed, and sunflower oils are different in their fatty acid patterns, the effectiveness of each added tocopherol was similar: γ-tocopherol stabilized best, followed by δ-tocopherol and α-tocopherol. At an equivalent concentration of 0.02% in lard at 110°C, the antioxidant activity of natural tocopherols was $\gamma > \delta > \alpha$. Pongracz et al. (1995) observed that the activity of tocopherols at 120°C does not increase in a linear manner; with increasing amounts of additive, the evaluated antioxidant activity was $\gamma > \delta > \beta > \alpha$.

TABLE 7.2
Shelf Life of Plant Oils (h) after Enrichment with 0.1% of Single Tocopherols (Toc) at 120°C (γ-Toc>δ-Toc>α-Toc)

Fat Source	Control Oil	α-Toc	γ-Toc	δ-Toc
Olive oil	6.1	7.1	14.1	13.7
Linseed oil	1.1	1.2	1.7	1.3
Sunflower oil	2.3	2.5	3.2	2.8
Rapeseed oil	4.3	3.9	5.2	4.3
Corn oil	4.4	4.0	5.0	4.7

The antioxidative potential of γ-tocopherol increases according to the concentration (Figure 7.9). A combined addition of γ-tocopherol with 0.02% of ascorbyl palmitate (AP) shows a further increase in oil stability at all tested concentrations.

A combination of 0.1% δ-tocopherol and 0.02% AP was most effective in protecting linseed oil against autoxidation. γ-Tocopherol that was found to be highly

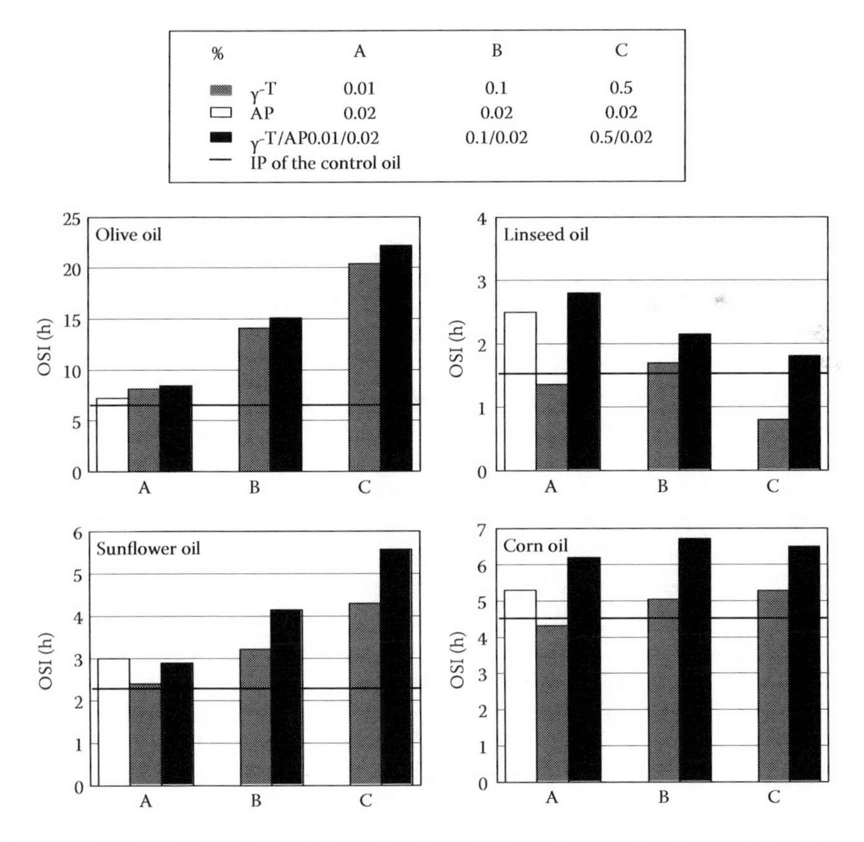

FIGURE 7.9 Oil stability (h) of plant oils after addition of γ-tocopherol, ascorbyl palmitate (AP), and the mixtures of both.

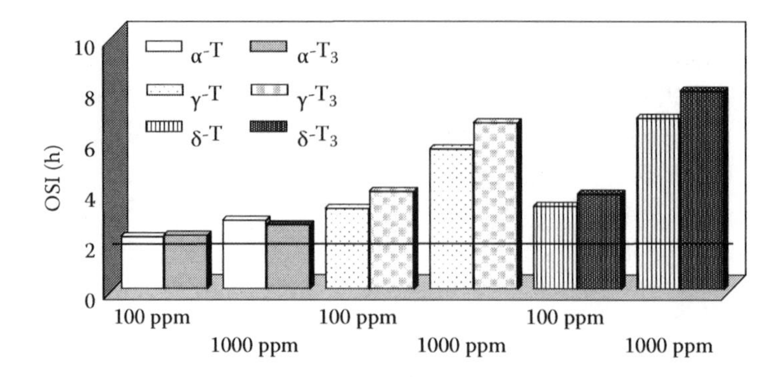

FIGURE 7.10 Shelf life of coconut fat after addition of tocopherol (T) or tocotrienols (T_3) at 160°C.

effective in olive and sunflower oil was ineffective as antioxidant in linseed oil added at 0.5%. The addition of 0.14% of γ-tocopherol reduced the induction period (IP) to 0.7 h, a decrease of 37.2%. The same effects, but less pronounced, were shown using a combination of γ-tocopherol/AP, because AP in single addition protected very well against oxidation. Tocotrienols have only rarely been investigated at higher temperatures. However, the few studies show the protective effect of γ-tocotrienol on shelf life superior to that of the tocopherols (Wagner et al., 2001; Schroeder et al., 2006) (Figure 7.10).

The addition of mixed tocopherols to palm oil considerably improved the Active Oxygen Method values before and after the treatment using frying and deep-frying conditions (Yuki and Morimoto, 1984). In the 1980s, not only tocopherols were tested for their improvement of the antioxidant activity under frying conditions but also some synergists for tocopherol such as citric acid and sodium pentapolyphosphate or *Penicillium herquei* (Nomura et al., 1987).

The storage stability of instant noodles (ramyon) fried in palm oil or beef tallow was preferred after addition of tocopherols to the frying fat. δ-Tocopherol protected better than α-tocopherol; ascorbyl palmitate acted as synergist for both tocopherols (Yang et al., 1988). The addition of α-tocopherol to the frying oil was not able to protect potato chips against autoxidation during a 7-weeks storage period at 60°C, but the addition of γ- and δ-tocopherol to the frying oil resulted in a lower peroxide value after 7 weeks (Aoyama et al., 1987). Using the formation of oxidized fatty acids during deep frying of potato crisps in rapeseed oil as quality criteria, no positive effects on the fat stability were shown when 0.1% α-tocopherol was added (Pazola et al., 1987).

The best protection provided by γ-tocopherol was reported by Shing-Huey et al. (1996) and Rho and Seib (1990), who tested the quality of different oils and instant noodles at frying temperatures. The observation of the higher protection of γ-tocopherol on plant oils has been proved in many studies at low (e.g., Isnardy et al., 2004; Wagner et al., 2004b) and higher (e.g., Warner et al., 2003; Neff et al., 2003) temperatures. Also, δ-tocopherol was recently shown to be much more efficient in protecting a commercial frying fat at 160°C than α-tocopherol (Nogala-Kalucka et al., 2005). The use of tocopherols reduced the formation of mutagenic/carcinogenic/

heterocyclic amines during frying of beef burgers in six different frying fats and oils (Johansson et al., 1995).

α-Tocopherol is oxidized much faster in oils used for deep-fat frying of potatoes than the other tocopherol homologues, with a reduction of 50% after 4–5 frying operations compared with 7 and 7–8 frying operations for β- and γ-tocopherol.

Since α-tocopherol is a more efficient hydrogen donor and radical scavenger than γ-tocopherol, it participates more readily in side reactions leading to partial loss of its antioxidant activity, which is often shown in food lipids.

Again, the presence 0.02% AP increased the number of frying operations before the tocopherol level decreased to 50% for α-, β-, and γ-tocopherol up to 8, 10, and 10–11 operations, respectively (Gordon and Kourimska, 1995). 0.02% of both α-tocopherol and 0.02% AP was able to delay lipid oxidation in canola oil when fried at 170°C. Based on the smoke point development, the use of the oil enriched with these two substances was at least prolonged by 25% (Önal and Ergin, 2002). In a more recent study, the stability of tocochromanols at frying conditions was correlated with the radical scavenging activity. Most stable was α-tocopherol; the least stable γ-tocotrienol (Rossi et al., 2006). Similar findings were made by Aladedunye and Przybylski, who also found γ-tocopherol less stable than α-tocopherol at frying at 185°C (Aladedunye and Przybylski, 2009).

Ascorbyl palmitate is highly effective in protecting frying and deep-frying fats and oils as well as the fried products. Stability tests at 180°C evaluated AP as the most efficient protector. Bharucha et al. (1980) found a higher activity for AP in comparison to acetals of ascorbic and erythorbic acid in reducing nitrosamine formation in bacon under household frying conditions after addition of AP to the frying pan. Addition of 0.02% AP reduced color development of the frying fat (animal fat/plant oil shortening) and plant oil (partially hydrogenated soybean oil). It also reduced peroxide values, the formation of conjugated diene hydroperoxides, and their degradation to volatile compounds, indicating that AP has the ability to inhibit thermal degradation of frying fats and oils (Gwo et al., 1985). A daily addition of 200 ppm AP into fresh, partially hydrogenated soybean oil shortening used to fry french fries, retarded fatty acid development, but increased color development and dielectric constant (Mancini-Filho et al., 1986). Augustin et al. (1987) showed that AP did not retard the rate of oil deterioration during frying of prawn crackers. AP was found to be a better antioxidant than rosemary extract, BHT, BHA, and δ-tocopherol in rapeseed oil during heating at 80°C and deep frying (Gordon and Kourimska, 1995). Ibrahim et al. (1991) found no significant effect on oil deterioration after addition of 200 mg/kg of AP to palm oil during frying.

At a temperature of 120°C, AP turned out to be a better protector of linseed oil and rapeseed oil against oxidation than tocopherols and their mixtures. The addition of 0.02%, 0.1%, and 0.2% of AP increased the IP of linseed oil from 1.1 h (control oil) to 2.5 h, 4.1 h, and 5.1 h, respectively. In olive oil, only γ-tocopherol in single addition was more effective than AP; the combination of both increased the IP most effectively. Oxidation of coconut fat at 200°C with the addition of 0.02% AP and the same amount of γ-tocopherol was significantly delayed (Wagner et al., 2003).

0.02% AP significantly reduced lipid oxidation in peanut oil after continuous frying of potato slices (Satyanarayana et al. 2000), and it had a significant effect on different unsaturated plant oils, resulting in a higher retention of natural antioxidants and a lower peroxide value (Masson et al., 2002). Further, the storage ability (at 63°C) of chips fried at 180°C was significantly increased after adding AP (Allam and El-Sayed, 2004).

It can be concluded from above that γ-tocopherol enrichment leads to the best protection of oils and the fried foods. Ascorbyl palmitate can also be very efficient and mainly acts as a synergist. It is able to improve the effectiveness of tocopherols at high temperatures in industrial processes as well as under household conditions.

7.3.2 Ubiquinone 50, Phylloquinone, Carotenoids, and Oil Stability

Experiments including intact animals, humans, and food systems support the view that ubiquinone, mainly in its reduced form ubiquinol, acts as an antioxidant (Beyer, 1990; Frei et al., 1990; Mukai et al., 1993; Petillo and Hultin, 2008). In addition to its role in energy metabolism, ubiquinone, as well as its reduced forms ubiquinols and ubichromenol, have been implicated as biological antioxidants (Ernster and Dallner, 1995). Especially, ubiquinols are highly effective in inhibiting lipid peroxidation in solutions, liposomes, mitochondria, microsomes, low-density lipoprotein, and *in vivo* systems. A kinetic study of free-radical-scavenging action showed a similar rate constant as the reduced form of ubiquinone 50, and of α-tocopherol (Mukai et al., 1993). Lambelet et al. (1992) observed no antioxidant activity of coenzyme Q10 in food systems after investigations with fish oils at 37°C. Also forming the corresponding semiquinone radical by reaction with peroxidizing lipid, ubiquinol Q10 was evaluated as a poor food antioxidant, because it rapidly reacts to its corresponding inactive quinone in air. Elmadfa and Wagner used the Rancimat at 120°C and tested the potential of ubiquinone Q10 at high temperature. The addition of 0.01% to five plant oils increased their susceptibility to lipid oxidation, thereby inducing a prooxidant effect of ubiquinone Q10 (data not published).

The enrichment of ubiquinone Q10 up to a concentration of 0.1% did not protect the oil. Apart from its action as an *in vivo* antioxidant, coenzyme Q10 is not recommended to be used as a food antioxidant at higher temperatures.

The first investigation of phylloquinone with plant oils was done by Parteshko et al. (1979). They found an increased stability of sunflower oil after the addition of 0.01%–0.1% of a phyllochromanol analog, which is similar to native phylloquinone. A kinetic study of the free radical–scavenging action of different hydroquinones was performed by Mukai et al. (1993). The inhibiting activity of a substituted phenoxy radical for the hydroquinones of phylloquinone and tocopherols was higher than it was shown for α-tocopherol.

Rancimat investigations in our laboratory at 120°C indicated mainly a prooxidative effect; a phylloquinone addition from 0.01% to 0.2% reduced the IP of linseed and rapeseed oil. The same concentrations did not affect the stability of olive, corn, and sunflower oil (data not published). Menadione (vitamin K_3) added to different plant oils was not able to protect tocopherols from undergoing oxidation (Kupczyk and Gogolewski, 2001; Kupczyk and Gogolewski, 2003).

The use of phylloquinone as a food antioxidant in general, but also at higher temperatures, cannot be recommended.

Carotenoids, and especially β-carotene, oxygen quenchers, and protector, against oxidation *in vivo*. In fats and oils, palm oil is the richest source of β-carotene, thereby also protecting the oil from lipid oxidation during storage. At frying conditions, it undergoes rapid degradation, due to its heat and light sensitivity (Khan et al., 2008). The first studies with β-carotene were conducted by Meara and Weir (1976). It was found that 800 ppm of β-carotene reduced the stability of palm oil at 100°C from 48 h for the crude oil to 0.6 h. The addition of 1000 ppm of β-carotene to lard, using the same conditions, reduced the IP from 16 h to 12 min. Also, in purified pine nut oil β-carotene acted as a prooxidant at concentrations of 0.01% (Kim et al., 1995).

The antioxidative properties of different carotenoids in oil model systems tested at 75°C–95°C were as follows: all-*trans*-β-carotene was less stable, followed by lutein, 9-*cis*-β-carotene, and lycopene (Henry and Schwartz, 1996). Levin and Mokady (1994) suggested higher antioxidant potency for 9-*cis*- β-carotene than for the all-*trans* isomer.

Investigations by Wagner and Elmadfa (not published) on β-carotene enrichment of different oils in concentrations of 0.01%–0.2% heated at 120°C resulted in a reduced shelf life of olive oil and rapeseed oil. No effects were found when β-carotene was added to linseed, corn, and sunflower oil.

β-Carotene was also not effective in protecting oils during heating in experiments carried out by Kajimoto et al. (1992). The effects of heating and frying operations on the retention of β-carotene in crude palm oil, a rich source of carotenoids in oils, was studied by Manorama and Rukimi (1992). Repeated deep-frying operations, using the same oil five times, resulted in a total loss of β-carotene by the fourth frying operation. Also, Schroeder et al. (2006) found no antioxidative effects of β-carotene in red palm oil frying and Rancimat heating tests.

This leads to the conclusion that none of the substances discussed in this part of the chapter, although they are important for maintaining human health, are able to protect oils from lipid oxidation at heating or frying temperatures.

7.4 SUMMARY

The antioxidative effects of nutritive substances in fats and oils at cooking and frying temperatures are influenced by a certain number of parameters including the test system, working temperature, light presence, manufacturing conditions, and availability of synergists and their amounts added. Evidence in supporting the role of tocopherols and ascorbyl palmitate as antioxidants at frying and heating conditions has been reviewed. γ-Tocopherol improves the oil stability more than other tocopherols. Ascorbyl palmitate is highly effective as a synergistic substance but, due to its structure, it can also act as a hydrogen donator. Phylloquinone, due to its quinone structure, is unable to protect oils at high temperatures but has some effects as an antioxidant at room temperature. Coenzyme Q10 can only slightly protect heated oils. More research is needed to understand the role of its reduced form (ubiqinol).

β-Carotene and related carotenoids function mainly as oxygen quenchers *in vivo* at low partial pressure. In food systems, especially in heated and fried oils, carotenes are decomposed faster than more stable antioxidants such as tocopherols. Mainly due to the high partial pressure and the sensitivity to oxidative decomposition on exposure to air, β-carotene is a poor food antioxidant.

REFERENCES

AOCS (1998). *Official Method of the American Oil Chemists' Society*, 5th ed., Cd 12b–92; Cd 12–57; Cd 8–53; Ti 1a–64; Cd 18–90.

Aladedunye, F.A. and Przybylski, R. 2009. Degradation and nutritional quality changes of oil during frying, *J. Am. Oil Chem. Soc.* 86:149–156.

Allam, S.S.M. and El-Sayed, F.E. 2004. Fortification of fried potato chips with antioxidant vitamins to enhance their nutritional value and storage ability, *Grasas y Aceites* 55:434–443.

Aoyama, M., Maruyama, T. Niiya, I., and Akatsuka, S. 1987. Antioxidant effects of tocopherols on palm oil by frying tests, *J. Jap. Soc. Food Sci. Technol.* 34:714–719.

Augustin, M.A., Chua, C.J.J., and Heng, L.K. 1987. Effects of silicone and ascorbyl palmitate on the quality of palm olein used for frying of prawn crackers, *J. Sci. Food Agric.* 40:87–93.

Beyer, R.E. 1990. The participation of coenzyme Q in free radical production and antioxidation, *Free Radic. Biol. Med.* 8:545–565.

Bharucha, K.R. Cross, C.K., and Rubin, L.H. 1980. Long-chain acetals of ascorbic and erythorbic acids as antinitrosamine agents for bacon, *J. Agric. Food Chem.* 28:1274–1281.

Belitz, H.-D. and Grosch, W. 1992. *Handbook of Food Chemistry*. 4th ed., Springer-Verlag, Berlin, Heidelberg, New York.

Britton, G. 1995. Structure and properties of carotenoids in relation to function, *FASEB. J.* 9:1551–1558.

Burton, G.W. and Ingold, K.U. 1984. Beta-Carotene: An unusual type of lipid antioxidant, *Science* 224:569–73.

Canfield, L.M., Davy, L.A., and Thomas, G.L. 1985. Anti-oxidant/pro-oxidant reactions of vitamin K, *Biochem. Biophys. Res. Commun.* 128:211–219.

Crane, F.L. 2001. Biochemical functions of coenzyme Q10, *J. Am. Coll. Nutr.* 20:591–598.

Ebermann, R. and Elmadfa, I. 2008. *Lehrbuch Lebensmittelchemie und Ernaehrung*. Springer-Verlag, New York.

Elmadfa, I. and Leitzmann, C. 2004. *Ernaehrung des Menschen*. 4th ed., Eugen Ulmer, Stuttgart.

Elmadfa, I. and Wagner, K-H. 1997. Vitamin E and stability of plant oils, *Fett/Lipid* 99:234–238.

Ernster, L. and Dallner, G. 1995. Biochemical, physiological and medical aspects of ubiquinone function, *Biochim. Biophys. Acta* 24:195–204.

Farhoosh, R., Esmaeilzadeh Kenari, R., and Poorazrang H. 2009. Frying stability of canola oil blended with palm olein, olive, and corn oils, *J. Am. Oil Chem. Soc.* 86:71–76.

Forsmark-Andree, P., Dallner, G., and Ernster, L. 1995. Endogenous ubiquinol prevents protein modification accompanying lipid peroxidation in beef heart submitochondrial particles, *Free. Radic. Biol. Med.* 19:749–757.

Frankel, E.N. 2005. *Lipid Oxidation*, 2nd ed., The Oily Press, Bridgwater, U.K.

Frankel, E.N. 2007. *Antioxidants in Food and Biology*, The Oily Press, Bridgwater, U.K.

Frankel, E.N. 1984. Chemistry of free radical and singlet oxidation of lipids, *Prog. Lipid Res.* 23:197–221.

Frei, B., Kim, M.C., and Ames, B.N. 1990. Ubiquinol-10 is an effective lipid-soluble antioxidant at physiological concentrations, *Proc. Natl. Acad. Sci.* 87:4879–4883.

Gordon, M.H. and Kourimska, L. 1995. Effect of antioxidants on losses of tocopherols during deep-fat frying, *Food Chem.* 52:175–177.

Gordon, M.H. and Mursi, E. 1994. A comparison on oil stability based on the Metrohm Rancimat with storage at 20°C, *J. Am. Oil Chem. Soc.* 71:649–651.

Gottstein, T. and Grosch, W. 1990. Model study of different antioxidant properties of α- and γ-tocopherol in fats, *Fat. Sci. Technol.* 92:139–145.

Gwo, Y.Y., Flick, G.J. Dupuy, H.P. Ory, R.L., and Baran W.L. 1985. Effect of ascorbyl palmitate on the quality of frying fats for deep fat frying operations, *J. Am. Oil Chem. Soc.* 62:1666–1671.

Handelman G.J. 1996. Carotenoids as scavengers of active oxygen species. In *Handbook of Antioxidants,* ed. E. Cadenas and L. Packer, 259–314. New York: Marcel Dekker.

Hasenhuettl, G.L. and Wan, P.J. 1992. Temperature effects on the determination of oxidative stability with the Metrohm Rancimat, *J. Am. Oil Chem. Soc.* 69:525–527.

Henry, L.J. and Schwartz, S.K. 1996. Thermal and oxidative stability of beta-carotene, lycopene and lutein, IFT Annual Meeting: book abstracts, p. 123.

Hill, S.E. 1994. Comparisons: Measuring oxidative stability, *Inform.* 5:104–109.

Huang, S.W., Frankel, E.N., and German J.B. 1995. Effects of individual tocopherols and tocopherol mixtures on the oxidative stability of corn oil triglycerides, *J. Agric. Food Chem.* 43:2345–2350.

Ibrahim, K., Augustin, M.A., and Ong, A.S.H. 1991. Effects of ascorbyl palmitate and silicone on frying performance of palm olein, *Pertanika* 14:53–57.

Isnardy, B., Wagner, K.H., and Elmadfa, I. 2004. Effects of alpha- gamma- and delta-tocopherols on the autoxidation of rapeseed oil triglycerides in a system containing low oxygen, *J. Agric. Food Chem.* 51:7775–7780.

Jadhav, S.J., Nimbalkar, S.S, Kulkarni, A.D., and Madhavi, D.L. 1996. Lipid oxidation in biological and food systems. In *Food Antioxidants*, ed. D.L Madhavi, S.S. Deshpande, and D.K. Salunkhe, 1–25, New York: Marcel Dekker.

Johansson, M.A.E., Fredholm, L, Bjerne, I., and Jaegerstad, M. 1995. Influence of frying fat on the formation of heterocyclic amines in fried beef burgers and pan residues, *Food Chem. Toxicol.* 33:993–1004.

Jorgensen, K. and Skibsted, L.H. 1993. Carotenoid scavenging of radicals, *Z. Lebensm. Unters. Forsch.* 196:423–429.

Kagan, V.E., Nohl, H., and Quinn, P.J. 1996. Coenzyme Q: Its role in scavenging and generation of radicals in membranes. In *Handbook of Antioxidants,* ed. E. Cadenas and L. Packer, 174–180. New York: Marcel Dekker.

Khan, M.I., Asha, M.R., Bhat, K.K., and Kathoon, S. 2008. Studies on coconut oil blends after frying potato chips, *J. Am. Oil Chem. Soc.* 85:1165–1172.

Kajimoto, G., Kanomy, Y., Kawakami, H., and Hamatani, M. 1992. Effects of antioxidants on the oxidative stability of oil and of tocopherol in oil during heating, *J. Jap. Soc. Nutr. Food Sci.* 45:291–295.

Kim, M., Rhee, S.H., and Cheig, H.S. 1995. Effect of tocopherols and carotene on the oxidation of purified pinenut oil in the model system, *J. Korean Soc. Food Nutr.* 24:60–66.

Kupczyk, B. and Gogolewski, M. 2001. Influence of added menadione (vitamin K3) on dissolution and dimerization of tocopherols and autoxidation of triacylglycerols during storage of plant oils, *Nahrung/Food* 45:9–14.

Kupczyk, B. and Gogolewski, M. 2003. Effect of menadione (vitamin K_3) addition on lipid oxidation and tocopherols content in plant oils, *Nahrung/Food* 47:11–16.

Lambelet, P., Loeliger, J., Saucy, F., and Bracco, U. 1992. Antioxidant properties of coenzyme Q10 in food systems, *J. Agric. Food Chem.* 40: 581–584.

Levin, G. and Mokady, S. 1994. Antioxidant activity of 9-*cis* compared to all-*trans* beta-carotene in vitro, *Free Radic. Biol. Med.* 17:77–82.

Mancini-Filho, J. Smith, L.M., Creveling, R.K., and Shaikh, H.F. 1986. Effects of selected chemical treatments on quality of fats used for deep frying, *J. Am. Oil Chem. Soc.* 63:1452–1456.

Manorama, R. and Rukimi, C. 1992. Crude palm oil as a source of beta-carotene, *Nutr. Res.* 12:223–232.

Masson, L., Robert, P., Dobarganes, M.C., Urra, C., Romero, N., Ortiz, J., Goicoechea, E., Pérez, P., Salamé, M., and Torres, R. 2002. Stability of potato chips fried in vegetable oils with different degree of unsaturation. Effect of ascorbyl palmitate during storage, *Grasas y Aceites* 53:190–198.

Matthäus, B. 2006. Utilization of high-oleic rapeseed oil for deep-fat frying of french fries compared to other commonly used oils, *Eur. J. Lipid Sci. Technol.* 108:200–211.

Meara, M.L. and Weir, G.S.D. 1976. The effect of beta-carotene on the stability of palm oil, *Riv. Ital. Sostanze Grasse* 53:178–180.

Min, D.B. and Kamal-Eldin A. 2008. *Lipid Oxidation Pathways*, AOCS Press, Champaign, IL.

Mukai, K., Morimoto, H., Kikuchi, S., and Nagaoka, S. 1993. Kinetic study of free-radical-scavenging action of biological hydroquinones (reduced forms of ubiquinone, vitamin K and tocopherol quinone) in solution, *Biochim. Biophys. Acta* 11:313–317.

Neff, W.E., Warner, K., and Eller, F. 2003. Effect of γ–tocopherol on formation of nonvolatile lipid degradation products during frying of potato chips in triolein, *J. Am. Oil Chem. Soc.* 80:801–806.

Nogala-Kalucka, M., Korczak, J., Elmadfa, I., and Wagner, K.-H. 2005. Effect of α- and δ-tocopherol on the oxidative stability of a mixed hydrogenated fat under frying conditions, *Eur. Food Res. Technol.* 221:291–297.

Nohl, H., Gille, L., and Staniek, K. 1998. The biochemical, pathophysiological, and medical aspects of ubiquinone function, *Ann. New York Acad. Sci.* 854:394–409.

Nomura, Y., Yasuda, A., Yamamoto, M., and Sugisawa, K. 1987. Synergistic effect of tocopherol, citric acid, and sodium pentapolyphosphate on the thermal oxidation of edible fats and oils, *Yukagaku* 36:117–119.

Normand, L., Eskin, N.A.M., and Przybylski, R. 2006. Comparison of the frying stability of regular and high-oleic acid sunflower oils, *J. Am. Oil Chem. Soc.* 83:331–334.

Önal, B. and Ergin, G. 2002. Antioxidative effects of α-tocopherol and ascorbyl palmitate on thermal oxidation of canola oil, *Nahrung/Food* 46:420–426.

Parteshko, V.G., Voitko, I.I., Makovetskii, V.P., Malinovskii, V.V., and Svishchuk, A.A. 1979. Antioxidant activity of the phyllochromanol analog in oxidizing sunflower oil, *Appl. Biochem. Microbiol.* 15:343–345.

Petillo, D. and Hultin, H.O. 2008. Ubiquinone-10 as antioxidant, *J. Food Biochem.* 32:173–181.

Pazola, Z., Buchowsi, M., Korczak, J., and Grzeskowiak, B. 1987. Effect of some antioxidants on fat stability during deep frying and storage of fried potato products. I. Potato crisps (part 1), *Ernährung/Nutrition* 11:481–487.

Pongracz, G., Weiser, H., and Matzinger, D. 1995. Tocopherols—antioxidants in nature, *Fett Wissenschaft Technol.* 97:90–104.

Rho, K.L. and Seib, P.A. 1990. Antioxidant effect of tocopherols in instant fried noodles (Ramyon), *ASEAN Food J.* 5:12–16.

Rice-Evans, C.A., Sampson, J., Bramley, P.M., and Holloway, D.E. 1997. Why do we expect carotenoids to be antioxidants in vivo? *Free. Radic. Res.* 26:381–398.

Rossell, J.B. 1983. Measurement of rancidity. In *Rancidity in Foods*, ed. J.J.C. Allen and J. Hamilton, 21–45, Applied Science Publishers, London.

Rossi, M., Alamprese, C., and Ratti, S. 2006. Tocopherols and tocotrienols as free radical-scavengers in refined vegetable oils and their stability during deep-fat frying, *Food Chem.* 102:812–817.

Satyanarayana, A., Giridhar, N., Joshi, G.J., and Rao, D.G. 2000. Ascorbyl palmitate as an antioxidant for deep frying of potato chips in peanut oil, *J. Food Lip.* 7:1–10.

Schroeder, M.T., Becker, E.M., and Skibsted L.H. 2006. Molecular mechanisms of antioxidant synergism of tocotrienols and carotenoids in palm oil, *J. Agric. Food Chem.* 54:3445–3453.

Schuler, P. 1990. Natural antioxidants exploited commercially. In *Food Antioxidants*, ed. B.J.F. Hudson, 100–112, Elsevier Applied Science, Amsterdam.

Shing-Huey, L,. Jeng-Leun, M., and Hau-Yang, T. 1996. Application of natural vitamin E oil for the use of frying oils and instant noodle processing, *Food Sci. Taiwan* 23:377–383.

St. Angelo, A.J. 1996. Lipid oxidation on foods, *Crit. Rev. Food Sci. Nutr.* 36:175–224.

Wagner, K.H. and Elmadfa, I. 2000. Effects of tocopherols and their mixtures on the oxidative stability of olive oil and linseed oil under heating, *Eur. J. Lipid Sci. Technol.* 102:624–629.

Wagner, K.H., Wotruba, F., and Elmadfa, I. 2001. Antioxidative potential of tocotrienols and tocopherols in coconut fat considering oxidation temperature, *Eur. J. Lipid Sci. Technol.* 103:746–751.

Wagner, K.H., Feigl, P., and Elmadfa, I. 2003. Ascorbyl palmitate and its synergism to tocopherols, *Forum Nutr.* 56: 347–348.

Wagner, K.H., Kamal-Eldin, A., and Elmadfa, I. 2004a. Gamma-tocopherol—the underestimated vitamin, *Ann. Nutr. Met.* 48:169–188.

Wagner, K.H., Isnardy, B., and Elmadfa, I. 2004b. Effects of alpha- gamma- and delta-tocopherols on the autoxidation of a 10% rapeseed oil triglyceride-in-water emulsion, *Eur. J. Lipid Sci. Technol.* 106:44–51.

Warner, K., Neff, W.E., and Eller, F.J. 2003. Enhancing quality and oxidative stability of aged fried food with γ-tocopherol, *J. Agric. Food Chem.* 51:623–627.

Yang, J.H., Chang, Y.S., and Shin, H.S. 1988. Relative effectiveness of some antioxidants on storage stability of instant noodle (Ramyon) fried in palm oil and beef tallow, *Kor. J. Food Sci. Technol.* 20:569–575.

Yanishlieva, N.V., Kamal-Eldin, A., Marinova, E.M., and Toneva, G.M. 2002. Kinetics of antioxidant action of α- and γ-tocopherols in sunflower and soybean triacylglycerols, *Eur. J. Lipid Sci. Technol.* 104:262–270.

Yuki, E. and Morimoto, K. 1984. Frying qualities of palm oil and palm olein under simulated deep-fat frying conditions, *Yakugaku* 33:265–269.

8 Nonnutrient Antioxidants and Stability of Frying Oils

Dimitrios Boskou

CONTENTS

8.1 INTRODUCTION

The use of natural antioxidants in food is receiving special attention because of the worldwide trend to avoid or minimize synthetic food additives. Great attention is also being paid to natural antioxidants because they prevent lipid peroxidation and oxidative stress, which is proposed to be closely related to aging, mutation, cancer, and other diseases. This intense interest of consumers in natural food products has broadened the market for naturally derived antioxidants.

Frying is one of the most commonly used cooking procedures in households, restaurants, and industrial food production. It is not easy to stabilize oils under frying conditions because the rate of initiation of oxidation is very high and the antioxidants added at acceptable mass fractions (0.01%–0.02%) decompose rapidly. Besides, most antioxidants tend to be somewhat volatile and undergo steam distillation at frying conditions. In addition to oxidation, other degradation reactions take place. As compounds formed from such reactions increase in number and quantity, the nutrition value of the oil is diminished, the sensory quality becomes unacceptable and, when total polar compounds reach a level of 20%–28%, the oil has to be discarded.

The effects of dietary-oxidized fats have been broadly studied. Feeding experiments using rats, mice, and other experimental animals revealed that ingestion of heated-oxidized fats, compared with fresh fats, may provoke an array of adverse biological effects such as oxidative stress and lipid metabolism changes mediated by alterations in gene expression (Ringseis and Eder, 2008). These harmful degradation products, however, are absent or present in minute quantities when good frying practices are applied and the oil is protected either by the conditions of frying or the presence of antioxidants. From the plethora of reviews and articles found in the literature, it is reasonable to assume that, when proper monitoring of discarding is practiced, the formation of substances that are toxic is of no practical significance. Nutritional and toxicological effects were detected only in discarded frying oils or extensively fried oils. A moderate consumption of fried food prepared under normal culinary practices is completely safe. The above indicates that knowledge of the stability of frying oils and the role of antioxidants is very important, and this knowledge should be made known not only to oil processors but also to consumers. So far, methods to stabilize heated oils have been based on hydrogenation and alteration of fatty acid composition, and on addition of primary and secondary antioxidants and silicone oil. Normally, less stable oils are hydrogenated to enhance their oxidative stability for deep-frying uses. However, considerable amounts of geometrical and positional isomers of fatty acids are formed, which are nutritionally undesirable. Recently, Criado et al. (2008) proposed enzymatic interesterification of olive oil with fully hydrogenated palm oil as a means of preparing frying fats with good stability and physical properties, devoid of geometrical isomers. The stability of frying oils can be also increased by reducing the unsaturation genetically or by blending polyunsaturated oils with less unsaturated oils. Such oils have been proposed as healthful alternatives to hydrogenated oils. (Warner, 2008).

Most approved antioxidants added to foods are synthetic. Methyl silicone and TBHQ are the compounds mostly used by the industry to improve performance of oils during frying (Berger, 2005). Antioxidants isolated from plant sources may provide an alternative to the current choices of effective, high-temperature oxidation inhibitors but, for the moment, there are few applications. These few applications and the results of studies conducted in the last decade provide evidence that deeper insight in the structure–reactivity relationships and proper formulation of antioxidants and synergists to achieve high stability might give a real alternative to hydrogenation. Novel natural antioxidants seem to be the landmark for future exploration in search of healthier, more stable frying oils. The industry is in need of such oils to satisfy a large number of consumers of chips and french fries, and also to reduce used oil discarding, which is not only a waste of money but a serious environmental problem.

This chapter discusses the stability of heated oils in the presence of both synthetic and natural antioxidants. Emphasis is given to nonnutrient natural antioxidants, mainly extracts obtained from the plants of the Lamiaceae family, which are rich in active phenols, and to lignans such as those present in sesame seed oil. Two important classes of natural antioxidants, tocopherols and sterols, that have an impact on the stability and performance of an oil at elevated temperatures are not discussed in detail because they are examined elsewhere in this book.

8.2 STABILITY OF FRYING OILS AND ANTIOXIDANTS

Frying is extensively used both at home and at commercial scale to enhance palatability and organoleptic properties of foods. During frying, the oil is exposed to atmospheric oxygen and moisture at high temperatures. As a result, many chemical changes occur both in the frying medium and in the fried product. These changes are due to hydrolytic oxidation, fission, and polymerization reactions, which are accelerated by an increase in temperature. The formation of breakdown and polymeric products affects the quality of the oil and the fried food (Chang et al., 1978; Cuesta et al., 1988; Dobarganes and Marquez-Ruiz, 1996; Perkins, 1996; Choe and Min, 2007). In a used frying medium, the distinguishable fractions are

1. Volatile breakdown products
2. Nonvolatile breakdown products
3. Original triacylglycerols
4. Polar triacylglycerols
5. Polar polymers
6. Nonpolar polymers

The presence of these reaction compounds is manifested during the frying process by color darkening; an increase in viscosity, free acidity, and foaming; and a decrease in smoke point (formation of volatiles).

For the protection of frying oils against thermal and oxidative deterioration blending, partial hydrogenation and the addition of various compounds or mixtures have been used. Hydrogenation forms *trans* fats, which are not wanted for nutritional reasons. Legislation requires labeling of *trans* fats on nutrition labels, and this has created a push to reformulate deep-fat frying oils since the consumers have generally a preference for trans-free oils. Antioxidants are effective during storage of fats, but there is a difference of opinion among scientists for their effectiveness in retarding deterioration of oils during frying (Augustin and Berry, 1983; Warner et al., 1985; Asap and Augustin, 1986; Yanishlieva et al., 1988; Akoh, 1994; Tyagi and Vasishtha, 1996; Berger, 2005; Farhoosh et al., 2009). This is partly because using actual frying conditions to study antioxidant effectiveness is not always easy, and extrapolation of data on the action of antioxidants from storage or heating experiments in model systems may be misleading. The rate of the initiation reactions of autoxidation is very high at frying temperatures; and antioxidants, added usually at levels 0.01%–0.02%, are rapidly decomposed. Besides, degradation reactions take place, which are not encountered when the oil is autoxidized at room temperature. Certain additives such as butylated hydroxytoluene (BHT) and butylated hydroxyanisole (BHA) are lost through volatilization, and the desired level of the additive is not maintained during the frying operation.

8.2.1 Synthetic Antioxidants and Secondary Oxidation Inhibitors

Among the various synthetic antioxidants used for the protection of frying oils against thermal oxidation, the most extensively studied are butylated hydroxyanisole (BHA), butylated hydroxytoluene (BHT), tertiary butyl hydroquinone (TBHQ), and propyl gallate (PG) (Peled et al., 1975; Augustin and Berry, 1983; Warner et al., 1985; Asap and Augustin, 1986; Tyagi and Vasishtha, 1996; Berger, 2005). The addition of TBHQ has been claimed to have a good "carry through" into potato chips deep-fried in oil containing 0.02% antioxidant. Asap and Augustin (1986) reported that an undesired effect of adding TBHQ is that it may darken the oil. Recently, Li Jiayi et al. (2006) proposed a new antioxidant, 1,1-di-(2′,5′–dihydroxy-4′-*tert*-butylphenyl)ethane(DHPE), prepared from TBHQ and acetaldehyde. Compared to the other known antioxidants, this compound was found the most potent as an antioxidant at room temperature but also in deep-fat frying of lard and soybean oil. DHPE is a strong radical scavenger, as shown by the 2,2-diphenyl-1-picrylhydrazyl (DPPH) radical assay.

Other substances that have been tested for their effect on the stability of frying oils are tocopherols, vitamin C, citric acid, mixtures of tocopherols and melanoidins, silicone oil, and polymeric antioxidants. Silicone oil was studied by Sims et al. (1972), Freeman et al. (1973), Rhee (1978), Ohta and Kusaka (1979), Frankel et al. (1985), and Kusaka et al. (1985). Freeman and his co-investigators (1973) heated sunflower oil containing 1 mg/kg of methyl silicone, and they reported a significant effect in preventing oxidation. Rhee (1978) studied the effect of methylsilicone, TBHQ, and BHA/BHT on the frying stability of salad oils. His tests suggested that methyl silicone improves frying stability, while TBHQ improves storage stability. The author recommends that methylsilicone (1.0 ± 0.5 mg/kg) and TBHQ (200 mg/kg) be added to vegetable oil to improve frying and storage stability.

Frankel and his collaborators in 1985 supplied evidence that the performance of partially hydrogenated soybean oil as a cooking medium was improved by adding TBHQ with methyl silicone. In the same year, Kusaka and others (1985) supported the view that silicone oil slows down the convection currents of the frying oil and, thus, thermal deterioration is suppressed. This supports an earlier suggestion by Freeman et al. (1973).

Secondary antioxidants and synergists such as citric acid, tartaric acid, EDTA, and ascorbyl palmitate were studied by Sherwin et al. (1974), Chun et al. (1984), Gwo et al. (1985), Warner et al. (1985), and others. Sherwin et al. (1974) used a mixture of phenolic antioxidants with silicone and esters of ethylene diamine tetraacetic acid (EDTA). Gwo and his co-workers (1985) found that the addition of 0.02% ascorbyl palmitate reduced color development of a frying fat consisting of an animal fat and vegetable oil and also of heated partially hydrogenated soybean oil. Warner and his co-investigators (1985) examined the stability of soybean oil heated at 190°C for various time periods. The combination of citric acid and methyl silicone had the greatest effect in lowering the intensity of odor of the heated oil. Chu (1991) reported a reduced increase in viscosity, refractive index, color, and total polar compounds when TBHQ and ascorbyl palmitate or propyl gallate and ascorbyl palmitate were added to heated soybean oil. The reduction in iodine value during 80 h of heating was also minimized.

The effect of a polymeric antioxidant (Anoxomer, approximately 4000 Da) on the fat stability during deep-frying of potato chips in low erucic acid rapeseed oil was investigated by Pazola et al. (1987). The oil containing the antioxidant suppressed oxidation as judged by measurement of petroleum ether insoluble fatty acids, dielectric constant, and color reactions. The same nonabsorbable antioxidant was examined by Vaisey-Genser and Ylimaki (1985) for its ability to limit the development of a heated oil odor in canola oil products heated to a frying temperature (190°C). While Anoxomer caused a modest improvement in the odor of heated solid frying shortenings, no consistent advantage was evident in heated cooking oils or in liquid shortenings used for frying.

Warner and his co-workers (1985) studied the effect of various additives, including a polymeric antioxidant, on the odor characteristics of heated soybean oil. The addition of the polymeric antioxidant, in various combinations with citric acid, did not affect the odor intensity as compared to the intensity of the control oils. Good results were obtained only with combinations of methyl silicone with citric acid and TBHQ.

Volcanic ash—A food-grade volcanic ash, MirOil Life Powder, is claimed to enhance the properties of frying oils (Rossell, 1996). When the powder is added to the oil, the citric acid and moisture absorbed in the ash react with soaps and a precipitate is formed, which can be removed by filtration. There are claims, however, that the powder has very little or no beneficial effect in large-scale industrial frying operations (Rossell, 1996).

Ethyl ferulate—Frying oils containing ethyl ferulate were found to have an effect similar to TBHQ when frying tests were conducted with frying oils. The latter were analyzed for retention of the additives and total polar compounds as indices of oil fry life (Warner and Laszlo, 2005). It has to be mentioned that ferulic acid is a known antioxidant phenol. It is part of the molecule of oryzanol, an oxidation and antipolymerization inhibitor of frying oils, present in rice bran oil (see Chapter 9).

8.2.2 Stabilization of Frying Oils with Natural Antioxidants

In the last three decades, there has been an increasing trend toward natural ingredients in food and many attempts have been made to replace chemicals in the stabilization of lipids (Bracco et al., 1981; Schuler, 1986; Djarmati et al., 1991; Nakatami, 1994; Tsimidou et al., 1995; Cuppett et al., 1997). The emphasis given to natural antioxidants is due to the toxicity concerns about some synthetic antioxidants but mainly to findings that relate active ingredients in the diet and free radical mechanisms in the human body. Defenses against oxidants, the role of both nutrient and nonnutrient antioxidants, and disease prevention have been extensively discussed in many reviews and books (Namiki, 1990; Diplock, 1992; Langseth, 1995; Kumpulainen and Salonen, 1996; Shahidi, 1997; Stanner et al., 2004; Kroon and Williamson, 2005 ; Scalbert et al., 2005; Boskou, 2006).

Researchers interested in the properties and occurrence of natural antioxidants concentrate on tocopherols and tocotrienols, ascorbic acid, carotenoids, phenolic acids, flavonoids, sesame lignans, phytosterols, extracts from the leaves of the *Lamiaceae* plants, olive tree leaves, tea leaves, phosphatides, oryzanols, propolis, and other natural sources. The effect of natural antioxidants on the stability of frying oils has been discussed by Shukla et al. (1997), White and Xing (1997), Chen and Ho (1997), Boskou (1999), Blekas and Boskou (1999), Kochhar (2000a, 200b), Jaswir et al. (2000), Berger (2005), Nogala-Kalucka et al., 2005, Farag et al. (2006), Choe and Min (2007), El Anany (2007), Chiou et al. (2008), and others.

Among the various natural antioxidants, extracts of herbs such as rosemary and sage have played an important role. Next to them, other members of the Lamiaceae family (including oregano, thyme, marjoram, dittany, and savory) and olive tree products show a stabilizing effect in heated oils (Chipault et al., 1952; Inatami et al., 1983; Houlihan et al., 1984; Farag et al., 2006; Economou et al., 1991; Chen et al., 1992; Lagouri and Boskou, 1995; Yanishlieva and Marinova, 1995; Nakatami, 1997; Choe and Min, 2007; Ramalho and Jorge, 2008). Recently, it was reported that pomegranate peels extract (El Anany Ayman, 2007) increases the stability of sunflower oil during deep-fat frying. The natural antioxidant used have been generally claimed to provide an antioxidation action equal to or higher than that of synthetic antioxidant. Che Man and Tan (1999) found the following order of effectiveness when palm olein was used for frying: oleoresin rosemary>BHA>sage extract>BHT>control. Jaswir et al. (2006) found a synergistic antioxidative effect of rosemary extract, sage extract, and citric acid in heated oils.

Studies related to the stabilizing effect of natural compounds on fats subjected to elevated temperatures are listed in Table 8.1.

8.2.2.1 Rosemary and Lemon Balm

The leaves of the plant *Rosemarinus officinalis* are mainly known as a spice and a flavoring material. Antioxidants are now extracted from this plant (Pratt and Hudson, 1990; Loliger and Wille, 1993). The advantage of rosemary is its availability as a natural source and the high antioxidant potential of its crude and refined extracts. The latter are easily found in the market in various forms.

Rosemary has been extensively studied as a means to stabilize fats at room temperature, but there are limited reports for its effect on frying oils. Pazola and his co-workers (1987) used rosemary and ethanol extracts from rosemary to study their effectiveness in retarding oxidation of heated low erucic acid rapeseed oil. The quality of potato chips fried in the same oil was also investigated. The changes in the frying oil were monitored by the determination of oxidized fatty acids insoluble in petroleum ether, measurement of dielectric constant, sensory evaluation, and color reactions with bromothymol blue and 2,6-dichlorophenol-indophenol. The stability of fried potato crisps during storage for 8 weeks was examined by determination of peroxide values, thiobarbituric acid indices, and sensory evaluation. The best results were obtained when rosemary was added to the potato dough.

Chu (1991) carried out a study that focused on soybean oil stability after the addition of various combinations of additives, including a mixture of rosemary concentrate and ascorbyl palmitate. When added at a level of 500 ppm each, the two additives provided a significant increase in the stability of the frying oil. This was demonstrated by measurements of refractive index, dielectric constant, viscosity,

TABLE 8.1
Chemical Compounds or Mixtures Tested for Their Effect on the Stability of Heated Oils

Active Material	Frying Oil/Temperature	Reference
Tocopherol mixture	Refined olive oil, high oleic acid sunflower oil/180°C	Kajimoto et al. (1985)
α-Tocopherol	Low erucic acid rapeseed oil/190°C	Palozza and Krinsky (1992)
α-Tocopherol	Soybean oil/185°C	Chu (1991)
γ-Tocopherol	High oleic and palmitic acid sunflower oil	Marmesat et al. (2008)
α- and γ-Tocopherols	Frying fat, 160°C	Nogala-Kalucka et al. (2005)
Rosemary extract (in a mixture with ascorbyl palmitate)	Soybean oil/185°C	Chu (1991)
Rosemary and sage extracts	Palm olein, frying potatoes	Che Man and Jaswir (2000)
Rosemary, sage extracts, BHA, BHT	Palm olein, frying	Che Man and Tan (1999)
Rosemary extract	Low erucic acid rape seed oil/190°C	Pazola et al. (1987)
Rosemary extracts	Soybean oil, 180°C	Ramalho and Jorge (2008)
Rosemary, sage extracts	Flaxseed oil, frying potatoes	Jaswir et al. (2006)
Summer savory (*Satureja hortensis*), ethanol extract	Sunflower oil/180°C	Yanishlieva et al. (1997)
Lavender and thyme	Sunflower oil, frying temperatures	Bensmira et al. (2007)
Sage extracts	Vegetable oils heated at frying temperatures	Kalantzakis and Blekas (2006)
Herb spices	Frying oil	Kimura and Kanamori (1982)
Olive leaf juice	Sunflower oil/180°C	Farag et al. (2006)
Burdock extracts	Lard	Kim and Choe (2004)
Cassia oil	Beef in frying oil	Hongxia Du and Hongjun Li (2008)
Leafy extracts	Sunflower oil, frying temperatures	Shyamala et al. (2005)
4α-Methylsterols from corn oil	Cotton seed oil/179°C	Chang and Mone (1969)
Sterols from wheat, germ oil and *Vermonia anthelminthica* oil	Safflower oil/180°C	Sims et al. (1972)
Olive oil sterols, Δ^5-avenasterol	Cotton seed oil/180°C	Boskou and Morton (1976)
Olive oil triterpene alcohols	Cotton seed oil/180°C	Boskou and Katsikas (1979)
Δ^5-Avenasterol, fucosterol	Glycerol trioleate/180°C	Gordon and Magos (1983)
Sunflower 4α-methylsterols (free and esterified)	Cotton seed oil/180°C	Blekas and Boskou (1986)
Oat sterol	Soybean oil, cotton seed oil/180°C	White and Armstrong (1986)
β-Sitosterol, stigmasterol	Purified triacylglycerols/180°C	Blekas and Boskou (1986)

(Continued)

TABLE 8.1 (Continued)
Chemical Compounds or Mixtures Tested for Their Effect on the Stability of Heated Oils

Active Material	Frying Oil/Temperature	Reference
Methanol oat extract	Soybean oil, cotton seed oil/180°C	Tian and White (1994)
Pomegranate peel extract	Sunflower oil/180°C	El Anany Ayman (2007)
Old tea extracts	Frying oil, chips 180°C	Zandi and Gordon (1999)
Citric acid-based antioxidant	Soybean oil	Warner and Gehring (2009)
Squalene	Safflower oil/180°C	Sims et al. (1972)
Squalene	Rapeseed oil	Malecka (2006)
Olive oil hydrocarbons	Cottonseed oil/180°C	Boskou and Katsikas (1979)
Sesamol	Corn oil/180°C	Fukuda et al. (1986a,b)
Propolis extract	Sunflower oil/180°C	Yanishlieva et al. (1988)
Phospholipids	Soybean oil, 185°C	Chu (1991)
Phospholipids	Olive oil, 170°C	Kourimska et al. (1993,1995)
Phospholipids	Paraffin, 180°C	Yuki et al. (1978)

color, and determinations of polar compounds, acid value, iodine value, anisidine value, and carbonyl value.

Michael Gordon and Lenka Kourimska (1995) studied the effect of an extract from rosemary on the losses of tocopherols during deep-fat frying of potatoes at 162°C. A commercial extract of the spice was added to refined low erucic acid rapeseed oil at a level of 0.1%. Potato chips were prepared by deep-frying in a laboratory scale, and the oil was allowed to cool at room temperature after each frying operation. The changes in tocopherols were followed by high-performance liquid chromatography determinations. It was found that alpha-tocopherol is lost much faster in comparison to the beta-, gamma-, and delta-homologues. After 4–5 frying operations, a 50% reduction in alpha-tocopherol was observed.

Rosemary extracts used to provide product integrity in meat and savory applications may have a strong taste, and are not applicable to more delicate applications such as sweet or neutral bakery goods and snacks. In order to overcome this challenge, selected varieties of lemon balm (*Melissa officinalis*), which is both mild in taste and effective, were tested. An active ingredient of lemon balm is rosmarinic acid (Figure 8.2).

8.2.2.2 Other Plants of the *Lamiaceae* Family

Nedyalka Yanishlieva and her coinvestigators (1997) reported on the stabilizing effect of an ethanol extract obtained from summer savory (*Satureja hortensis* L.). This extract was previously studied by Yanishlieva and Marinova (1995) for its effect on sunflower oil autoxidation and compared to similar extracts of other plants of the same family (*Mentha piperita, Mentha spicata, Melissa officinalis, Origanum vulgare, Ocimum vasilicum*). Since the extract from *Satureja hortensis* was found to be the most effective in retarding peroxide accumulation during oxidation of sunflower oil, Yanishlieva and her research team attempted to study the

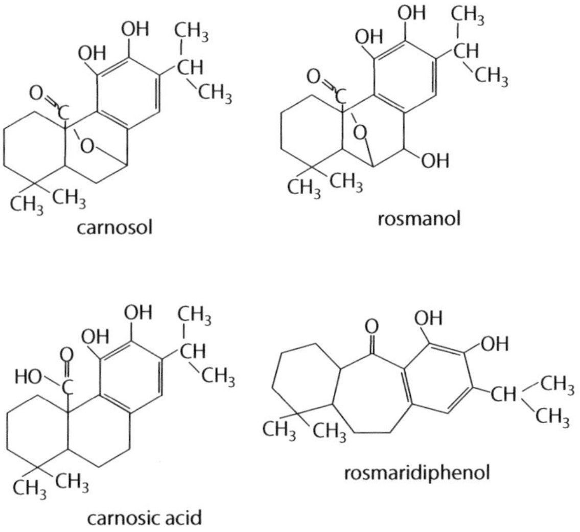

FIGURE 8.1 The main phenolic antioxidants of rosemary.

potentiality of the active extract in simulated deep-frying of sunflower oil. Control lipid samples containing 0.1% and 0.5% of the extract were placed into glass vials and heated continuously at 180 ± 5°C in an oven for 8 h each day for 7 days. At the end of each heating period, the oxidative stability of the samples at 100°C and the concentration of total unchanged triacylglycerols (TAG) were determined. The rate of decrease of TAG was found to be much lower in the first 20 h of heating for the samples containing 0.1% and 0.5% ethanol extract from *Satureja hortensis*. The oil not containing the additives reached a level of 73% of TAG in 30 h, while the oil containing the extract at a concentration 0.5% was far more stable (44 h to reach a level of 73% TAG). The limit of 73% TAG that corresponds to 27% polar materials was selected because, at this concentration of polar compounds, the oil is regarded as deteriorated and consequently unacceptable (Guhr et al., 1981; Andrikopoulos, 1989).

The effect of the savory extract seems to be very impressive. It is not known, however, if the flavor of the extract is compatible with the various types of food, which may be fried in sunflower oil. No attempt has been made to correlate the antioxidative and antipolymerization activity with specific phenolic compounds that may be present in the extract. It is known from the literature that hydrocinnamic acids (e.g., caffeic acid) and their derivatives (e.g., rosmarinic acid) are found in oregano plants at high concentrations (Reschke, 1983; Nakatami, 1994; Bertelsen et al., 1995; Boskou et al., 2006). These phenols may contribute significantly to the stabilizing effect of the extract obtained from summer savory.

As already mentioned, the benefits of antioxidants as part of a healthy diet have been known for some time, and health and nutrition experts recommend various sources of natural antioxidants such as tea, spinach, berries, olives, broccoli, and many others. The study of the composition of herbs and spices that have antioxidant effects allows today for other foods to be enriched with substances that make them "healthier." Based on this principle, Houhoula et al. (2003) suggested the addition of oregano (the ground spice) or oregano extracts (obtained with organic solvents) to cottonseed oil used for the preparation of potato chips. The addition of the spice decreased the rate of peroxide formation in the oil absorbed in the stored chips. This increase of stability was attributed to the presence of antioxidant phenols. According to the authors, "oregano may have multivariate benefits if used in frying oils." The deterioration of dietary-important unsaturated fatty acids is retarded, and the formation of harmful radicals is restricted. On the other hand, the flavonoids and other antioxidant components of oregano have antimicrobial, anti-thrombotic, antimutagenic, and anticarcinogenic benefits within the human body. Antioxidant compounds present in oregano extracts are not fully investigated. Research based on chromatographic and spectrometric techniques (Vekiari et al., 1993) and especially on high-performance liquid chromatography and combined advanced NMR methodologies (proton–proton double quantum filter correlation, proton–carbon heteronuclear multiple quantum coherence, and on line solid-phase extraction in LC-NMR) confirmed the presence of flavonols, flavones, flavanones, dihydroflavonol, and rosmarinic acid (Exarchou et al., 2002, 2003) (Figure 8.2).

8.2.2.3 Other Extracts

Shyamala et al. undertook in 2005 an investigation with the objective of evaluating the antioxidant capacity of extracts obtained from leafy vegetables in sunflower and groundnut oils heated to frying temperatures. The leafy vegetables examined were cabbage (*Brassica oleracia var capitata*), coriander leaves (*Coriandrum sativum*), honggone (*Alternanthera sessilis*), and spinach (*Spinacia oleracea*). The extracts were found to confer a protective effect on peroxide formation on storage of heated oils and, according to the authors, can serve as substitutes for synthetic antioxidants. The addition of spinach powder in flour-rolled dough as a natural antioxidant before frying at 160°C was also studied by Lee et al. (2002). Rhee et al. (1992) examined the effect of polyphenols from cottonseed to cottonseed oil heated at 180°C.

8.2.2.4 Lignans

A stable cooking oil composed of all natural components was launched in 1986 in the United States and in Europe (Haumann, 1996). This was approved by German health authorities for marketing as dietetic food, and contains mainly high oleic acid sunflower oil, in addition to rice bran oil and sesame seed oil. The latter is known to be the most resistant to oxidative rancidity among vegetable oils (Kikugawa et al., 1983; Kamal-Eldin and Appelquist, 1994).

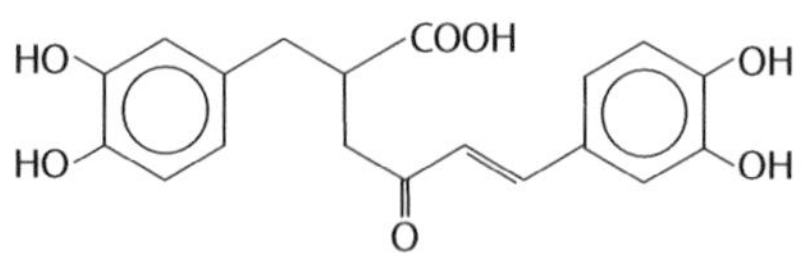

FIGURE 8.2 Rosmarinic acid.

The stability of sesame oil has been attributed to tocopherols, ethylidene side-chain sterols, and lignans such as sesamol, sesamolin, and sesaminol and sesamol dimers (Fukuda et al., 1986a,b; Fukuda et al., 1994; Kamal-Eldin and Appelqvist, 1994). Sesamolin is a precursor of sesamol and sesamol dimers. Sesamol (3,4-methylene-dioxy phenol) is present in small amounts in crude sesame oil. As demonstrated by Kikugawa et al. (1983), these levels are insufficient to explain the high stability of the oil. Sesamolin [2-(3,4-methylenedioxy phenyl)-6-(3,4-methylenedioxyphenoxy)-*cis*-3,7-dioxabicyclo [3.3.0]octane], on the contrary, is a major constituent of the unsaponifiable fraction of sesame oil, and it liberates sesamol and sesaminol [2-(3,4-methylenedioxy-6-hydroxyphenyl)-6-(3,4-methylenedioxyphenyl)-*cis*-3,7-dioxabicyclo-[3.3.0] octane] during refining and frying (Fukuda et al., 1986a; Mohamed and Awatif, 1998). In this way, sesamolin, although not a strong antioxidant itself, is a precursor to two strong antioxidants that are responsible for the stability of processed sesame seed oil. Sesamol dimer is also a potent antioxidant comparable to the parent compound (Kikugawa et al., 1983; Fukuda et al., 1986b). To prove the strong antioxidant effect of sesamol, Fukuda and his coinvestigators (1986a) carried out a model experiment. Corn oil was supplemented with 0.2% sesamol and heated at 180°C for 3 h (a 100% reduction of sesamol content). Thereafter, the oil was kept at 60°C for 7 days, and the extent of oxidation was examined. The system, which contained corn oil with the additive, proved to be significantly more stable than the control not containing sesamol. Today, palm olein, sunflower oil, and other oils containing, at low levels, specially refined sesame seed and rice bran oils are routinely applied in industrial-scale frying of potato chips to obtain improved flavor stability.

Structures of antioxidant lignans are given in the following text. Two lesser-known lignans were identified in oils from certain wild Sesamum species, 2-episesalatin and sesangolin (Kamal-Eldin and Appelqvist, 1994). Their role as stabilizing agents is not known (Figure 8.3).

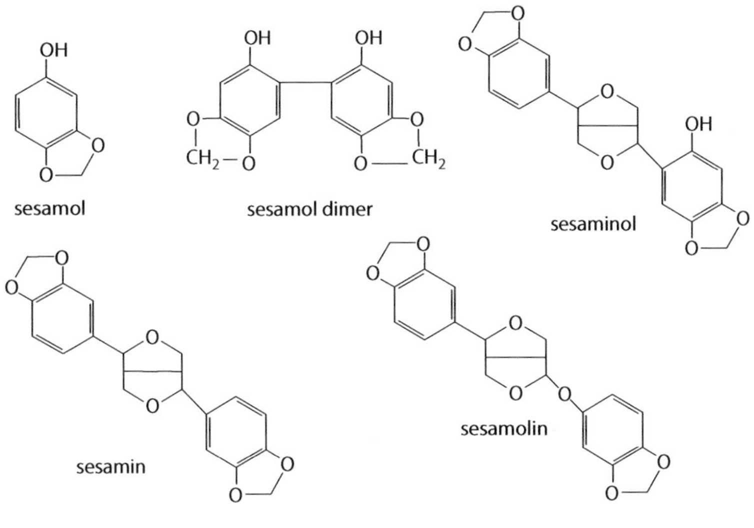

FIGURE 8.3 Structures of sesamol and related lignans.

8.2.2.5 Propolis

Propolis (bee glue) is a resinous material collected by bees from different plant exudates and used as a general-purpose sealer in the construction of hives. It is known to possess a broad spectrum of biological activities (antibacterial, fungicidal, anti-inflammatory, etc.). Its chemical composition is very complex: at least 150 compounds have been identified, most of them phenolics such as hydrobenzoic acids, hydroxycinnamic acids (*p*-coumaric, ferrulic, caffeic), esters of hydroxycinnamic acids, and flavonoids (flavones, flavonols, chalcones, methoxy flavanones) (Bankova et al., 1992, 1996).

Yanishlieva and her co-workers (1988) studied the thermal stability of sunflower oil heated at 180°C with free access of oxygen after the addition of 0.05% of a natural antioxidant obtained from propolis by ethanol extraction. The additive caused a 6 times increase in the stability of the fried oil as judged by the accumulation of polar compounds in the first 25 h of heating.

8.2.2.6 Phospholipids

Phospholipids are known to possess moderate antioxidant activity, which is enhanced in the presence of phenolic antioxidants. The mode of their action is not completely known. According to Pokorny and his collaborators, the improvement of oxidative stability is related to the ability of phosphatides to form chelate compounds with trace metals and to inhibit peroxide formation (Pokorny et al., 1992; Kourimska et al., 1993).

There is rather scant information concerning the use of lecithins to improve the stability of frying oils. Chu (1991) added 1000 ppm of lecithin in combination with 500 ppm ascorbyl palmitate to soybean oil used for frying for 16 h each day. Although lecithin had an adverse effect on oil color, a significant antioxidant effect was observed from the lower total polar compounds percentage, refractive index, viscosity, and dielectric constant values. Kourimska et al. (1995) studied the effect of soybean and rapeseed lecithins on a mixture of refined and virgin olive oil. The oil was selected because of its low polyunsaturated fatty acids content and the presence of natural antioxidants. Natural lecithin was prepared by degumming crude protein with the addition of water and phosphoric acid. The oil was added to a household fryer where potatoes were fried at 170°C. The level of lecithin added was 0.1%. Small additions of lecithins were not found to increase foaming and did not exhibit any noticeable adverse effect on the quality of french fries. The authors claimed that the addition of lecithin reduced the amount of antinutritional compounds such as hydroperoxides, dimers, and conjugated fatty acids and, therefore, the addition could be recommended. The mechanism of this action of lecithins is rather a synergism with tocopherols or other phenolic antioxidants naturally present in the oil.

Yuki and his co-workers (1978) studied the effect of lecithin on the thermal oxidation of tocopherols in liquid paraffin. Treatment at 180°C in the presence of 0.2%–2.0% lecithin demonstrated a significant inhibition of tocopherol oxidation and a reduced amount of dimers. This effect was attributed to an interaction and the formation of a hydrogen bond between the choline moiety of the phosphatide and tocopherols.

8.2.2.7 Squalene

Squalene is a highly unsaturated hydrocarbon ($C_{30}H_{50}$) with important biological properties. It is a metabolic precursor of cholesterol and other sterols; as an oxygen carrier, it has been extensively researched and found to play a key role in maintaining health. Today, many experts claim that squalene can enhance the quality of life, if taken continuously, and that its consumption is beneficial for patients with heart disease, diabetes, arthritis, hepatitis, and other diseases.

Squalene is found in large quantities in shark liver oil and occurs also in small amounts in olive oil, wheat germ oil, bran oil, and yeast. The presence of squalene in olive oil probably makes a significant contribution to the health effects of the latter.

Squalene has been reported to possess moderate antioxidant properties in methyl oleate and methyl linoleate at 63°C (Govind Rao and Achaya, 1968). Psomiadou and her co-workers (1998) also reported a moderate antioxidant activity in olive oil triacylglycerols at levels ranging from 7,000 to 12,000 ppm. These concentrations were selected because the hydrocarbon accounts for more than 50% of the nonsaponifiable fraction of virgin olive oil and its concentration may be as high as 12,000 mg/kg oil. According to Manzi and her collaborators (1998), squalene and tocopherol act as chain-breaking antioxidants. Tocopherol scavenges radicals faster, but the tocopheryl radical formed is reduced by squalene and tocopherol is regenerated. Thus, squalene is consumed faster than tocopherol.

Concerning the effect of squalene at frying temperatures, there is rather scant information. According to Sims and his co-workers (1972), the addition of 0.5% squalene retards the degradation of unsaturated fatty acids in safflower oil heated at 185°C. The effect is marked but not so strong as that of sterols with an ethylidene side chain (citrostadienol, Δ^5-avenasterol). Similar results were reported by Malecka (2006), who studied heated rapeseed oil containing squalene at levels from 0.1% to 1.0%.

The hydrocarbon fraction of olive oil was tested as a polymerization inhibitor in cottonseed oil heated at 180°C for 60 h (Boskou and Katsikas, 1979). The fraction acted as an antioxidant, but its effect was not pronounced. Since the main constituent of the hydrocarbon fraction of olive oil is squalene, it can be concluded that the moderate antioxidant effect observed is due to this polyunsaturated hydrocarbon.

The effect of unsaponifiable matter from olive oil deodorization distillates on the stability of sunflower oil during frying and on the quality of potato chips was studied by Abdalla (1999). A protective effect against oxidation was found, which was attributed to squalene, delta-avenasterol, and tocopherols.

8.2.2.8 Citric-Acid-Based Natural Antioxidants

In a recent publication, Warner and Gehring (2009) proposed a new antioxidant based on citric acid, which is claimed to extend the frying life of regular soy bean oil and the shelf life of food fried in the oil. This additive limits the need for stable *trans* fat containing hydrogenated oils or modified oils rich in monounsaturated and low in polyunsaturated fatty acids.

8.3 FRYING OILS AS SOURCES OF NATURAL ANTIOXIDANTS

8.3.1 Olive Oil as a Frying Medium

Virgin olive oil shows a remarkable resistance during domestic deep-frying of potatoes or in other uses at frying temperatures (Andrikopoulos et al., 1989; Romero et al., 1995; Romero et al., 1996; Aggelousis and Lalas, 1997). Compared to other vegetable oils such as sunflower, cottonseed, corn, and soybean oils, olive oil has a significantly lower rate of alteration, as demonstrated by measurements of viscosity, total polar compounds, and loss of tocopherols. This increased stability to thermal oxidation explains why this oil can be used for repeated frying operations before reaching the rejection point of approximately 27% total polar compounds (Andrikopoulos et al., 1989; Garcia Mesa et al., 1996; Aggelousis and Lalas, 1987; Kalantzakis et al., 2006).

One possible explanation for the resistance of olive oil to rapid deterioration at elevated temperatures might be the presence of natural antioxidants in the unsaponifiables. These antioxidants are α-tocopherol, squalene, and mainly Δ^5-avenasterol. The latter is a known antipolymerization factor (Boskou and Morton, 1976; Gordon and Magos, 1983).

The polar antioxidants found in virgin olive oil may also make a contribution to the increased resistance to oxidation and polymerization (Beltran Maza et al., 1996). This polar fraction of potent antioxidants occurring in olive oil in ppm quantities comprises a complex mixture of simple phenols and *o*-diphenols such as tyrosol and hydroxytyrosol and phenolic acids, mainly derivatives of hydroxycinnamic, *p*-hydroxybenzoic, and *p*-hydroxyphenylacetic acids (Boskou, 2006). Tyrosol and hydroxytyrosol are encountered as free alcohols but mainly as aglycones of their glycosides. Traces of the latter, for example, oleuropein, are also found. Among the various phenols present in olive oil, *o*-diphenols such as hydroxytyrosol and caffeic acid have been demonstrated to have a strong antioxidant activity (Papadopoulos and Boskou, 1991). Other phenols present in olive oil in trace amounts are lignans (such as pinoresinol), flavonoids (mainly apigenin and luteolin), and hydroxyl-isochromans (Boskou, 2009) (Figure 8.4).

Beltram Maza and his collaborators (1996) attempted to study the protective role of phenolic compounds and α-tocopherol in olive oil degradation during frying of potatoes. Their approach was based on monitoring the concentration of these compounds by HPLC after each frying and for 25 consequent 5 min fryings at 190°C. From the degradation rate measured, the authors concluded that hydroxytyrosol and α-tocopherol make a significant contribution to the stability of the oil.

The concentration of several classes of compounds (simple and complex phenols, lignans, and phenolic acids) was determined in olive oil samples heated at 180°C (Brenes et al., 2002; Carrasco-Pancorbo et al., 2007). Hydroxytyrosol and some oleuropein aglycons were found to be more sensitive to heating in relation to other phenols, especially in relation to lignans and hydroxyl-tyrosol acetate.

These properties of olive oil have been taken into consideration in studies related to canning of fried fish. Cuesta and his collaborators (1996) and also Bastida et al. (1996) suggested that sardines be fried in olive oil, which is used for only a few fryings. In this way, the fat intake profile is improved; besides, nutritional benefits of

p-hydroxybenzoic acid *p*-hydroxyphenyl acetic acid syringic acid

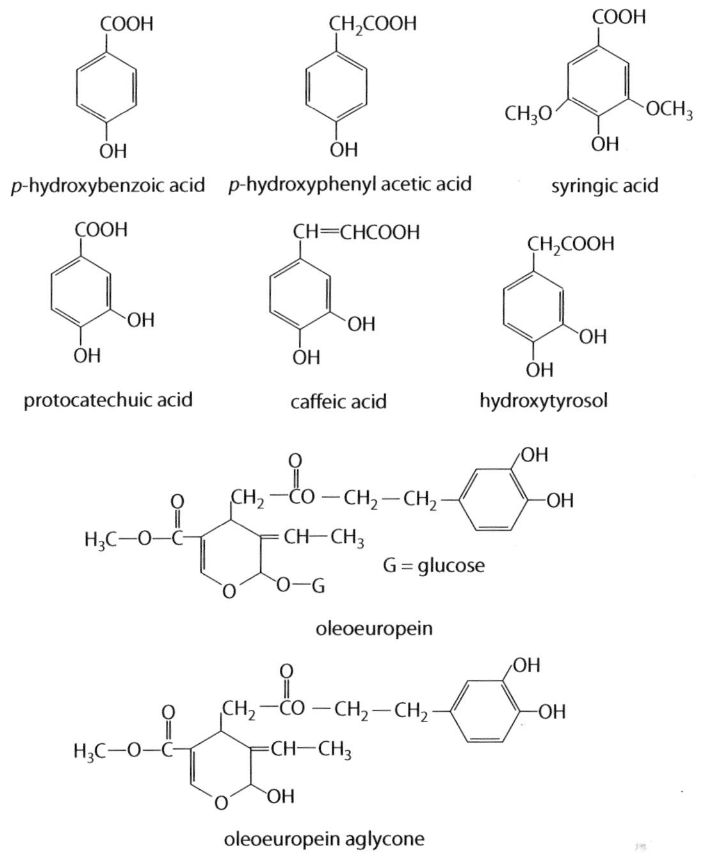

protocatechuic acid caffeic acid hydroxytyrosol

G = glucose

oleoeuropein

oleoeuropein aglycone

FIGURE 8.4 Antioxidant phenolic compounds reported to be present in olive oil. (From Boskou, 2009. In *Olive Oil, Minor Constituents and Health,* ed. D. Boskou, 12–36, Boca Raton, FL: CRC Press.)

olive oil and omega-3 fatty acids can be combined. This practice seems to be in line with prevailing theories concerning the nutritional role of olive oil and PUFA n-3 fatty acids, and their possible protective effect against various forms of cancer and cardiovascular diseases. (Simopoulos and Robinson, 1998).

Retention of antioxidant activity. Andrikopoulos et al. (2002) studied the rate of deterioration of natural antioxidants of virgin olive oils, sunflower oil, and a vegetable shortening during the domestic deep-frying and pan-frying of potatoes. The retention of total phenolic compounds ranged from 70% to 80% (first frying) to 20% to 30% (eighth frying). The retention of α-, β-, and γ-tocopherols during the eight fryings was from 85% to 90% (first frying) to 15% to 40% (eighth frying). Tocopherols of sunflower oil disappeared completely after the sixth frying operation.

Brenes et al. (2002), Carrasco-Pancorbo et al. (2007), and Allouche et al. (2007) studied the effect of heating on extra virgin olive oil quality indices and phenol composition. Tocopherols and polyphenols were the most affected classes of compounds. Hydroxytyrosol (3,4–dihydroxy-phenethyl alcohol; 3,4-DHPEA) and certain secoiridoid derivatives with elenolic acid were more sensitive in relation to lignans, oleic acid, sterols, squalene, and triterpenic alcohols; (erythrodiol and uvaol). Triterpenic acids (oleanolic and maslinic) were quite constant, exhibiting a high stability against oxidation during frying.

Losses of antioxidant activity of olive oil due to heating, measured by ABTS and DPPH radical decolorization, electron spin resonance, and other methods, were reported by Carlos-Espin et al. (2000), Pellegrini et al. (2001), Quiles et al. (2002), Gomez-Alonso et al. (2003), Valavanidis et al. (2004), and Kalantzakis et al. (2006). Virgin olive oil has a remarkable thermal stability but, on the other hand, it should not be seen only as a good frying medium. If health effects are expected from the phytochemicals present, the number of heating operations should be restricted to a minimum.

In recent research, Kalogeropoulos et al. (2007) used vegetables such as potatoes, green peppers, zucchinis, and eggplants for shallow frying in olive oil according to the Mediterranean traditional culinary practice, and measured the level of bioactive ingredients (alpha-tocopherol, phenolics, and triterpene acids). The overall retention of the antioxidants and triterpene acids (maslinic, oleanolic, ursolic) in the oil and the food ranged from 32% to 64% for alpha-tocopherol, 25% to 70% for phenolic compounds, and 35% to 83% for triterpene acids.

8.3.2 Other Stable Oils

S. P. Kochhar (2000a,b) explained the concept of a stable frying oil in his article "Stable and Healthful Frying Oil for the 21st Century." The idea is not to use hydrogenation to obtain high solid fat frying oil but a high oleic acid oil whose stability is further enriched by sesame seed, rice bran, and olive oil. These oils are rich in sterols such as delta-5-avenasterol and citrostadienol that are antioxidant at elevated temperatures. A range of other antioxidants present in these oils are tocopherols, oryzanol (sterol ferulate), sesamolin (a precursor of an antioxidant), sesaminol, sesamol dimers, and others. Boskou., 1999, Blekas and Boskou., 1999, Nasirullah and Rangaswamy, 2005, Sharif et al. (2009) indicated that bene hull oil, a species of pistachio growing wild in Iran, has a strong antioxidant activity higher than those of sesame seed and rice bran oil during the frying process.

Fractions obtained by solvents of varying polarity from nontraditional tree seed oils (sal, mango, and mahua) were also shown to increase the stability when incorporated into base frying oils such as refined soybean and rice bran oils (Nasirullah, 2001). In a study published in 2006 Warner and Dunlap evaluated the stability of expeller-pressed soybean oil used in potato frying. This oil had a slightly higher content of tocopherols and phytosterols compared to the refined oil.

The reduction in the antiradical power of soybean oil during frying was determined using the 2,2-diphenyl-1-picrylhydrazyl radical (DPPH) (Van Loon et al., 2006). Kalantzakis et al. (2006) studied the effect of heating on the antioxidant activity of virgin olive oil, refined olive oil, and other vegetable oils by measuring the radical scavenging activity toward the 1,1-diphenyl-2-picrylhydrazyl radical

(DPPH). It was observed that olive oil lost its radical scavenging activity at a shorter time of heating in relation to other vegetable oils much richer in tocopherols, but it reached the level of 25% total polar content (rejection point for a heated fat) after prolonged heating; all the other oils reached this upper limit in shorter time periods.

8.4 ANTIOXIDANTS AND ACRYLAMIDE FORMATION

Acrylamide has been classified as a possible carcinogen by various scientific societies and agencies. It is formed through Maillard reactions between asparagines and reducing sugars in many starch-rich foods. Particularly high levels of acrylamide have been found in potato crisps, potato chips, and also crisp bread, breakfast cereals, various bakery products, and coffee. Oxidized vegetable oils seem to promote acrylamide formation. Oxidative processes, free radical intermediates, and the effect of natural antioxidants have been studied by Hadegaard et al. (2008). The presence of antioxidants such as ascorbic acid, ferulic acid, and others reduces the amount of acrylamide in potato chips. Napolitano et al. (2008) demonstrated that virgin olive oil with a high content of *o*-diphenols is able to efficiently inhibit acrylamide formation in crisps in domestic deep-frying for the preparation of potato crisps.

Commercial rosemary preparations are sold to reduce acrylamide in fried food.

Other strategies that aim at reducing acrylamide formation in potato products have also been proposed. These are related to the selection of potato varieties, storage conditions, and time and temperature of frying. Mitigation strategies proposed include the addition of enzymes The industry's efforts now focus on the use of some acrylamide-reducing enzymes (asparigenases) that convert the precursor asparagine into other amino acids that do not form acrylamide (Konings et al., 2007).

8.5 CONCLUSION

Frying processes utilizing cooking oil at elevated temperatures can cause various degradation effects in the oil, including oxidation, hydrolysis, and/or polymerization. The oxidation products from heated fats used for frying foodstuffs are probably not detrimental to the metabolic processes in the human body at the concentrations normally encountered in food. However, it is difficult to fully study all the changes that take place in the fatty acids of fats and oils during the frying process. It is therefore important to have methods and compositions that provide beneficial and cost-effective improvements in the cooking performance of oil used at high temperatures.

The approach to preparing stable frying oils by using mixtures rich in natural antioxidants combines two functions of phytochemicals: increase in resistance to oxidative deterioration and medicinal effects. Thus, the method, apart from being stabilizing, similar to the traditional hydrogenation or fatty acid modification by breeding, may help food manufacturers and oil processors prepare good-quality products that may also provide health benefits. For the moment, the whole approach for the enhancement of the stability of frying oils, based on the beneficial effect of minor constituents including phytochemicals, appears to be a successful alternative to existing methods of stabilization. It is, however, necessary to understand better the role of natural antioxidants present in oils that are used for frying and to identify new sources of novel antioxidants. Thermal decomposition of antioxidant phenols must also be thoroughly investigated to check interactions with other food components or the possible formation of compounds of toxicological concern.

REFERENCES

Abdalla, A.E.M. 1999. Antioxidative effect of olive oil deodorizer distillate on frying oil and quality of potato chips, *Fett/Lipid*, 101: 57–63.

Aggelousis, G. and Lalas, S. 1997. Quality changes of selected vegetable oils during frying of doughnuts, *Riv. Ital. Sostanze Grasse*, 74: 559–562.

Akoh, C.C. 1994. Oxidative stability of fat substitutes and vegetable oils by the oxidative stability index method, *J. Am. Oil Chem. Soc.*, 71: 211–216.

Allouche, Y., Jimenez, A., Gaforio, J.J., Uceda, M., and Beltran, G. 2007. How heating affects extra virgin olive oil quality indexes and chemical composition, *J. Agric. Food Chem.*, 55: 9646–9654.

Andrikopoulos, N.K., Tzamtzis, V.A., Giannopoulos, G.A., Kalantzopoulos, G.K., and Demopoulos, C.A. 1989. Deterioration of some vegetable oils. I: During heating or frying of several foods, *Rev. Franc. Corps Gras.*, 36: 127–129.

Andrikopoulos, N., Dedousis, G., Falirea, A., Kalogeropoulos, N., and Hatzinikola, H. 2002. Deterioration of natural antioxidants of vegetable edible oils during the domestic deep-frying and pan frying of potatoes, *Int. J. Food Sci. Technol.*, 53: 351–363.

Asap, T. and Augustin, M.A., 1986. Effect of TBHQ on quality characteristics of RBD olein during frying, *J. Am. Oil Chem. Soc.*, 63: 1169–1172.

Augustin, M.A. and Berry, S.K. 1983. Efficacy of the antioxidants BHA and BHT in palm olein during frying, *J. Am. Oil Chem. Soc.*, 63: 1109–1172.

Bankova, V., Christov, R., Stoev, G., and Popov, S. 1992. Determination of phenolics from propolis by capillary gas chromatography, *J. Chromatogr.*, 607: 150–153.

Bankova, V., Nikolova, N., and Marcucci, M. 1996. A new lignan from Brazilian propolis, *Naturforsch*, 51b: 735–737.

Bastida, S., Sanchez-Muniz, F.J., Cava, F., Viejo, J.M., and Marcos, A., 1996. Benefits of the consumption of olive oil-fried sardines in the prevention of hypercholesterolemia in rats. Effects on some serum. lipids and cell-damage marker enzymes, *2nd International Symposium on Frying of Food,* Abstracts of posters, Fundacion Espanola de la Nutrition, Madrid, 1998, *Grassas y Aceites*, 49: 370.

Beltran Maza, G., Jimenez Marquez, A., Garcia Mesa, J.A., and Friaz Ruiz, L. 1996. Evolution of extra virgin olive oil natural antioxidants during continuous frying, *2nd International Symposium on Frying of Food, Abstracts of Posters, Fundacion Espanola de la Nutrition*, Madrid, 1998, *Grassas y Aceites*, 49: 372.

Bensmira, M., Jiang, B., Nsabiurma, C., and Jian, T. 2007. Effect of lavender and thyme incorporation in sunflower oil on its resistance to frying temperatures, *Food Res. Intern.,* 40: 341–346.

Berger, K.G. 2005. The use of palm oil in frying, *Frying Oil Series, Malaysia Palm Oil Promotion Council,* Publication 665.3.

Bertelsen, G., Christophersen, C., Nielsen P.H., Madsen, H.L., and Stadel, P. 1995. Chromatographic isolation of antioxidants guided by a methyl linoleate assay, *J. Agric. Food Chem.,* 43: 1272–1275.

Blekas, G. and Boskou, D. 1986. Effect of esterified sterols on the stability of heated oils, In: *The Shelf Life of Food and Beverages*, ed. G. Charalambous, 641–645, Amsterdam: Elsevier.

Blekas, G. and Boskou, D. 1999. Phytosterols and stability of frying oils. In: *Frying of Food,* eds. Boskou, D. and Elmadfa, I., 205–222, Lancaster, PA: Technomic Publishing Co.

Boskou, D. 1999. Non nutrient antioxidants and stability of frying oils. In: *Frying of Food*, eds. Boskou, D. and Elmadfa, I., 183–204, Lancaster, PA: Technomic Publishing Co.

Boskou, D. 2006. Sources of natural antioxidant phenols, *Trends Food Sci. Technol.*, 17: 505–512.

Boskou, D. 2009. Phenolic compounds in olives and olive oil. In: *Olive Oil, Minor Constituents and Health,* ed. D. Boskou, 12–36, Boca Raton, FL: CRC Press.

Boskou, D. and Morton, I.D. 1976. Effect of plant sterols on the rate of deterioration of heated oils, *J. Sci. Food Agric.*, 27: 928–932.

Boskou, D. and Katsikas, H. 1979. Effect of olive oil hydrocarbons and triterpene alcohols on the stability of heated cotton seed oil, *Acta Alimentaria*, 8: 317–320.

Boskou, D., Exarchou, V., and Kefalas, P. 2006. Antioxidant phenols in aromatic plants. In *Natural Antioxidant Phenols,* eds. Boskou,D., Gerothanassis, I.P., and Kefalas, P., 157–174, Kerala, India: Research Signpost.

Bracco, U., Loliger, J., and Viret, J.L. 1981. Production and use of natural antioxidants, *J. Am. Oil Chem. Soc.*, 58: 686–690.

Brenes, M., Garcia, A., and Dobarganes, C. 2002. Influence of thermal treatment simulating cooking processes on the polyphenol content in virgin olive oil, *J. Agric. Food Chem.*, 50: 5962–5067.

Carlos-Espin, J., Soler-Rivers, C., and Wichers, J., 2000. Characterization of the total free radical scavenging capacity of vegetable oils and oil fractions using 2,2-diphenyl-1-picrylhydrazyl radical, *J. Agric. Food Chem.*, 48: 648–656.

Carrasco-Pancorbo, A., Cerretani, L., Bendini, A., Segura-Carretero, A., Lercker, G., and Fernandez-Gutierrez, A. 2007. Evaluation of the influence of thermal oxidation on the phenolic composition and on the antioxidant activity of extra-virgin olive oils, *J. Agric. Food Chem.*, 55: 4771–4780.

Chang, S.S. and Morton, P. E. 1969. *Breakdown Inhibitors in Food Processing*, *Review No 5.* ed. Rednarcyk, 209–211., Park Ridge, NJ: Noyes Development Corporation.

Chang, S.S., Peterson, R.J., and Ho, C.T. 1978. Chemical reactions involved in deep fat frying of food, *J. Am. Oil Chem. Soc.*, 55: 718–723.

Che Man, Y. and Tan, C.P. 1999. Effects of natural and synthetic antioxidants on changes in refined, bleached and deodorized palm olein during deep-fat frying of potato chips, *J. Am. Oil Chem. Soc.*, 76: 331–330.

Che Man, Y.B. and Jaswir, I. 2000. Effect of rosemary and sage extracts on frying performance of refined bleached and deodorized palm olein during deep-fat frying, *Food Chem.*, 69: 301–307.

Chen, J.H. and Ho, C.-T. 1997. Antioxidant activities of caffeic acid and its related hydroxy-cinnamic acid compounds, *J. Agric. Food Chem.*, 45: 2374–2378.

Chen, Q, Shi, H., and Ho, C-T. 1992. Effects of rosemary extracts and major constituents on lipid oxidation and soybean lipoxygenase activity, *J. Am. Oil Chem. Soc.*, 69: 999–1002.

Chiou, A., Karpathiou, V., Ghioxari, A., and Andrikopoulos, N.K. 2008. Oleuropein in french fries during the successive deep-frying in sunflower oil enriched with olive leaf extract, *6th Eurofed Lipid Congress*, Athens 2008, *Book of Abstracts*, FRY-01.

Chipault, J.R., Mizuno, G.R., Hawkins, J.M., and Lundberg, W.O. 1952. The antioxidant properties of natural spices, *Food Res.*, 17: 46–55.

Choe, E. and Min, D.B. 2007. Chemistry of deep frying, *J. Food Sci.*, 72: R77–R86.

Chu, Yan-Hwa 1991. A comparative study of analytical methods for evaluation of soybean oil quality, *J. Am. Oil Chem. Soc.*, 68: 379–383.

Chun, J.H., Kim, S.K., and Lee, E.H. 1984. Effect of BHA, TBHQ, silicone and citric acid on the thermal deterioration of soybean oil, *Pusan Susan Taehok, Yongu Pogo Chayon Kwahak*, 24: 113–120, *Chem. Abstr.*, 1985, 104: 86720.

Criado, M., Hernandez-Martin, E., Lopez-Hernandez, A., and Otero, C. 2008. Enzymatic interesterification of olive oil with fully hydrogenated palm oil: Characterization of fats, *Eur. J. Lipid Sci. Technol.*, 110: 714–724.

Cuesta, C., Sanchez-Muniz, F.J., and Varela, G. 1988. Nutritive value of frying fats. In: *Frying of Food, Principles, Changes, New Appro*aches, eds. G. Varela, A.E. Bender, and Morton I.D., 112–129. Chichester, England: Ellis Harwood.

Cuesta, I., Perez, M., Ruiz-Roso, B., and Varela, G. 1996. Comparative study of the effect on the sardine fatty acids composition during deep-frying and canning in olive oil, *2nd International Symposium on Frying of Food,* Abstracts of posters, Fundacion Espanola de la Nutrition, Madrid, 1998, *Grassas y Aceites*, 49: 371.

Cuppett, S., Schnepf, M., and Hall, C. III. 1997. Natural antioxidants—are they a reality. In: *Natural Antioxidants, Chemistry, Health Effects and Applications,* ed. F. Shahidi, 12–23, Champaign, IL: AOCS Press.

Diplock, A.T. 1992. The role of antioxidant nutrients in disease, health and nutrition, *Inform*, 3: 1214–1217.

Djarmati, Z., Jankov, R.M., and Schwirtlich, E. 1991. High antioxidant activity of extracts obtained from sage by supercritical CO_2 extraction, *J. Am. Oil Chem. Soc.*, 68: 731–734.

Dobarganes, M.C. and Marquez-Ruiz, G., 1996. Dimeric and higher oligomeric triglycerides. In: *Deep Frying*, eds. E.G. Perkins and M.D. Erickson, 89–111, Champaign, IL: AOCS Press.

Economou, K.D., Oreopoulos, V., and Thomopoulos, C.D. 1991. Antioxidant properties of some plant extracts of the Labiatae family, *J. Am. Oil Chem. Soc.*, 68: 109–113.

El Anany Ayman, M. 2007. Influence of pomegranate (*Punica granatum*) peel extract on the stability of sunflower oil during deep-fat frying process, *Elect. J Food Plants Chem.,* 2: 14–19.

Exarchou, V., Nenadis, N., Tsimidou, M., Gerothanasis, I.P., Troganis, A., and Boskou, D. 2002. Antioxidant activities and phenolic composition of extracts from Greek oregano, Greek sage and summer savory, *J. Agric. Food Chem.*, 50: 5294–5299.

Exarchou, V., Godejohann, M., Van Beek, T.A., Gerothanasis, I.P., and Vervoort, J., 2003. LC-UC-solid-phase extraction-NMR-Ms combined with a cryogenic flow probe and its application to the identification of compounds present in Greek oregano, *Anal. Chem.*, 75: 6288–6294.

Farag, R.S., Badei, A.Z.M.A., Hewedi, F.M., and El-Baroty, G.S.A. 1989. Antioxidant activity of some spice essential oils on linoleic acid oxidation in aqueous media, *J. Am. Oil Chem. Soc.*, 66: 792–799.

Farag, R.S., Mahmoud, E.A., and Basuny, A.M. 2006. Use of olive leaf juice as a natural antioxidant for the stability of sunflower oil during heating, *Int. J. Food Sci. Technol.*, 42: 107–115.

Farhoosh, R., Kenari, E., and Poorazrang, H. 2009. Frying stability of conola oil blend with palm olein, olive oil and corn oils, *J. Am. Oil Chem. Soc.,* 86: 71–76.

Frankel, E.N., Warner, K., and Moulton, S.R. 1985. Effect of hydrogenation and additives on cooking performance of soybean oil, *J. Am. Oil Chem. Soc.*, 62: 1354–1358.

Freeman, I.P., Padley, F.B., and Sheppard, W.L. 1973. Use of silicones in frying oils, *J. Am. Oil Chem. Soc.,* 50: 101–103.

Fukuda, Y., Nagata, M., Osawa, T., and Namiki, M. 1986a. Chemical aspects of the antioxidative activity of roasted sesame seed oil and the effect of using the oil for frying, *Agric. Biol. Chem.*, 50: 857–862.

Fukuda, Y, Nagata, M., Osawa, T., and Namiki, M. 1986b. Contribution of lignan analogues to antioxidative activity of refined unroasted sesame seed oil, *J. Am. Oil Chem. Soc.*, 63: 1027–1031.

Fukuda, Y., Osawa, I., Kawakishi, S., and Namiki, M. 1994. Chemistry of lignan antioxidants in sesame seed oil. In: *Food Phytochemicals for Cancer Prevention*, eds. Ho Chi-Tang, T. Osawa, Mon-Tuan Huang, and R.T. Rosen, 264–273, Washington, DC: American Chemical Society.

Garcia Mesa, J.A., Jimenez Marquez, A., Beltran Maza, G., and Friuz Ruiz, L. 1996. Thermoxidation of different vegetable oils used in deep frying of potatoes. Abstract of posters, *2nd International Symposium on Frying of Food*, Fundacion Espanola de la Nutrition, Madrid, 1998, *Grassas Aceites*, 49: 375.

Gomez-Alonso, S., Fregapane, G., and Salvador Desamparados, M., 2003, Changes in phenolic composition and antioxidant activity of virgin olive oil during frying, *J. Agric. Food Chem.*, 51: 667–672.

Gordon, H.M. and Kourimska, L.C. 1995. On losses of tocopherols during deep-fat frying, *Food Chem.*, 52: 175–177.

Gordon, M.H. and Magos, P. 1983. The effect of sterols on the oxidation of edible oils, *Food Chem.*, 10: 141–147.

Govind Rao, M.K. and Achaya, K.T. 1968. Antioxidant activity of squalene, *J. Am. Oil Chem. Soc.*, 45: 296.

Guhr, G., Gerz, C., Waibel, J., and Arens, H. 1981. Bestimmung der Polaren Anteile in Fritierfetten, *Fetten Seifen Anstrichm.*, 83: 373–376.

Gwo, Y.Y., Flick, G.J., Dupuy, H.P., Ory, R.L., and Barn, W.L. 1985. Effect of ascorbyl palmitate on the quality of frying fats for deep frying operations, *J. Am. Oil Chem. Soc.*, 62: 1666–1671.

Hadegaard, R.V., Granby, K., Frandsen, H., Thygensen, J., and Skibsed, L.H. 2008. Acrylamide in bread. Effect of prooxidants and antioxidants, *Eur. Food Res. Technol.*, 227: 519–525.

Haumann, B.F. 1996. Frying fats, *Inform*, 7: 321–334.

Hongxia Du and Hongjun Li, 2008, Antioxidant effect of Cassia essential oil on deep-fried beef during the frying process., *Meat Sci.*, 78: 641–648.

Houlihan, C.M., Ho, C.-T., and Chang, S.S. 1984. Elucidation of the chemical structure of a novel antioxidant, rosmaridiphenol, isolated from rosemary, *J. Am. Oil Chem. Soc.*, 61: 1036–1039.

Houhoula, D.P., Oreopoulou, V., and Tzia, C. 2003,Antioxidant efficiency during frying and storage of potato chips, *J. Sci. Food Agric.*, 83: 1499–1503.

Inatami, R., Nakatami, N., and Fuwa, H.1983. Antioxidant effect of the constituents of rosemary and their derivatives, *Agric. Biol. Chem.*, 47: 521–528.

Jaswir, I., Che Man, Y.B., and Kitts, D. 2000. Use of natural antioxidants in refined palm olein during repeated deep-fat frying, *Food Res. Int.*, 33: 501–508.

Jaswir, I., Che Man, B., and Kitts, D.D. 2000a. Effect of natural antioxidants in controlling alkaline contaminant material (ACM) in heated palm olein, *Food Res. Int.*, 33: 75–81.

Jaswir, I., Kitts, D.D., Che Man, Y., and Hassan, T. 2006. Synergistic effect of rosemary, sage and citric acid on fatty acid retention of palm olein during deep frying, *J. Am. Oil Chem. Soc.*, 77: 527–533.

Kajimoto, G., Yoshida, H., and Shibanara, H. 1985. A role of tocopherol on the heat stability of vegetable oils, *Nippon Eiyo, Shokuryo Gakkaishi*, 38: 301–307, C.A., 1985, 103: 159318b.

Kalantzakis, G. and Blekas, G. 2006. Effect of Greek sage and summer savory extracts on vegetable oil thermal stability, *Eur. J. Lipid Sci. Technol.*, 108: 842–847.

Kalantzakis, G., Blekas, G., Peglidou, K., and Boskou, D. 2006, Stability and radical-scavenging activity of heated olive oil and other vegetable oils, *Eur. J. Lipid Sci. Technol.*, 108: 329–335.

Kalogeropoulos, N., Mylona, A., Chiou, A., Ioannou, M., and Andrikopoulos, N. 2007. Retention of natural antioxidants after shallow frying of vegetables in olive oil, *Lebemsm-Wissensch. Technol.*, 40: 1008–1017.

Kamal-Eldin, A. and Appelqvist, L.A. 1994. Variations in the composition of sterols, tocopherols and lignans in seed oils from four Sesamum species, *J. Am. Oil Chem. Soc.*, 71: 149–156.

Kikugawa, K., Arai, M., and Kurechi, T. 1983. Participation of sesamol in stability of sesame oil, *J. Am. Oil Chem. Soc.*, 60: 1528–1533.

Kim, M. and Choe, E. 2004. Effects of burdock (*Arctium lappa* L) extracts on autoxidation and thermal oxidation of lard, *Food Sci. Biotechnol.*, 13: 460–466.

Kimura, Y. and Kanamori, T. 1982. Method of frying foods in the presence of a spice antioxidant, U.S. Patent US 4 363 823, *FSTA*, 1983, 9T: 536.

Kochhar, S.P. 2000a. Stable and healthful frying oils for the 21st century, *Inform*, 11: 642–647.

Kochhar, S.P., 2000b. Stabilisation of frying oils with natural antioxidative components, *Eur. J. Lipid Sci. Technol.*, 102: 552–559.

Konings, E.M., Ashby, P., Hamlet, C.G., and Thompson, G.A.K. 2007. Acrylamide in cereal and cereal products: A review on progress in level reduction, *Food Addit. Contam.*, 24 Suppl.: 47–59.

Kourimska, Z., Reblova, J., and Pokorny, J. 1993. Stabilization of dietetic oils containing phospholipids against oxidative rancidity. In: *Phospholipids: Characterization, Metabolism and Novel Biological Applications*, eds. G. Leve and R. Paltauf, 372–377, Champaign, IL: AOCS Press.

Kourimska, L., Pokorny J., and Reblova, Z. 1995. Phospholipids as inhibitors of oxidation during food storage and frying, *Prehrambeno-Technol. Biotechnol. Rev.*, 32: 91–94.

Kroon, P. and Williamson, G. 2005. Polyphenols: Dietary components with established benefits to health? *J. Sci. Food Agric.*, 85:1239–1240.

Kumpulainen, J.T. and Salonen, J.T. 1996. Natural antioxidants and food quality, In: *Atherosclerosis and Cancer Prevention*, The Royal Society of Chemistry, Cambridge, U.K.: Thomas Graham House.

Kusaka, H.S., Nagano, S., and Ohta, S. 1985. On functions of silicone oil in frying oil. Influence of silicone oil on convection of frying oil, *Yukagaku*, 34: 187–191, *Chem. Abstracts*, 102: 165461.

Lagouri, V. and Boskou, D. 1995. Screening for antioxidant activity of essential oils obtained from spices. In: *Food Flavors: Generation, Analysis and Process Influence*, ed. G. Charalambous, 869–879, amsterdam: Elsevier Science.

Langseth, L. 1995. *Oxidants, Antioxidants and Disease Prevention*, 1–24, *Ilsi Europe* Press.

Lee, J., Lee, S., Lee, H., Park, K., and Xhoe, E. 2002. Spinach (*Spinacea oleracea*) powder as a natural food-grade antioxidant in deep-fat fried products, *J. Agric. Food Chem.*, 50: 5664–5669.

Li, J., Wang, T., Wu, H., Ho, C.T., and Weng, X. 2006. 1,1-Di-(2′,5′–dihydroxy-4′-*tert*-butylphenyl)ethane: A novel antioxidant, *J. Food Lipids*,13: 331–340.

Loliger, L. and Wille, J. 1993. Natural antioxidants, *Oils Fats Int.* 9: 18–22.

Malecka, M. 2006. The effect of squalene on the thermostability of rapeseed oil, *Food/Nahrung*, 38: 135–140.

Manzi, P., Panfili, G., Esti, M., and Pizzoferrato, L. 1998. Natural antioxidants in the unsaponifiable fraction of virgin olive oils from different cultivars, *J. Sci. Food Agric.*, 77: 115–120.

Marmesat, S., Velasco, L., Ruiz-Mendez, M.R., Fernandez-Martinez, J.M., and Dobarganes, M.C. 2008. Thermostability of genetically modified sunflower oil differing in fatty acid and tocopherol compositions, *Eur. J. Lipid Sci. Technol.*, 110: 776–782.

Mohamed, H.M.A. and Awatif, I.I. 1998. The use of sesame oil unsaponifiable matter as a natural antioxidant, *Food Chem.*, 62: 269–276.

Nakatami, N. 1994. Antioxidative and antimicrobial constituents of herbs and spices. In: *Spices, Herbs, and Edible Fungi*, ed. G. Charalambous, 251–284, Amsterdam, B.V.: Elsevier Science.

Nakatami, N. 1997. Antioxidants from spices and herbs. In: *Natural Antioxidants, Chemistry, Health Effects and Applications*, ed. F. Shahidi, 64–75, Champaign, IL: AOCS Press.

Namiki, M. 1990. Antioxidants, antimutagens in food, *Crit. Rev. Food Sci. Nutr.*, 29: 273–300.

Napolitano, A., Morales, F., Sacchi, R., and Fogliano, V. 2008, Relationship between virgin olive oil phenolic compounds and acrylamide formation in fried crisps, *J. Agric. Food Chem.*, 56: 2034–2040.

Nasirullah, 2001, Development of deep frying edible vegetable oils, *J. Food Lip.*, 8: 295–304.

Nasirullah, N. and Rangaswamy, B.L., 2005, Oxidative stability of healthful frying medium and uptake of inherent nutraceuticals during deep frying, *J. Am. Oil Chem. Soc.*, 82:753–757.

Nogala-Kalucka, M., Korczak, J., Elmadfa, I., and Wagner, K-H., 2005. Effect of alpha-and delta-tocopherol on the oxidative stability of a mixed hydrogenated fat under frying conditions *Eur. J. Food Sci. Technol.*, 221: 292–297.

Ohta, S. and Kusaka, H. 1979. The function of silicone oil in frying oil, *J. Am. Oil Chem. Soc.*, 56: 202A.

Palozza, P. and Krinsky, N.I. 1992. β-Carotene and α-tocopherol are synergistic antioxidants, *Arch. Biochem. Biophys.*, 297: 184–187.

Papadopoulos, G. and Boskou, D. 1991. Antioxidant effect of natural phenols on olive oil, *J. Am. Oil Chem. Soc.*, 68: 669–672.

Pazola, Z., Buchowski, M., Korczak, J., and Gzeskowiak, B. 1987. Einflub Einiger Antioxidantien auf die Fettstabilisierung Wahrend des Fritierens und der Lagerung von Fritierten Kartoffelprodukten. I. Kartoffelkrispies, *Ernahrung*, 11: 481–487.

Peled, M., Gutfinger, T., and Letan, A. 1975. Effect of water and BHT on stability of cottonseed oil during frying, *J. Sci. Food Agric.*, 26: 1655–1666.

Pellegrini, N., Visioli, F., Burrati, S., and .Brigheti, F. 2001. Direct analysis of total antioxidant activity of olive oil and studies on the influence of heating., *J. Agric. Food Chem.,* 49: 2532–2538.

Perkins, E.G. 1996. Volatile odor and flavor components formed in deep frying, In: *Deep Frying*, eds. E.G. Perkins and M.D. Erickson, 43–48. Champaign, IL: AOCS Press.

Pokorny, J., Reblova, Z., Ranny, M., Kanova, J., Panek, J., and Davidek, J. 1992. Natural lecithins and phosphorylated acylglycerols as inhibitors of autoxidation of fats and oils, *Nahrung*, 36: 461–465.

Pratt, D.E. and Hudson, P.J.F. 1990. Natural antioxidants not exploited commercially. In: *Food Antioxidants*, ed. B.J.F. Hudson, 171–191. London: Elsevier.

Psomiadou, E., Panagiotopoulou, P.M., and Tsimidou, M. 1998. Squalene and olive oil stability, *1st International Conference of the Chemical Societies of the SouthEast European Countries*, Halkidiki, Greece, Abstract p. PO 426.

Quiles, L.J., Ramirez-Tortoza, M.C., Gomez, J.A, Huertas, R.J., and Mataix, J. 2002. Role of vitamin E and phenolic compounds in the antioxidant capacity measured by ESR of virgin olive oil and sunflower oils after frying., *Food Chem.*, 76: 461–468.

Ramalho, V.C. and Jorge, N. 2008. Antioxidant action of rosemary extract in soybean oil submitted to thermoxidation, *Grasas Aceites*, 59: 128–131.

Rhee, J.S. 1978. Effect of methyl silicone, TBHQ and BHA/BHT on frying and storage stabilities of the vegetable salad oil in high density polyethylene bottles, *Korean J. Food Sci. Technol.*, 10: 250–257.

Rhee, K.S., Housson, S.E., and Ziprin, Y.A. 1992. Enhancement of frying oil stability by a natural antioxidant ingredient in the coating system of fried meat nuggets, *J. Food Sci.*, 57: 789–791.

Ringseis, R. and Eder, K. 2008. Effects of dietary oxidized fats on gene expression in mammals, *Inform*, 19,657–659.

Reschke, A. 1983. Capillargaschromatographische Bestimmung der Rosmarinsaűre in Blattgewűrzen, *Z. Lebensm. Unters Forsch*, 176: 116–119.

Romero, A., Cuesta, C., and Sanchez-Muniz, F.J. 1996. Behaviour of extra virgin olive oil in potato frying. Thermoxidative alteration of the fat content in the fried product, Abstract of posters, *2nd International Symposium on Frying of Food, Fundacion Espanola de la Nutrition*, Madrid, 1998, *Grassas Aceites*, 49: 370.

Romero, A., Cuesta, C., and Sanchez-Muniz, F.J., 1995. Quantitation and distribution of polar compounds in an extra virgin olive oil used in frying with turnover of fresh oil, *Fat Sci. Technol.*, 97: 403–407.

Rossell, J.B. 1996. Industrial frying process, *2nd International Symposium on Frying of Food*, 135, Fundacion Espanola de la Nutrition, Madrid.

Scalbert, A., Manach, C., Morand, C., and Jimenez, L. 2005. Dietary polyphenols and the prevention of disease, *Crit. Rev. Food Sci. Nutr.*, 45: 287–306.

Shahidi, F. 1997. *Natural Antioxidants, Chemistry, Health Effects and Applications*, Champaign, IL: AOCS Press.

Sharif, A., Farhoosh, R., Khodaparast, M.H.H., and Kafrani, M.H.T. 2009. Antioxidant activity of Bene hull oil compared with sesame and rice bran oils during the frying process of sunflower oil, *J. Food Lip.,* 16: 394–406.

Sherwin, E.R., Luckado, B.M., and Freeman, G.J. 1974. Antioxidant compositions, defensive publications, T. 918 003, *J. Am. Oil Chem. Soc.*, 51: 364A.

Shyamala, B.N., Gupta, S., Lakshmi, A.J., and Prakash, J., 2005. Leafy vegetable extracts-antioxidant activity and effect on storage stability of heated oils, *Innov. Food Sci. Emerg. Technol.*, 6: 239–245.

Shukla, V.K.S., Wanasundara, P.K.J.P.D., and Shahidi, F. 1997. Natural antioxidants from oilseeds. In: *Natural Antioxidants, Chemistry, Health Effects and Applications*, ed. F. Shahidi, 97–132, Champaign, IL: AOCS Press.

Simopoulos, A.P., and Robinson, J. 1998. *The Omega Plan*, New York: Harper Collins Publishers.

Sims, R.J., Fioriti, J.A., and Kanuk, M.J. 1972. Sterol additives as polymerization inhibitors for frying oils, *J. Am. Oil Chem. Soc.*, 49: 298–301.

Stanner, S.A., Hughes, J., Kelly, C.N., and Butriss, J. 2004. A review of epidemiological evidence for the antioxidant hypothesis, *Public Health Nutr.*, 7: 407–422.

Tian, L.L. and White, P.J. 1994. Antipolymerization activity of oat extract in soybean and cotton seed oils under frying conditions, *J. Am. Oil Chem. Soc.*, 71: 1087–1090.

Tsimidou, M., Papavergou, E., and Boskou, D. 1995. Evaluation of oregano antioxidant activity in mackerel oil, *Food Res. Int.*, 28: 431–433.

Tyagi, V.K. and Vasishtha, A.K. 1996. Changes in the characteristics and composition of oils during deep-fat frying, *J. Am. Oil Chem. Soc.,* 73: 499–506.

Valavanidis, A.C., Nisiotou, C., Papageorgiou, Y., Kremli, I., Satravelas, N., and Zinieris, N. 2004. Comparison of the radical scavenging potential of polar and lipidic fractions of olive oil and other vegetable oils under normal conditions and after thermal treatment, *J. Agric. Food Chem.*, 52: 2358–2365.

Van Loon, W.A.M., Linssen, J.P.H., Leader, A., and Voragen, G.J., 2006. Anti-radical power gives insight into lipid oxidation events during frying, *J. Sci. Food. Agric.*, 86: 1446–1451.

Vaisey-Genser, M. and Ylimaki, G. 1985. Effect of a nonabsorbable antioxidants on canola oil stability to accelerated storage and to frying temperatures, *Can. Inst. Food Sci. Technol. J.*, 18: 67–71.

Vekiari, S.A., Oreopoulou, V., Tzia, C., and Thomopoulos, C.D., 1993. Oregano flavonoids as lipid antioxidants, *J. Am. Oil Chem. Soc.*, 70: 483–487.

Warner, K., Mounts, T.L., and Kwolek, W.F. 1985. Effects of antioxidants, methyl silicone and hydrogenation on room odor of soybean cooking oils, *J. Am. Oil Chem. Soc.*, 62: 1483–1486.

Warner, K.A. and Laszlo, J.A. 2005. Addition of ferulic acid, ferulate, and feruloylated monoacyl- and diacylglycerols to salad and frying oils, *J. Am. Oil Chem. Soc.*, 82: 647–652.

Warner, K. and Dunlap, C. 2006. Effects of expeller-pressed physically refined soybean oil on frying oil stability and flavor of french-fried potatoes, *J. Am. Oil Chem. Soc.,* 83: 435–441.

Warner, K. 2008. Mid-oleic/ultra low linolenic acid soybean oil—a healthful new alternative to hydrogenated oils for frying, *J. Am. Oil Chem. Soc.,* 85: 945–951.

Warner, K. and Gehring, M.M. 2009. High-temperature natural antioxidant improves soy oil for frying, *J. Food. Sci.*, 74: C500–5.

White, P.J. and Armstrong, L.S. 1986. Effect of selected oat sterols on the deterioration of heated soybean oil, *J. Am. Oil Chem. Soc.*, 63: 525–529.

White, P.J. and Xing, Y. 1997. Antioxidants from cereals and legumes. In: *Natural Antioxidants, Chemistry, Health Effects and Applications*, ed. F. Shahidi, 25–63, Champaign, IL: AOCS Press.

Yanishlieva, N., Stefanov, K., Marinova, E., and Seizova, K. 1988. Composition and oxidation changes in sunflower seed lipids, *Bulgarian Acad. Sci.*, 21: 112–119.

Yanishlieva, N.V. and Marinova, E., 1995. Antioxidant activity of selected species of the family Lamiaceae grown in Bulgaria, *Die Nahrung*, 34: 458–463.

Yanishlieva, N.V., Marinova, E.M., Marekov, I.N., and Gordon, M.H. 1997. Effect of an ethanol extract from summer savory (*Satureja hortensis*) on the stability of sunflower oil at frying temperatures, *J. Sci. Food Agric.*, 74: 524–530.

Yuki, E., Morimoto, K., Ishikawa, Y., and Noguchi, H. 1978. Inhibition effect of lecithin on the thermal oxidation of tocopherols, *J. Japan Oil Chem. Soc.*, 27: 425–430.

Zandi, P. and Gordon, M.H., 1999, Antioxidant activity of extracts from old tea leaves, *Food Chem.*, 64: 285–288.

9 Phytosterols and Frying Oils

Georgios Blekas and Dimitrios Boskou

CONTENTS

9.1 INTRODUCTION

Stabilization of cooking oils with natural antioxidants has not been extensively studied, and industrial applications are limited. The technique, which still serves industry in the production of frying oils, is mainly modification of fatty acid composition and addition of silicone oil and/or synthetic antioxidants.

In the last 20 years, consideration of the significance of certain minor constituents of oils and fats for the stabilization of heated oils has contributed to the conclusion that natural antioxidants are a very promising alternative to existing technologies. This approach is based on the use of powerful antioxidants, which occur naturally, and has an additional advantage. Natural mixtures of antioxidants act not only as free radical scavengers and antipolymerization agents in frying oils but they may have functional properties as well (improvement of cholesterol balance, prevention of lipid peroxidation).

This chapter discusses the chemistry of phytosterols with antipolymerization properties, undertakes further in-depth analysis to highlight the significance of certain structural characteristics for a stabilizing effect, and gives information about phytosterol oxidation products found in oils and fried foods heated at frying temperatures.

9.2 OCCURRENCE AND BIOLOGICAL IMPORTANCE OF PHYTOSTEROLS

Sterols are 3-hydroxy derivatives of cyclopentano-perhydrophenanthrene, a condensed four-ring system with 17 carbon atoms. The presence of methyl groups at C-4 differentiates 4-methylsterols and 4,4-dimethylsterols (triterpene alcohols) from 4-desmethylsterols. The substitution at C-17 by side-chain alkyl groups together with some unsaturation, commonly in the B-ring, is responsible for the main differences between the compounds of the same class. Unsaturated sterols (mostly Δ^5) are conventionally designated as stenols, while their saturated analogues are called stanols.

Sterols are minor constituents of oils and fats. They make up the major portion of the unsaponifiable fraction of most edible fats and oils. They are widely distributed in nature, occurring both in the free form or combined (most frequently as esters of higher aliphatic acids and as glycosides). Sterols, sterol esters, and sterol glycosides are, to different extents, soluble in fats but are completely insoluble in water. Free sterols and sterol glycosides, with polar hydroxy groups and apolar condensed ring elements, have a strong affinity both to water and to fats. Therefore, they have the ability to reduce the interface tension of water–oil mixtures. In vegetable fats and oils, sterols containing lipoproteins with a strong lipophilic character were also found (Homberg and Bielefeld, 1985).

Sterols can be classified according to their origin, as animal sterols or as plant sterols. The latter can be subdivided into phytosterols (higher plant sterols) and mycosterols (lower plant sterols present in the lipid fraction of yeast and fungi). Cholesterol, with a C27 carbon skeleton, is the characteristic sterol of higher animals. It is important as a component of all tissues and as a precursor of steroid hormones and bile acids. Cholesterol occurs also in plants, usually in very small quantities, and marine red algae (Brooks, 1970). β-Sitosterol, stigmastenol, stigmasterol, and avenasterols, with a C29 carbon skeleton, as well as campesterol and brassicasterol with a C28 carbon skeleton, are major phytosterols present in vegetable oils at significantly higher levels than cholesterol in animal fats (Itoh et al., 1973a; Weihrauch and Gardner, 1978; Kornfeldt and Croon, 1981; Homberg and Bielefeld, 1985; Homberg and Bielefeld, 1989).

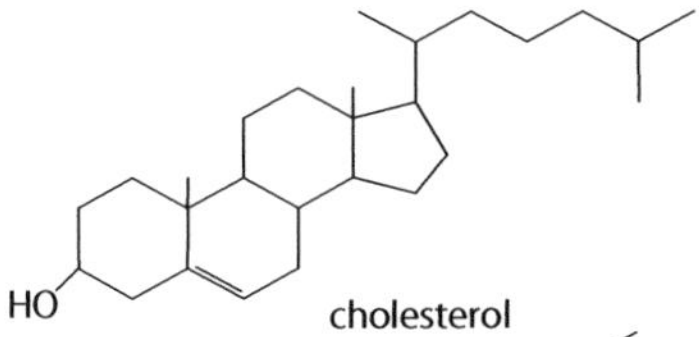

cholesterol 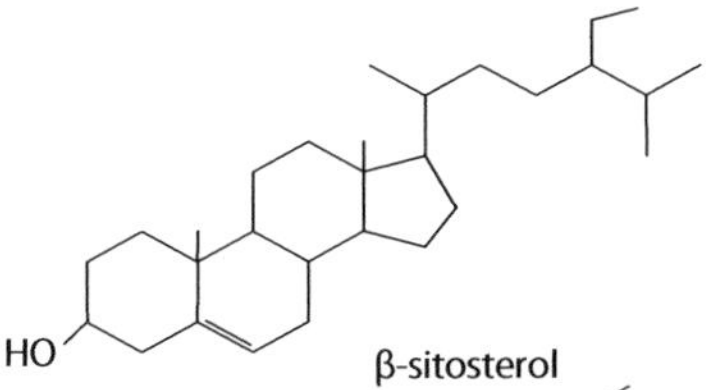 β-sitosterol

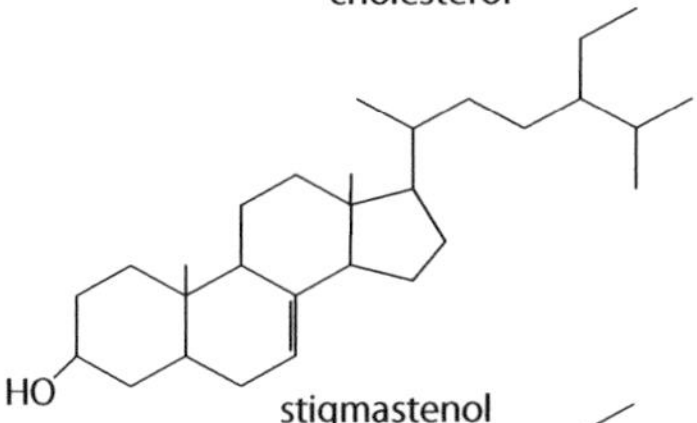

stigmastenol 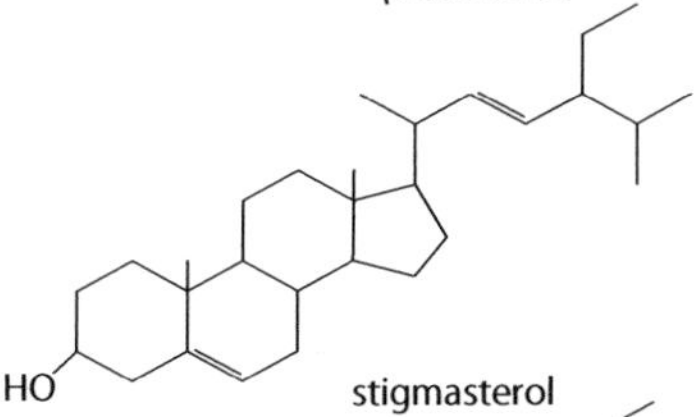 stigmasterol

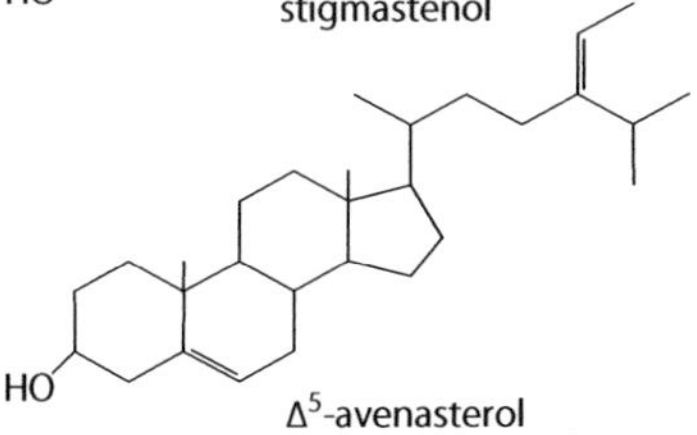

Δ^5-avenasterol 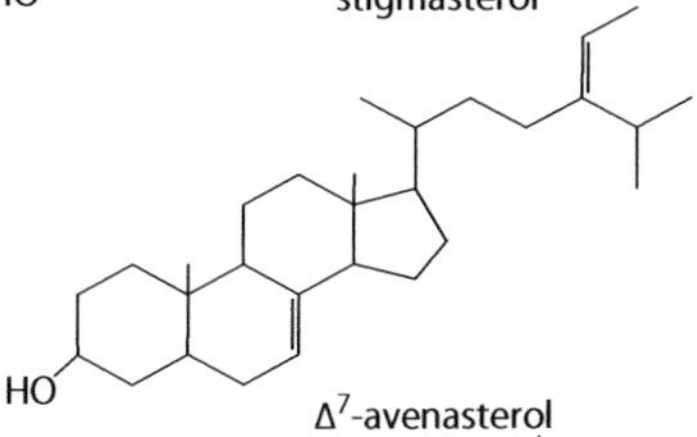 Δ^7-avenasterol

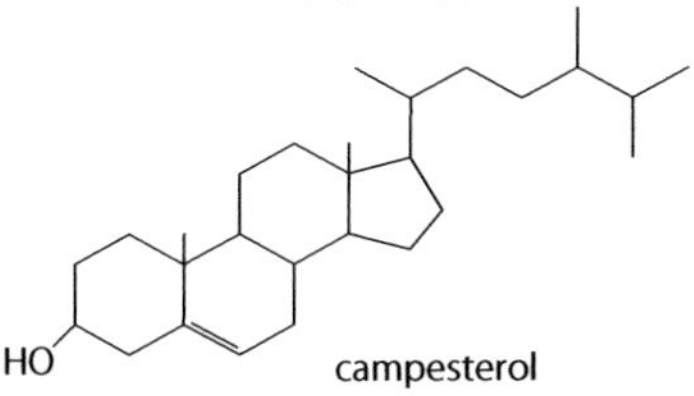

campesterol 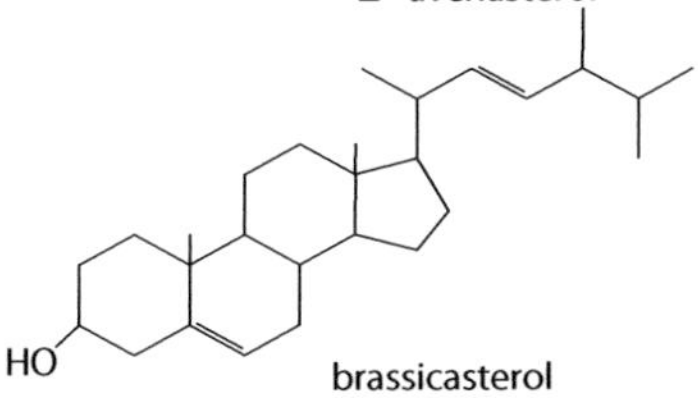 brassicasterol

4-Methylsterols are intermediates in sterol biosynthesis, and they are always present in fats and oils accompanying them. The predominant 4-methylsterols are citrostadienol, obtusifoliol, gramisterol, and cycloeucalenol (Itoh et al., 1973b; Kronfeldt and Croon, 1981).

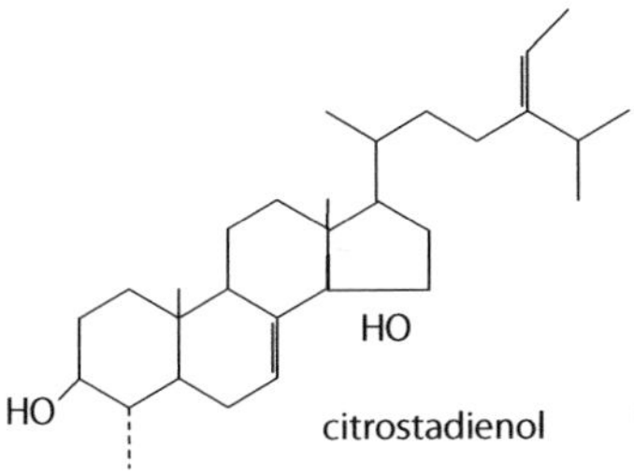

citrostadienol 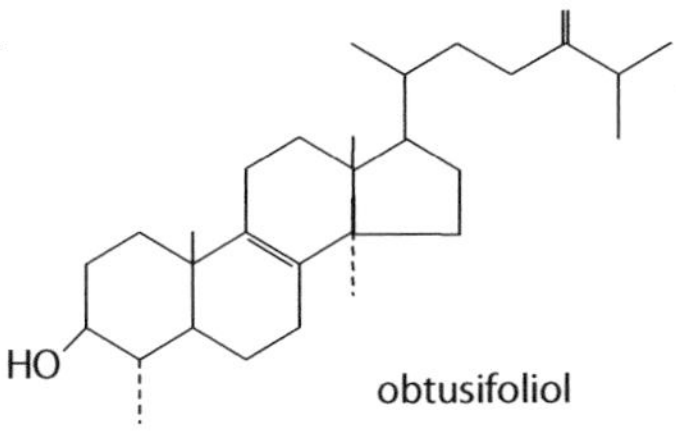 obtusifoliol

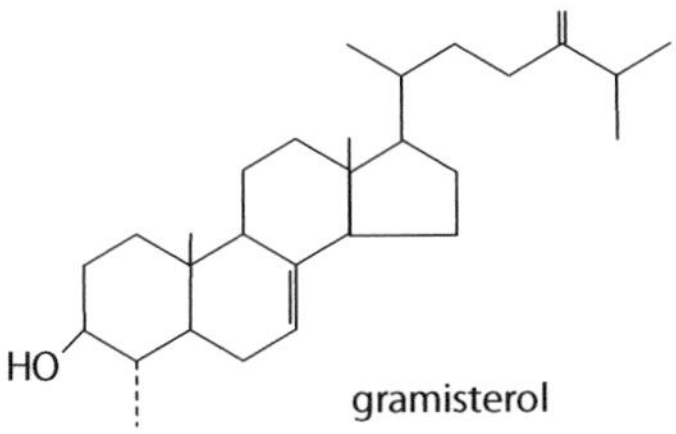

gramisterol 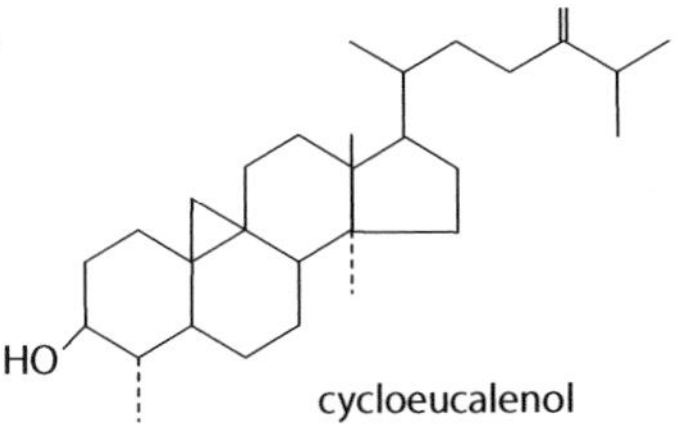 cycloeucalenol

Phytosterols are functional ingredients. They have an absorption level 20 times lower than that of cholesterol: the latter is absorbed in the human digestive system in significant amounts (Vuovisto and Miettinen, 1994). This difference in absorption affects the availability of phytosterols and cholesterol, and has some nutritional implications. It has been known since the 1950s that phytosterols inhibit the absorption of cholesterol in the body during digestion. The first observation was made by Peterson et al. (1951), who found that serum cholesterol of chicks, when elevated by dietary cholesterol, may be lowered by increasing β-sitosterol in the diet. Kudchodkar and his co-investigators (1976) reported that treatment with plant sterols (a mixture of 65% β-sitosterol, 30% campesterol, and 5% stigmasterol) increased the fecal excretion of metabolites of endogenous cholesterol by 35%–73% over the control values and also decreased the absorption of dietary cholesterol. Erickson (1971) prepared the first hypocholesteremic product. It was a shortening suitable as a cooking, frying, and salad oil, which contained an oil enriched with 1.5%–3.0% β-sitosterol or stigmasterol oleate. However, it was Igmar Wester, the head of the research and development department of a Finnish company, who devised a method for manufacturing sitostanol esters (Heasman and Mellentin, 1998). These esters are fat soluble and can be added to spreads such as margarines. Such margarines have a cholesterol-lowering effect, which was demonstrated by clinical trials, and they are claimed to be a successful type of functional food (Heasman and Mellentin, 1998). Weststrate and Meijer (1998) demonstrated that a margarine containing sterol esters from soybean, mainly esters of β-sitosterol, campesterol, and stigmasterol, is as effective as a margarine containing sitostanol esters in lowering blood total and LDL cholesterol levels. Today, many hypocholesteremic spreads containing phytosterols and sitostanols are commercially available and are broadly used because consumers believe that the obtained small decrease in plasma cholesterol level is beneficial for their health. There are also claims for the preparation of deep-frying oils combined with free or esterified phytosterols, which provide fried food with enhanced health benefits (Hayes et al., 2004; Nakhasi et al., 2007).

9.3 STRUCTURAL FEATURES OF PHYTOSTEROLS AND ANTIPOLYMERIZATION ACTIVITY

Some phytosterols are potent antipolymerization agents. Chang and Mone (1960) were the first to suggest that the addition of a relatively small amount of naturally occurring phytosterols to a cooking fat gives it improved resistance to darkening and polymerization upon prolonged treatment at elevated temperatures. These investigators used as additives "α-sitosterols," a mixture of sterols isolated from corn oil by steam distillation under vacuum at 160°C or by extraction of wheat germ with the fatty material in which it was to be added. α-Sitosterols were added to winterized cottonseed oil at levels ranging from 0.001% to 2%. The samples, with and without the additives, were heated in deep-fat fryers at 180°C ± 1°C for 144 h. α-Sitosterols were found to strongly inhibit polymerization, which was monitored by the decrease in linoleic acid content, the increase in viscosity, and by measurements of color and nutritional value.

Sims and his co-workers (1972) carried out additional studies using the unsaponifiables of olive, corn, wheat germ, and *Vermonia anthelmintica* oils, as well as pure sterols. The additives were dispersed in safflower oil, and the samples were heated in beakers on electric hot plates at 180°C ± 5°C for 7 h periods with overnight cooling to room temperature. The rate of destruction of polyunsaturated fatty acids was followed by measurement of iodine value, refractive index, viscosity, and fatty acid composition. It was found that the unsaponifiable fraction of olive, corn, and *Vermonia anthelmintica* oil and the sterols vernosterol, fucosterol, and Δ^7-avenasterol were effective as antipolymerization agents, whereas ergosterol, lanosterol, spinasterol, β-sitosterol, stigmasterol, and cholesterol were ineffective or slightly prooxidant. The wheat germ oil unsaponifiables appeared only moderately effective in retarding oxidation, although they were reported to contain a relatively high level of α-sitosterols. When this fraction was further purified to increase the level of 4-methyl-sterols, its effect on the heated safflower oil was more pronounced. Gas chromatographic analysis indicated the presence of citrostadienol as the compound largely responsible for the antioxidant activity of wheat germ oil unsaponifiables. All sterols, which were effective as antipolymerization agents, contained an ethylidene group in the side chain, which was claimed to be an essential structural feature for this activity.

Boskou and Morton (1975) indicated that citrostadienol and Δ^5-avenasterol deteriorate more rapidly than the other components of the 4-methylsterol and 4-desmethylsterol fractions when virgin olive oil is held at 180°C ± 5°C for different periods of time. The same investigators (1976) studied the effect of sterols isolated from virgin olive oil and green algae on the rate of deterioration of cottonseed oil heated at frying temperatures. At the end of each 24 h heating period, aliquots were removed for measuring refractive index, viscosity, iodine value, and fatty acid composition. It was found that the olive oil sterol fraction containing Δ^5-avenasterol and the green algae sterol fraction, which consisted exclusively of Δ^5-avenasterol, showed antipolymerization properties in contrast to the cottonseed oil sterol fraction consisting mainly of β-sitosterol. The sterol fraction isolated from brown algae containing only fucosterol, the *trans*-isomer of Δ^5-avenasterol, was also tested. This sterol minimized deterioration of heated cottonseed oil, but its effect was not equal to that of Δ^5-avenasterol. This cannot be explained by the difference in the geometry of the side-chain double bond. The presence of Δ^5-avenasterol and other ethylidene side-chain sterols in olive oil may contribute to the remarkable stability of this oil at frying temperatures. The aggregate data describing the behavior of virgin olive oil during repeated frying operations indicate that this oil is more stable in comparison with other vegetable oils (Andrikopoulos et al., 1989; Andrikopoulos and Demopoulos, 1989; Romero et al., 1995; Aggelousis and Lalas, 1997). It is interesting to note that in olive oils from certain areas in the Mediterranean region, percentages of Δ^5-avenasterol in the sterol fraction are very high (up to 36%; Boskou, 2006).

Gordon and Magos (1983) investigated the effect of sterols on the oxidative polymerization of a triglyceride mixture, similar in composition to olive oil, by periodic measurements of iodine value and linoleic acid content of the substrate during heating at 180°C. Δ^5-Avenasterol and fucosterol were found to be effective antipolymerization agents, while other sterols, including cholesterol and stigmasterol, were found to be ineffective. It was also shown that the antioxidant effect of Δ^5-avenasterol increased with concentration in the range 0.01%–0.1%. The same investigators attempted to explain the effectiveness of sterols as natural antipolymerization agents. They postulated that lipid free radicals react rapidly with sterols at unhindered allylic carbon atoms to form a radical, which can isomerize to a relatively stable allylic tertiary free radical. The latter does not tend to react readily.

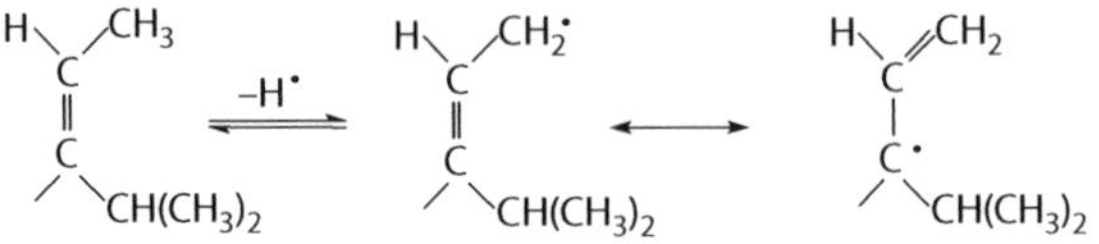

$$\mathrm{H(CH_3)C{=}C(CH(CH_3)_2)} \xrightarrow{-\mathrm{H}^{\bullet}} \mathrm{H(CH_2^{\bullet})C{=}C(CH(CH_3)_2)} \longleftrightarrow \mathrm{H(CH_2{=})C{-}C^{\bullet}(CH(CH_3)_2)}$$

An ethylidene group in the side chain seems to be a prerequisite for antipolymerization activity, but a further effect may arise from the presence and position of endocyclic double bonds. Vernosterol, a sterol with two endocyclic double bonds, was found to be more effective than Δ^7-avenasterol and fucosterol (Sims et al., 1972). These latter sterols have only one endocyclic double bond.

HO

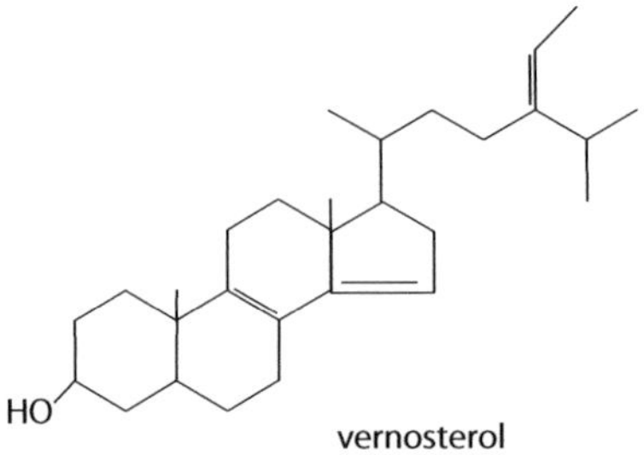

vernosterol

In vernosterol there is a rapid formation of free radicals at C29, but slower formation of free radicals at C11 or C16, which are stabilized by delocalization over two double bonds. This additional stabilization mechanism may also contribute to the antipolymerization activity of this sterol.

White and Armstrong (1986) carried out studies using oat sterol fractions containing Δ^5-avenasterol and pure β-sitosterol. The additives were added to soybean oil, and the samples were heated at 180°C. Changes in fatty acid percentages and formation of conjugated dienoic acids and high-molecular-weight compounds were monitored in all samples. The heated oils with added oat sterol fractions containing Δ^5-avenasterol deteriorated more slowly than did the controls. Heated oil with added pure β-sitosterol was altered at a rate similar to that of the control.

Blekas and Boskou (1986) demonstrated that both free sterols and steryl esters formed from the same sterols have similar effects in reducing the deterioration of heated oils. The sterol mixture was isolated from crude sunflower oil by saponification, fractional crystallization of the unsaponifiables, and preparative thin layer chromatography. 4-Methylsterols were used as such or as stearates, which were prepared by esterification with stearic acid chloride. The changes in the substrate (refined cottonseed oil) were monitored by measurements of linoleic acid content, iodine value, refractive index, and viscosity. Steryl esters were found to be equally effective with the free 4-methylsterols in minimizing oxidative polymerization in the heated oil. These findings are consistent with Sim's hypothesis that it is the ethylidene group in the side chain that is important for the antipolymerization properties of sterols and that the effectiveness of the 4-methylsterol fraction of sunflower oil is not influenced by the hydroxyl group at C3. The same investigators (1988) also studied the effect of β-sitosterol and stigmasterol on the rate of deterioration of heated triacylglycerols obtained from commercial triolein purified by column chromatography on activated silicic acid. Both sterols were found to have no significant effect on the stability of the heated triacylglycerol mixture. β-Sitosterol, the most common sterol, has been claimed to be prooxidant in heated sunflower oil (Yanishlieva, 1981) and tristearin (Yanishlieva and Schiller, 1983) in comparison to pure β-sitosterol. Sitosteryl stearate was found to exhibit a stronger prooxidative effect (Yanishlieva et al., 1982). It was pointed out that 6-hydroperoxy-stigmast-4-en-3-one and 7-hydroxy-stigmast-5-en-3β-ol, formed from β-sitosterol during heating, were mainly responsible for the prooxidative activity of β-sitosterol (Yanishlieva, 1981). These findings, however, are of little practical importance because the oxidation products used in these studies were prepared from pure β-sitosterol by blowing oxygen at 120°C and then added to the substrates at 100°C or 120°C. These conditions differ greatly from those in deep-fat frying. Ghavami and Morton (1984) have shown that sterols were lost from soybean oil during heating in a deep-fat fryer but they were unable to detect oxidation products in the unsaponifiable fraction of the oil. It is believed that when a large volume of oil is heated in a deep-fat fryer, oxygen supply is poor, and the reactions in which sterol radicals participate lead to polymerization rather than oxidation products.

Gordon and Williamson (1989) confirmed the decrease in avenasterol content of oils heated at 180°C and the ineffectiveness of this sterol as antioxidant at room temperature or under accelerated test conditions in a Rancimat apparatus at 100°C. They also confirmed the positive effect of avenasterol in retarding the loss of tocopherols from heated oils. Gordon (1990) reported that Δ^5-avenasterol does not stabilize oils heated in an oven at 100°C, 180°C, and 200°C, but is active in stabilizing oils heated on a hot plate at 100°C or 180°C. A possible explanation for these results is "that oil samples heated on a hot plate become oxygen deficient because deterioration of the oil is faster at the higher temperature near the base of the sample while transport of oxygen is limited at the lower temperature near the top of the sample." The mechanism proposed to explain the antipolymerization properties of Δ^5-avenasterol is similar to that suggested by Burton and Ingold (1984) for the antioxidant activity of β-carotene. At high partial pressures of oxygen, the reaction of the avenasterol-derived tertiary allylic radical with oxygen occurs at a reasonable rate to form the less hindered peroxy radical, which is ineffective in interrupting the autoxidation chain reaction (Gordon, 1990).

$$\xrightleftharpoons{O_2}$$

However, although several products arising from the oxidation of A and B rings of Δ^5-avenasterol were isolated, no oxidation products from the side chain were detected in an oil heated for 72 h at 180°C (Gordon and Magos, 1984). Obviously, more experimental work is needed to elucidate this mechanism.

Yan and White (1990) carried out additional studies to determine the ability of linalyl acetate and undecylenic acid to reduce oxidative changes in soybean oil held at frying temperatures. These compounds were used as additives because they have an ethylidene group in their structure.

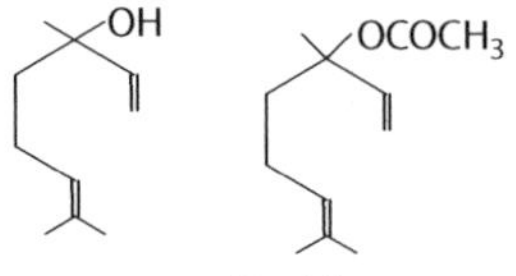

OH OCOCH$_3$

$H_2C{=}CH(CH_2)_8COOH$

undecylenic acid

linalool linalyl acetate

Linalyl acetate was prepared by acetylation of linalool. Pure linalool was also tested as additive, but proved to be slightly prooxidative when compared with the control. Acetylation of linalool to linalyl acetate caused the formation of many by-products, which were tested after separation and purification by thin layer chromatography and identification by GC-MS. These by-products were similar in structure to linalyl acetate (they retained the basic ethylidene double bond), and they were equally effective as antipolymerization agents. Compared to Δ^7-avenasterol, linalyl acetate was found to be less effective. At levels of 400 mg/kg, its effect was similar to that obtained by the addition of methyl silicone at levels 0.3 mg/kg. Undecylenic acid had some protective effect, but less pronounced than that of linalyl acetate. The latter was also found to provide some protection in heated lard enriched with cholesterol (Yan and White, 1991).

Recently, Winkler and Warner (2008b) conducted a study using soybean oil, which was heated at 180°C for 8 h. Indigenous tocopherols and phytosterols were previously removed by molecular distillation, and pure phytosterols were added to the stripped oil at levels ranging from 0.05 to 0.5 wt%. Measurement of the rate of polymerization was followed by measurements of fatty acid composition, residual phytosterol content, and triacylglycerol dimers and polymers. The authors concluded that under the conditions of the study, the degree of unsaturation of each phytosterol was more important for the antipolymerization effect than the presence of an ethylidene side group. These discrepancies in the literature concerning the importance of an ethylidene side chain may be due to the different models used, the conditions and duration of heating, and the lack of pure sterols with double bonds both in the main skeleton and the side chain (such as Δ5-avenasterol and citrostadienol) that could be used as model compounds in the heating experiments. Winkler and Warner (2008a) also indicated that the antipolymerization effect is more pronounced at higher concentrations. Phytosterols added at 1.0% and 2.5% significantly decreased the oxidative stability index (OSI) of high-oleic sunflower and soybean oils, which were stripped previously for the removal of indigenous tocopherols and phytosterols. The study showed that the addition of phytosterols at such high concentrations had a strong impact on the thermal and oxidative stability of oils.

9.4 APPLICATIONS TO FRYING OILS

Fried food is liked by consumers because of its flavor. Fats used in frying account for an important share of per-capita edible fat consumption. However, frying fats are not considered as good for human health as unheated ones, and they are not declared as such. Therefore, efforts are always under way to improve the quality and stability of fats used in frying and to manufacture products with a healthier profile.

Modified vegetable oils with a specific fatty acid composition, such as high-oleic canola, safflower, and sunflower oils, as well as low-linolenic canola and soybean oils, have been shown to have enhanced nutritional and stability characteristics. The key advantages of these modified oils are that they have a remarkable stability, obtained by little or no hydrogenation, and a low content in saturated fatty acids. Modified oils are blended with other fats and oils to improve their stability. An example is a fat, made from vegetable oil, and cholesterol-stripped animal fat that has been shown to lower serum cholesterol (Haumann, 1996). This product has neither the cholesterol found in usual animal fats nor the *trans* fatty acids due to partial hydrogenation. In frying applications, the modified fat mentioned earlier has been claimed to be superior to traditional beef fat and hydrogenated soybean oil. This superiority is manifested by a lower rate of dienes and trienes formation and a leveling up of thermal oxidation and polymerization products.

Another example of a stable and healthier frying oil is a high-oleic sunflower-oil-based product that incorporates a blend of rice bran oil and specially processed sesame seed oil (Haumann, 1996). High-oleic sunflower oil is obviously used because of its low saturated fatty acid and linolenic acid content. Haumann (1996) states that an ideal frying oil should have a high oleic acid content (more than 60%), a low content of saturated fatty acids (less than 15%), and a very low content of linolenic acid (less than 0.5%). High-oleic sunflower oil fulfills these criteria, but any other oil rich in oleic acid could also be used for the same purpose.

The presence of sesame seed and rice bran oils in such frying oils provides additional stability, which is due not only to the fatty acid composition but mainly to constituents with antioxidant and antipolymerization properties.

Sesame seed oil, which is known for its stability, has been studied for the presence of antioxidants by many investigators (Kikugawa et al., 1983; Fukuda et al., 1986a, 1986b, 1986c). Kamal-Eldin et al. (1992) and Kamal-Eldin and Appelqvist (1994) thoroughly investigated the minor constituents of sesame oil extracted from sesame seeds of different varieties of the cultivated *Sesamun indicum* L and from three related wild species (*S. alatum*, *S. angustifolium*, and *S. radiatum*). The three sterol fractions (4-desmethylsterols, 4-monomethylsterols, and 4,4-dimethylsterols), tocopherols and lignans were analyzed. The contents of the main antipolymerization sterols and tocopherols are given in Table 9.1.

The wild species were found to contain higher amounts of antipolymerization sterols in comparison to *Sesamum indicum* L., with the exception of *S. alatum*. The oils from the wild species were also richer in tocopherols. Percentages of individual tocopherols were more or less the same in all the oils. The predominant tocopherol was the γ-homologue (96%–99%), which is the most antioxidative form.

TABLE 9.1
Antipolymerization Sterol and Tocopherol Content (mg/kg) of Oils Obtained from Four Sesamum Species

Sample	Δ^5-Avenasterol	Δ^7-Avenasterol	Citrostadienol	Tocopherols*
S. indicum L.	590	50	150	585
S. alatum	2115	80	100	265
S. angustifolium	1670	80	215	745
S. radiatum	620	160	545	805

Source: Kamal-Eldin et al., 1992. Seed lipids of *Sesamum indicum* and related wild species in Sudan. The sterols, *J. Sci. Food Agric.*, 59: 327–334; Kamal-Eldin, A. and Appelqvist, L.-Å. 1994. Variations in the composition of sterols, tocopherols and lignans in seed oils from four Sesamum species, *J. Am. Oil. Chem. Soc.*, 71: 149–156.

* Mainly γ-tocopherol.

Concerning the rest of the antioxidants present in sesame seed oil, these are sesamolin, sesamol, and sesaminol (Fukuda et al., 1986a, 1986b; Kamal-Eldin and Appelqvist, 1994). The antioxidant activity of lignans is discussed in another chapter in this book (see Chapter 8). Here, some additional information is provided to complete the whole spectrum of strong antioxidants found in these so-called "stable and healthier frying oils." Sesamol and sesaminol, two strong antioxidants, are decomposition products of sesamolin. A mechanism for the formation of sesamol and sesaminol from sesamolin present in an oil by heating under anhydrous conditions in the presence of acid clay was proposed by Fukuda et al. (1986c).

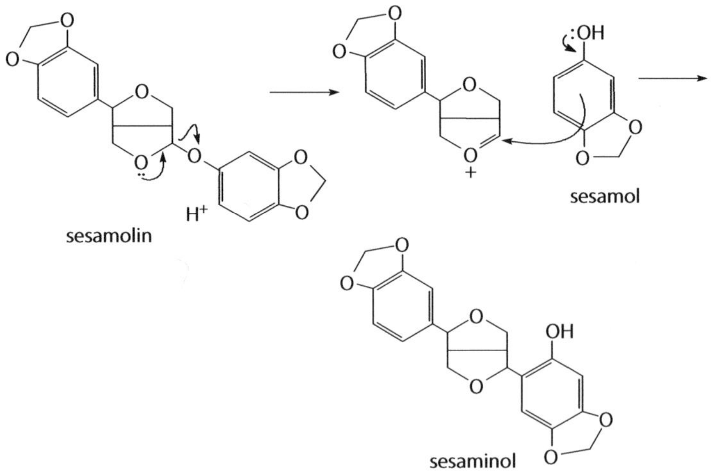

Four stereoisomers of sesaminol with somewhat different antioxidative activities have been found to be produced during the industrial bleaching process of unroasted sesame seed oil (Nagata et al., 1987). Among these isomers, sesaminol and two episesaminols with antioxidant activities comparable to that of sesamol and γ-tocopherol are present in commercial sesame seed oils (Fukuda et al., 1986a).

Sesaminol shows heat stability at frying temperatures. Sesaminol and sesamol were added to corn oil, which was heated at 180°C; 40.5% of the added sesaminol remained unaltered after 6 h heating, whereas sesamol was completely decomposed after heating for 4 h (Fukuda et al., 1994). The content of sesaminol in refined sesame seed oil obtained from fresh unroasted sesame seeds is about 500–1000 mg/kg (Fukuda et al., 1994). Crude sesame oil obtained from roasted sesame seeds was found to be more stable than the refined oil from unroasted seeds because it contains more sesamol formed from sesamolin and also browning products formed by Maillard reactions (Fukuda et al., 1988; Yen, 1990; Namiki, 1995; Fukuda et al., 1996). Products of nonenzymic browning show a synergistic effect with γ-tocopherol, sesamol, and combinations of γ-tocopherol with sesamol.

The other basic constituent of stable frying oil is rice bran oil. This is an oil characterized by its very high content in unsaponifiables. Itoh et al. (1973a) give a value of 4.2% for unsaponifiables and values as high as 1.8% and 0.42%, respectively, for 4-desmethylsterols and 4-methylsterols in the oil. Other investigators give values for the content in unsaponifiables from 4.4%–6% in raw and from 1.9%–3.5% in refined rice bran oils (Gaydou et al., 1980; Sah et al., 1983; Bhattacharyya and Bhattacharyya, 1987; Gupta, 1989).

Rice bran oil is rich in sterols known for their antipolymerization properties, such as Δ^5-avenasterol (about 900 mg/kg), Δ^7-avenasterol (about 360 mg/kg), and citrostadienol (about 1680 mg/kg) (Itoh et al., 1973a, 1973b). These sterols obviously increase the stability of sunflower oil at frying temperatures. Rice bran oil also contains tocopherols, mainly β- and γ-tocopherols (16–358 mg/kg), and β- and γ-tocotrienols (62–975 mg/kg) (Rogers et al., 1993).

Another group of compounds naturally occurring in rice bran oil are γ-oryzanols. These are esters of ferulic acid with 4-desmethylsterols, such as β-sitosterol (I) and campesterol (II), and 4,4-dimethylsterols, such as cycloartanol (III), cycloartenol (IV), and 24-methylenecycloartanol (V) (Rogers et al., 1993).

The levels of these compounds in several brands of edible rice bran oil were found to be 35–232 mg/kg for IV, 30–314 mg/kg for V, 39–342 mg/kg for II, and 0–84 mg/kg for III + I (Rogers et al., 1993).

The sterol part of γ-oryzanols lacks any structural characteristic, which would justify an antioxidant and antipolymerization effect at frying temperatures. Some antioxidant activity for γ-oryzanol has been reported (Diack and Saska, 1994), but this has to be substantiated by further research. It is possible that the antioxidant activity is due to ferulic acid esterified with the sterols in the γ-oryzanol molecules. Ferulic acid is a monomethylated *o*-diphenol, and its activity has been reported by many investigators (Yanishlieva and Marinova, 1992; Scott et al., 1993; Marinova and Yanishlieva, 1994; Lavanjinha et al., 1996; Chen and Ho, 1997). It is interesting to note that γ-oryzanols are functional constituents. They show physiological effects such as a decrease in cholesterol absorption, decrease in hepatic cholesterol biosynthesis, decrease in the levels of cholesterol in plasma, and an increase in fecal bile acid excretion (Rogers et al., 1993).

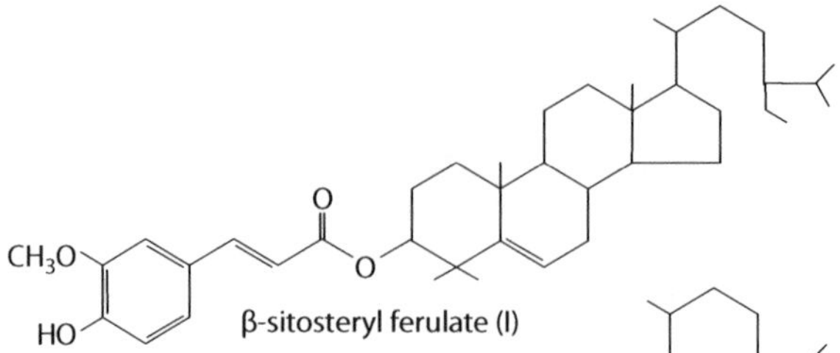

β-sitosteryl ferulate (I)

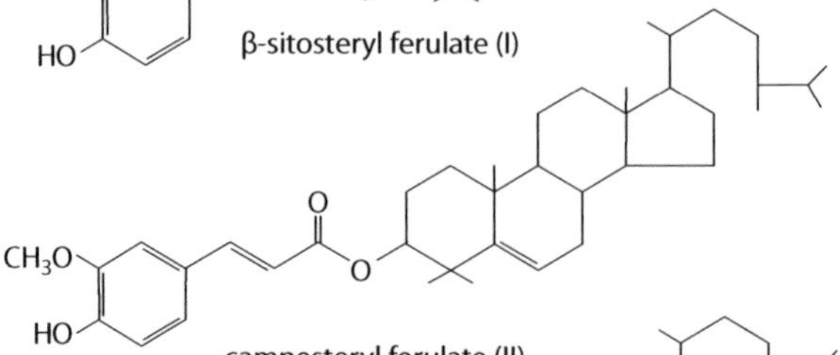

campesteryl ferulate (II)

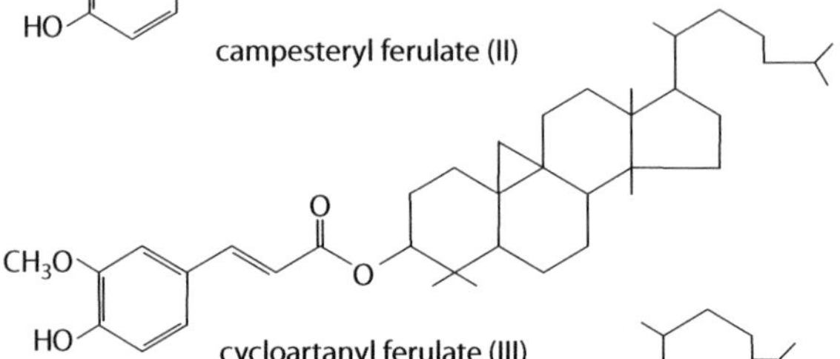

cycloartanyl ferulate (III)

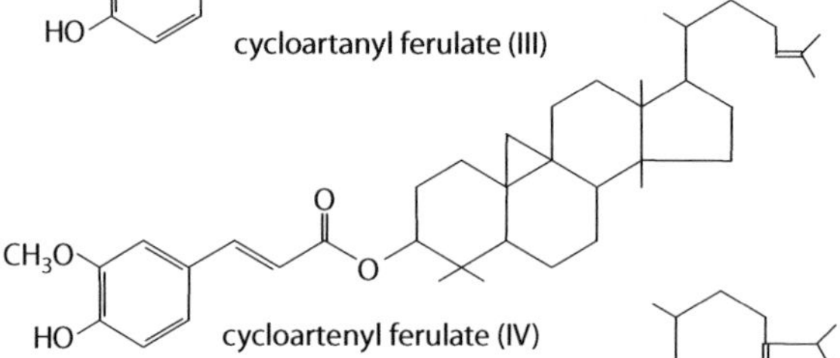

cycloartenyl ferulate (IV)

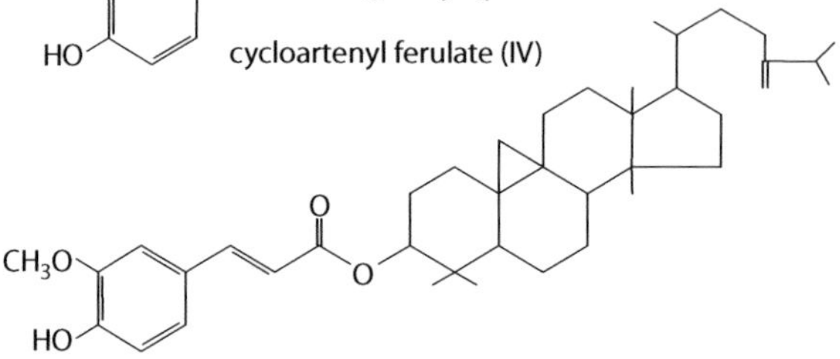

24-methylenecycloartanyl ferulate (V)

9.5 PHYTOSTEROL OXIDATION PRODUCTS

Phytosterols, like cholesterol, are oxidized mainly by a free radical mechanism (autoxidation), giving rise to the formation of compounds known as oxyphytosterols or phytosterol oxidation products (POPs). The oxygen attack occurs predominantly at C7 and also at the tertiary carbon atoms in the side chain. Initially, a radical is formed due to the abstraction of the sterol reactive allylic hydrogen at C7. This radical reacts further with molecular oxygen, yielding 3β-hydroxy-5-en-7-peroxyl radicals stabilized by hydrogen abstraction in the form of the more stable epimeric 7-hydroperoxides. During heating or storage, sterol 7-hydroperoxides can decompose to the 7-hydroxysterol epimers and 7-ketosterol. Important secondary oxidation products formed via a molecular interaction between sterol hydroperoxides and intact sterols are also the 5,6-epoxysterol epimers, which can be converted to their triol derivatives through hydration in an acidic environment (Tai et al., 1999), and POPs formed by thermal decomposition of phytosterol side-chain hydroperoxides (Lampi et al., 2002; Johnsson and Dutta, 2003). POPs with the same structures are formed by exposure of vegetable oils in sunlight or artificially generated light and by irradiation of vegetable oils (Zhang et al., 2006). Formation, occurrence, and biological effects of POPs are presented in detail in a recently published review (Ryan et al., 2009).

9.5.1 Identification of Phytosterol Oxidation Products

POP's identification is based on their gas chromatographic separation, using a mass detector, after isolation from the sample by applying several techniques (preparative and analytical TLC, column chromatography and analytical TLC, and SPE and analytical TLC) and preparation of TMS derivatives (Lampi et al., 2002; Johnsson and Dutta, 2003; Apprich and Ulberth, 2004; Rudzinska et al., 2009). Additional steps for lipid extraction (Nourooz-Zadeh and Appelqvist, 1992; Dutta and Appelqvist, 1997; Dutta, 1997; Zunin et al., 1998; Tabee et al., 2008a) and/or isolation of the unsaponifiable lipids (Dutta and Appelqvist, 1997; Lampi et al., 2002; Grandgirard et al., 2004; Zhang et al., 2005; Cercaci et al., 2007) are needed for POP's identification in fried foods and foods containing lipids of plant origin.

Primary and secondary POPs have been identified in unheated and heated phytosterol standards (Yanishlieva et al., 1980; Daly et al., 1983; Gordon and Magos, 1984; Lampi et al., 2002; Johnsson and Dutta, 2003; Johnsson et al., 2003; Säynäjoki et al., 2003; Apprich and Ulberth, 2004; Grandgirard et al., 2004; Johnsson and Dutta, 2006; Zhang et al., 2005; Rudzińska et al., 2009), purified triacylglycerols enriched with phytosterol standards (Yanishlieva and Schiller, 1983; Blekas and Boskou, 1989; Soupas et al., 2005), vegetable fats and oils enriched or nonenriched with phytosterol standards (Nourooz-Zadeh and Appelqvist, 1992; Dutta, 1997; Oehrl et al., 2001; Lampi et al., 2002; Giuffrida et al., 2004; Grandgirard et al., 2004; Zhang et al., 2005; Cercaci et al., 2007; Soupas et al., 2007), oil-in-water emulsions (Cercaci et al., 2007), and foods containing phytosterols (Lee et al., 1985; Nourooz-Zadeh and Appelqvist, 1992; Dutta and Appelqvist, 1997; Dutta, 1997; Zunin et al., 1998; Rudzińska et al., 2005; Zunin et al., 2006; Garcia-Llatas et al., 2008; Tabee et al., 2008a). Some of these products have been discussed in detail by Dutta et al. (2007).

9.5.1.1 Identified Thermoxidation Products of β-Sitosterol

The main β-sitosterol thermoxidation products are the following:

- (24R)-Ethylcholest-5-en-3β,7α-diol (7α-hydroxysitosterol)
- (24R)-Ethylcholest-5-en-3β,7β-diol (7β-hydroxysitosterol)
- (24R)-Ethylcholest-5-en-3β-ol-7-one (7-ketositosterol)
- (24R)-5α,6α-Epoxy-24-ethylcholastan-3β-ol (sitosterol-5α,6α-epoxide)
- (24R)-5β,6β -Epoxy-24-ethylcholastan-3β-ol (sitosterol-5β,6β-epoxide)
- (24R)-Ethylcholestan-3β, 5α, 6β-triol (sitostanetriol)

Minor β-sitosterol thermoxidation products are

- (24R)-Ethylcholest-5,24-dien-3β-ol (Yanishlieva et al., 1980)
- (24R)-Ethylcholest-5-en-3β,4β-diol (Grandgirard et al., 2004)
- (24R)-Ethylcholest-5-en-3β,6α-diol (Soupas et al., 2007)
- (24R)-Ethylcholest-5-en-3β,6β-diol (Grandgirard et al., 2004; Soupas et al., 2007)
- (24R)-Ethylcholest-4-en-3β,6β-diol (Grandgirard et al., 2004)
- (24R)-Ethylcholest-5-en-3-one (Daly et al., 1983)
- (24R)-Ethylcholest-4-en-3-one (Yanishlieva and Schiller, 1983; Daly et al., 1983)
- (24R)-Ethylcholest-3,5-dien-7-one (Yanishlieva and Schiller, 1983)
- (24R)-Ethylcholest-4,6-dien-3-one (Yanishlieva and Schiller, 1983)
- (24R)-Ethylcholest-4-en-6α-ol-3-one (Johnsson and Dutta, 2003)
- (24R)-Ethylcholest-4-en-6β-ol-3-one (Yanishlieva and Schiller, 1983; Johnsson and Dutta, 2003)
- (24R)-Ethylcholest-4-en-3,6-dione (Daly et al., 1983)
- (24R)-Ethylcholest-5-en-3β,24-diol (Lampi et al., 2002; Johnsson and Dutta, 2003)
- (24R)-Ethylcholest-5-en-3β,25-diol (Johnsson and Dutta, 2003; Soupas et al., 2007)

9.5.1.2 Identified Thermoxidation Products of Stigmasterol

The main thermoxidation products of stigmasterol, similar to those of β-sitosterol, are 7α-hydroxy-stigmasterol, 7β-hydroxy-stigmasterol, 7-ketostigmasterol, stigmasterol-5α,6α-epoxide, stigmasterol-5β,6β-epoxide, and stigmastentriol. Minor stigmasterol oxidation products are

- (24S)-Ethylcholest-4,22-dien-3-one (Blekas and Boskou, 1989)
- (24S)-Ethylcholest-4,22-dien-3β,4β-diol (Grandgirard et al., 2004)
- (24S)-Ethylcholest-4,22-dien-3β,6β-diol (Grandgirard et al., 2004; Soupas et al., 2007)
- (24S)-Ethylcholest-5,22-dien-3β,24-diol (Johnsson et al., 2003)
- (24S)-Ethylcholest-5,22-dien-3β,25-diol (Lampi et al., 2002; Johnsson et al., 2003)
- (24S)-Ethylcholest-5,22-dien-3β-ol-25-one (Johnsson et al., 2003)
- Pregn-5-en-3β-ol-20-one (Blekas and Boskou, 1989)

9.5.1.3 Identified Thermoxidation Products of Other Plant Sterols

Other phytosterols, such as campesterol and brassicasterol, which are present at high or low levels in edible fats and oils used often as a frying medium or as raw materials in the production of foods, are also subjected to thermoxidation. Their main oxidation products are similar to those of β-sitosterol and stigmasterol (7α- and 7β-hydroxycampesterol, 7-ketocampesterol, campesterol-5α,6α- and 5β,6β-epoxide, campestanetriol, 7α- and 7β-hydroxybrassicasterol, 7-ketobrassicasterol,

brassicasterol-5α,6α- and 5β,6β-epoxide, and brassicastanetriol). Minor campesterol and brassicasterol oxidation products are

- (24R)-Methylcholest-5-en-3β,6β-diol (Grandgirard et al., 2004; Soupas et al., 2007)
- (24R)-Methylcholest-5-en-3β,4β-diol (Grandgirard et al., 2004)
- (24R)-Methylcholest-5-en-3β,24-diol (Johnsson and Dutta, 2003)
- (24R)-Methylcholest-5-en-3β,25-diol (Lampi et al., 2002; Johnsson and Dutta, 2003)
- (24R)-Methylcholest-4-en-6α-ol-3-one (Johnsson and Dutta, 2003)
- (24R)-Methylcholest-4-en-6β-ol-3-one (Johnsson and Dutta, 2003)
- (24R)-Methylcholest-5,22-dien-3β,6β-diol (Grandgirard et al., 2004)
- (24R)-Methylcholest-5,22-dien-3β,4β-diol (Grandgirard et al., 2004)

9.5.2 Determination of Phytosterol Oxidation Products in Food

The methodologies applied for determination of POPs in foods are based on analytical techniques developed for analysis of cholesterol oxidation products. The most frequently employed technique is gas chromatography. The methods used are comprehensively reviewed by Guardiola et al. (2004). The main steps of analysis are lipid extraction, saponification of lipids and extraction of unsaponifiables, isolation and purification of oxidized sterols by preparative TLC, SPE, or liquid chromatography, derivatization of purified sterol oxides to TMS-ethers, and gas chromatographic analysis (Nourooz-Zadeh and Appelqvist, 1992; Dutta and Appelqvist, 1997; Dutta, 1997; Oehrl et al., 2001; Lampi et al., 2002; Grandgirard et al., 2004; Louter, 2004; Zhang et al., 2005; Conhillo et al., 2005; Zunin et al., 2006; Johnsson and Dutta, 2006; Cercaci et al., 2007; Soupas et al., 2007). An alternative methodology is based on transesterification of the extracted lipids and purification of the product obtained by SPE (Johnsson and Dutta, 2006; Tabee et al., 2008a). Silica cartridges are used to remove nonoxidized sterols and to obtain fractions enriched in sterol oxides that are generally present in small quantities. Gas chromatographic analysis is performed on a nonpolar capillary column, for example, DB-5, or on a combination of two capillary columns of the same dimensions combined with a press-fit connector. A capillary column coated with a medium-polarity stationary phase is connected to the on-column injector of the gas chromatograph, and a capillary column coated with a low-polarity stationary phase is connected to a flame ionization detector of the gas chromatograph or to a mass spectrometer (Johnsson and Dutta, 2006; Tabee et al., 2008a). GC/MS is used for quantification in selected ion monitoring (SIM) mode (Zhang et al., 2005; Zunin et al., 2006). As an internal standard, 19-hydroxycholesterol is mainly used.

French fries and potato chips (Dutta, 1997; Dutta and Appelqvist, 1997; Tabee et al., 2008a), vegetable oils (Lampi et al., 2002; Grandgirard et al., 2004; Zhang et al., 2005, 2006; Rudzińska et al., 2005), rapeseeds and peanuts (Rudzińska et al., 2005), phytosterol enriched and nonenriched margarines (Johnsson and Dutta, 2006; Zhang et al., 2005; Conchillo et al., 2005), and wheat flour (Nourooz-Zadeh and Appelqvist, 1992) have been found to contain POPs at moderate levels. POPs have also been identified as minor constituents in bakery products (Cercaci et al., 2006), infant formulas (Zunin et al., 1998), and liquid infant foods (Garcia-Llatas et al., 2008).

9.5.3 Factors Affecting the Level of Phytosterol Oxidation Products in Food

The concentration of POPs in food is dependent on levels of phytosterols in the food, phytosterol structure, degree of unsaturation, water content of the lipid matrix, heating time, and heating temperature. According to Soupas and her co-investigators (2005), the formation of sterol oxides, especially at 180°C, did not account for all the phytosterol losses. This indicates the presence of other oxidation products. In a study conducted by Rudzińska and her co-investigators (2009), it is mentioned that β-sitosterol, oxidized in the presence of air by heating at 60°C, 120°C, and 180°C for different time periods, contained intact β-sitosterol, degradation and oxidation products of β-sitosterol, volatiles formed by degradation of β-sitosterol oxides, and oligomers formed by condensation/polymerization of β-sitosterol oxides (at levels affected by heating temperature and time). The amount of oxidation products by heating of β-sitosterol at 180°C was found to be lower than that at 120°C and decreased when the heating time was increased. According to Soupas et al. (2004), both temperature and heating time, as well as lipid matrix and sterol structure, affect phytosterol oxidation, The interactions between lipid matrices and temperatures have a drastic effect on the level of the sterol oxides formed and also on pathways of oxidation reactions.

9.5.4 Retardation of Phytosterol Oxidation by Natural Antioxidants

Some efforts have been made by several investigators to retard phytosterol oxidation by using natural antioxidants in some foods, such as peanuts (Malecka et al., 2003), virgin olive oil (D' Evoli et al., 2006), and refined vegetable oils (Tabee et al., 2008b), or purified triacylglycerols from sunflower oil enriched with stigmasterol (Rudzińska et al., 2004). Malecka and her co-workers (2003) studied the effect of ethanolic extracts obtained from black currant, raspberry, and tomato seed on phytosterol oxide formation in peanut samples subjected to accelerated storage conditions at 60°C. Peanuts were treated with rapeseed oil enriched with the extracts. All extracts were found to retard the formation of POPs, but most effective was black currant seed extract. Rudzińska and her co-investigators (2004) studied the effect of ethanolic extracts of rosemary, green tea, α-tocopherol, and butylated hydroxytoluene (BHT) on stigmasterol oxidation in purified triacylglycerols obtained from sunflower oil incubated at 60°C for 3, 6, and 9 days. The total increase of the stigmasterol oxidation products was the lowest in samples containing α-tocopherol, but the content of stigmastentriol increased the most in this case. In all analyzed samples, stigmasterol 5α,6α-epoxide was found to be formed in the highest amounts among the analyzed stigmasterol oxidation products. Rosemary extract was also added to extra virgin olive oil heated at 180°C for different time periods with the aim of studying its influence on phytosterol oxidation (D' Evoli et al., 2006). The investigators indicated that after

6 h of heating, the levels of total sitosterol oxides formed when rosemary extract was added were 45% lower. The effect of α-tocopherol on phytosterol oxidation in refined olive oil during heating at 180°C for different time periods was studied by Tabee et al. (2008b). The levels of total POPs formed after heating were found to be significantly lower when α-tocopherol was added.

9.5.5 Refining and Formation of Phytosterol Oxidation Products

Disterylethers and steradienes are sterol derivatives formed by dehydration reactions during fat and oil refining process (Weber et al., 1992; Verleyen et al., 2002). Besides steradienes, other steroidal hydrocarbons, such as steratrienes, have been detected in refined oils (Memmie et al., 1994; Bortolomeazzi et al., 1996). Their formation from 7-hydroxy derivatives of β-sitosterol treated with bleaching earths was long ago reported by Niewiadomski and Sawicki (1964), Kaufmann and Hamza (1970), and Kaufmann et al. (1970). Bleaching experiments carried out by Bortolomeazzi and his co-investigators (2003) on a sample of sunflower oil bleached at 80°C for 1 h with 1% and 2% of both acidic and neutral earths showed a reduction of the 7-hydroxyphytosterols and partial formation of steratrienes with double bonds in the ring system at the 2-, 4-, and 6-positions. Deodorization experiments carried out on the same sunflower oil sample at 180°C under vacuum for 1 h showed that no dehydration products were formed and recovery of the formed 7-hydroxysterols was complete. The investigators also reported that no detectable levels of phytosterol oxides were found in crude palm or coconut oils, and that the levels of phytosterol oxides in lampante olive oils and crude peanut oils were lower than those in crude sunflower and maize oils. The main POPs quantified in crude oils as well as in refined sunflower oils were 7-ketositosterol and the 7-hydroxy isomers of β-sitosterol, campesterol, and stigmasterol. 7-Keto and both 7-hydroxy derivatives of brassicasterol were also the main POPs quantified in semirefined low-erucic-acid rapeseed oil (Lambelet et al., 2003). Their levels were not highly modified during deodorization at temperatures ranging from 200°C to 250°C. Rudzińska and her co-workers (2005) reported that the levels of POPs in rapeseed oil samples from three different industrial productions in the same factory ranged between 42 and 48 mg/kg when rapeseeds were pressed, and between 52 and 59 mg/kg when rapeseeds were extracted. The levels of POPs were found to increase gradually during the different refining steps (66–72 mg/kg in degummed oils, 75–78 mg/kg in neutralized oils, 90–99 mg/kg in bleached oils, and 100–110 mg/kg in deodorized oils).

9.5.6 Safety Aspects of Phytosterol Oxidation Products

Phytosterols, when consumed, block cholesterol absorption in the gut and help to lower blood cholesterol. A health claim has been approved by nutrition authorities for phytosterols in certain foods; thus, the demand for phytosterols and their subsequent addition to foods has been on the increase.

Some phytosterols have been found to inhibit oxidative degradation and polymerization of unsaturated fatty acids during high-temperature heating. Losses of phytosterols in various frying systems is generally low, indicating that these compounds are relatively stable to frying, and thus may confer some antidegradative activity after other antioxidant constituents have been destroyed. However, under adverse conditions, plant sterols can undergo oxidation similar to other unsaturated lipids. The products of oxidation at high temperatures are monomeric oxidation products but also dimers and polymers (Lampi et al., 2009).

There is accumulating evidence that cholesterol oxidation products may have adverse effects on human health. Relatively recent studies have also demonstrated that similar oxidation products are formed from cholesterol and phytosterols. Thus, the stability of these lipid compounds has been a matter of concern. The losses of phytosterols in frying oils with varying fatty acid composition were studied by Soupas et al. (2007), Winkler et al. (2007), and Tabee et al. (2008b). High-oleic sunflower oil, corn, hydrogenated soybean, expeller-pressed soybean, and expeller-pressed low-linolenic acid soybean oils were used by Winkler et al. (2007) for frying potato chips in a pilot-plant-scale continuous fryer. The same oils, and regular soybean oil, were also used in intermittent batch frying of tortilla chips. Soupas (2006) applied GC-MS to study the oxidation products of phytosterols from various processing methods. Phytosterols and phytostanols were found to be stable during processes such as spray-drying and UHT-type heating followed by storing. When, however, commercially available products were used in pan-frying, phytosterol oxidation products could be formed at nonacceptable levels. The stability of different phytosterol products during pan-frying in oil-based liquid margarines and butter oil was also evaluated by Soupas et al. (2007). The highest phytosterol oxide content was measured in saturated butter oil enriched with free phytosterols, whereas the highest matrix dimer and polymer contents were found in free phytosterol-enriched liquid margarine. Generally, common hypocholesteremic spreads should not be used for frying. Today, there are claims that scientists have succeeded in developing methods of adding free or esterified phytosterols to frying oils, but the information one can obtain from the published abstracts of the relative patents is very poor.

9.5.6.1 Toxicity Data

An extensive package of safety data has shown phytosterol esters to be safe for human use. However, even though phytosterols are very stable lipids, oxidation may occur at low levels under extreme heating conditions, resulting in phytosterol oxides. There is a plethora of publications related to the loss of phytosterols and the level of oxidized sterols in food, but there are only few real feeding or other toxicological experiments with oxidized phytosterols or human data. Lea et al. (2004) generated a phytosterol oxide concentrate (POC) by prolonged heating of phytosterol esters in the presence of oxygen. The genotoxicity and subchronic toxicity of POC, which contained approximately 30% phytosterol oxides, were assessed by a subchronic feeding study and a series of *in vitro* genotoxicity assays. According to the authors, the oxidized material has been shown to raise no obvious concerns for human safety. Based on the feeding study (phytosterol oxides were administered in the diet of the rat for 90 consecutive days), no obvious evidence of toxicity was observed. A NOEL (No Observed Adverse Effect Level) was established at an estimated dietary level of phytosterol oxides of 128 mg/kg/day for males and 144 mg/kg/day for females. Abramsson-Zetterberg et al. (2007) attempted a genotoxicity study *in vivo* using a flow cytometer-based micronucleus assay in mice. POPs such as triols and epoxides obtained from an oxidized mixture of sitosterol and campesterol gave no evidence of genotoxic effect.

The potential toxic effects of sitosterol oxides on U937 and other cell lines were studied by Maguire et al. (2003) and Ryan et al. (2005). Thermally oxidized β-sitosterol demonstrated similar cytotoxic effects as 7-β-hydroxycholesterol, but at higher concentrations. The cytotoxicity of cholesterol and sitosterol and their oxides was examined in culture-derived macrophage cell lines C57BL/6 by Adcox et al. (2001). The results of this study indicate that phytosterols oxidized during frying may cause cellular damage in an *in vitro* cell line similar to that of cholesterol oxides cause such a damage, although less severe.

Exogenous phytosterol oxides were shown to be well absorbed and to accumulate in the body. However, no promotion of atherosclerosis was observed in apo E–deficient mice by Tomoyori et al. (2004). Angiogenicity and atherogenicity studies are conducted more often with cholesterol oxidation products. Such oxidation products, which can also be produced endogenously in tissues through radical-induced reactions, have been shown to cause injury to vascular endothelial and smooth cells, and to induce experimental atherosclerosis alone or in combination with cholesterol (Peng et al., 2005).

9.6 CONCLUSIONS

The sterol composition of frying oils is important for the quality of the fried food. Complete information concerning the exact chemical composition of stable and healthy frying oils cannot be obtained. However, from the literature, one can conclude that future trends for the production of stable frying oils will be not only to modify fatty acid composition but also to add novel antioxidant mixtures that provide increased heat stability and benefit human health. Among the various classes of antioxidants that enhance the resistance of the heated fats to degradation and prolong their ability to be used in repeated frying operations, certain sterols seem to be significant from a practical point of view.

REFERENCES

Abramsson-Zetterberg, L., Svensson, M., and Johnsson, L. 2007. No evidence of genotoxic effect *in vivo* of the phytosterol oxidation products triols and epoxides, *Toxicol. Lett.*, 173: 132–139.

Adcox, C., Boyd, L., Oehrl, L., Allen, J., and Fenner, G. 2001. Comparative effects of phytosterol oxides and cholesterol oxides in cultured macrophage-derived cell lines, *J. Agric. Food Chem.*, 49: 2090–2096.

Aggelousis, G. and Lalas, S. 1997. Quality changes of selected vegetables oils during frying of doughnuts, *Riv. Ital. Sost. Grasse*, 74: 559–565.

Andrikopoulos, N.K. and Demopoulos, C.A. 1989. Deterioration of some vegetable oils. II. after two years storage of fried and non-fried samples, *Rev. Franc. Corps Gras,* 36: 213–215.

Andrikopoulos, N.K., Tzamtzis, V.A., Giannopoulos, G.A. Kalantzopoulos, G.K., and Demopoulos, C.A. 1989. Deterioration of some vegetable oils. I. During heating or frying of several foods, *Rev. Franc. Corps Gras*, 36: 127–129.

Apprich, S. and Ulberth, F. 2004. Gas chromatographic properties of common cholesterol and phytosterol oxidation products, *J Chromatogr. A*, 1055: 169–176.

Bhattacharyya, A.C. and Bhattacharyya, O.K. 1987. Deacidification of high free fatty acid rice bran oil by reesterification and alkali neutralization, *J. Am. Oil Chem. Soc.*, 64: 128–131.

Blekas, G. and Boskou, D. 1986. Effect of esterified sterols on the stability of heated oils. In: *The Shelf Life of Foods and Beverages*, ed. G. Charalambous, 641–645, Amsterdam, Elsevier.

Blekas, G. and Boskou, D. 1988. Effect of β-sitosterol and stigmasterol on the rate of deterioration of heated triacylglycerols. In: *Frontiers of Flavor*, ed. G. Charalambous, 403–408, Amsterdam, Elsevier.

Blekas, G. and Boskou, D. 1989. Oxidation of stigmasterol in heated triacylglycerols, *Food Chem.*, 33: 301–310.

Bortolomeazzi, R., Pizzale, L., Novelli, Novelli, L., and Conte L. S. 1996. Steroidal hydrocarbons formed by dehydration of oxidized sterols in refined oils, *Riv. Ital. Sostanze Grasse*, 73: 457–460.

Bortolomeazzi, R., Cordado, F., Pizzale, L., and Conte L. S. 2003. Presence of phytosterol oxides in crude vegetable oils and their fate during refining, *J. Agric. Food Chem.*, 51: 2394–2401.

Boskou, D. 2006. *Olive Oil-Chemistry and Technology*, 2nd ed., Champaign, IL, AOCS Press.

Boskou, D. and Morton, I.D. 1975. Changes in the sterol composition of olive oil on heating, *J. Sci. Food Agric.*, 26: 1149–1153.

Boskou, D. and Morton, I.D. 1976. Effect of plant sterols on the rate of deterioration of heated oils, *J. Sci. Food Agric.*, 27: 928–932.

Brooks, C.J.W. 1970. Steroids: Sterols and bile acids, In: *Rodd's Chemistry of Carbon Compounds*, Vol. II, Part D, ed. S. Coffey, 95–96, Amsterdam, Elsevier.

Burton, G.W. and Ingold, K.V. 1984. β-Carotene: An unusual type of lipid antioxidant, *Science*, 224: 569–573.

Cercaci, L., Conchillo, A., Rodriguez-Estrada, M., Ansorena, D., Astiasaran, I., and Lercker, G. 2006. Preliminary study on health-related lipid components of bakery products, *J. Food Protection*, 69: 1393–1401.

Cercaci, L., Rodriguez-Estrada, M., Lercker, G., and Decker, E. 2007. Phytosterol oxidation in oil-in-water emulsions and bulk oil, *Food Chem.*, 102: 161–167.

Chang, S.S. and Mone, P.E. 1960. Stabilization of Fats and Oils. U.S. Patent 2, 966, 413.

Chen, J.H. and Ho, C-T. 1997. Antioxidant activities of caffeic acid and related hydroxycinnamic acid compounds, *J. Agric. Food Chem.*, 45: 2374–2378.

Conchillo, A., Cercaci, L., Ansorena, D., Rodriguez-Estrada, M.T., Lercker, J., and Astiasarán, I. 2005. Levels of phytosterol oxides in enriched and nonenriched spreads: Application of a thin-layer chromatography-gas chromatography methodology, *J. Agric. Food Chem.*, 53: 7844–7850.

Daly, G.G., Finocchiaro E.T., and Richardson T. 1983. Characterization of some oxidation products of β-sitosterol, *J. Agric. Food Chem.*, 31: 46–50.

D' Evoli, L., Huikko, L., Lampi, A.-M., Lucarini, M., Lombardi-Boccia, G., Nicoli, S., and Piironen, V. 2006. Influence of rosemary (*Rosmarinus officinalis,* L.) on plant sterol oxidation in extra virgin olive oil, *Mol. Nutr. Food Res.*, 50: 818–823.

Diack, M. and Saska, M. 1994. Separation of vitamin E. and γ-oryzanols from rice bran by normal-phase chromatography, *J. Am. Oil Chem. Soc.*, 71: 1211–1217.

Dutta, P.C. and Appelqvist, L.-Å. 1997. Studies on phytosterol oxides. I. Effect of storage on the content in potato chips prepared in different vegetable oils, *J. Am. Oil Chem. Soc.*, 74: 647–657.

Dutta, P.C. 1997. Studies on phytosterols oxides. II. Content in some vegetable oils and in french-fries prepared in these oils, *J. Am. Oil Chem. Soc.*, 74: 659–666.

Dutta, P.C., Przybylski, R., Eskin, N.A., and Appelqvist, L.-Å. 2007. Formation and analysis and health effects of oxidized sterols in frying fat. In: *Deep Frying: Practices, Chemistry and Nutrition*, ed. M.D. Erickson, 125–178, Champaign, IL, AOCS Press.

Erickson, B.A. 1971. Hypocholesteremic shortening, *Ger. Offen.*, 2, 035, 069.

Fukuda, Y., Nagata, M., Osawa, T., and Namiki, M. 1986a. Contribution of lignan analogues to antioxidative activity of refined unroasted sesame seed oil, *J. Am. Oil Chem. Soc.*, 63: 1027–1031.

Fukuda, Y., Nagata, M., Osawa, T., and Namiki, M. 1986b. Chemical aspects of the antioxidative activity of roasted sesame seed oil and the effect of using the oil for frying, *Agric. Biol. Chem.*, 50: 857–862.

Fukuda, Y., Isobe, M., Nagata, M., Osawa, T., and Namiki, M. 1986c. Acidic transformation of sesamolin, the sesame-oil constituent, into an antioxidant bisepoxylignan, sesaminol, *Heterocycles*, 24: 923–926.

Fukuda, Y., Osawa, T., Kawagishi, S., and Namiki, M. 1988. Oxidative stability of foods fried with sesame oil, *J. Jap. Soc. Food Sci. Technol.*, 35: 28–32.

Fukuda, Y, Osawa, T. Kawagishi, S., and Namiki, M. 1994. Chemistry of lignan antioxidants in sesame seed and oil. In: *Food Phytochemicals for Cancer Prevention II*, eds. C.-T. Ho, T. Osawa, M.-T. Huang, and R.T. Rozen, 264–274, Washington, DC, American Chemical Society.

Fukuda, Y., Koizumi, Y., Ho, R., and Namiki, M. 1996. Synergistic action of the antioxidative components in roasted sesame seed oil, *J. Jap. Soc. Food Sci. Technol.*, 43: 1272–1277.

Garcia-Llatas, G., Cercaci, L., Rodriguez-Estrada, M.T., Jesús Lagarda, M., Farré, R., and Lercker, G. 2008. Sterol oxidation in ready-to-eat infant foods during storage, *J. Agric. Food Chem.*, 56: 469–475.

Gaydou, E.M., Raoniza Finimanana, R., and Bianchini, J.P. 1980. Quantitative analysis of fatty acids and sterols in malagasy rice bran oils, *J. Am. Oil Chem. Soc.*, 54: 141–142.

Ghavami, M. and Morton, I.D. 1984. Effect of heating at deep-fat frying temperature on the sterol content of soya bean oil, *J. Sci. Food Agric.*, 35: 569–572.

Giuffrida, F., Destaillats, F., Robert, F., Skibsted, L.H., and Dionisi, F. 2004. Formation and hydrolysis of triacylglycerol and sterol epoxides: Role of unsaturated triacylglycerol peroxyl radicals, *Free Radic. Biol. Med.*, 37: 104–114.

Gordon, M.H. 1990. Plant sterols as natural antipolymerization agents, In: I*nternational Symposium on New Aspects of Dietary Lipids. Benefits, Hazards and Use*, ed. 23–34. Royal Swedish Academy of Sciences, Göteborg.

Gordon, M.H. and Magos, P. 1983. The effect of sterols on the oxidation of edible oils, *Food Chem.*, 10: 141–147.

Gordon, M.H. and Magos, P. 1984. Products from the autoxidation of Δ^5-avenasterol, *Food Chem.*, 14: 295–301.

Gordon, M.H. and Williamson, E. 1989. A comparison of headspace analysis with other methods for assessing the oxidative deterioration of edible oils. In: *Trends in Food Science*, eds. W.S. Lien and C.W. Foo, 53–57, Singapore, Singapore Institute of Food Science and Technology.

Grandgirard, A., Martine, L., Joffre, C., Juaneda, P., and Berdeaux, O. 2004. Gas chromatographic separation and mass spectrometric identification of mixtures of oxyphytosterol and oxycholesterol derivatives: Application to a phytosterol-enriched food, *J. Chromatogr. A*, 1040: 239–250.

Guardiola, F., Bou, R., Boatella, J., and Codony, R. 2004. Analysis of sterol oxidation products in food, *J. AOAC Int.*, 87, 441–465.

Gupta, H.P. 1989. Rice bran offers India an oil sources, *J. Am. Oil Chem. Soc.*, 66: 620–623.

Haumann, B.F. 1996. The goal: Tastier and healthier fried foods, *Inform*, 7: 320–333.

Hayes, K., Pronczuk, A., and Perlman, D. 2004. Nonesterified phytosterols dissolved and recrystallized in oil reduce plasma cholesterol in gerbils and humans, *J. Nutr.*, 134: 1395–1399.

Heasman, M. and Mellentin, J. 1998. Single ingredients—global markets, *Int. Food Ingredients*, 1: 16–18.

Homberg, E. and Bielefeld, B. 1985. Freie und Gebundene Sterine in Pflanzenfetten, *Fette-Seifen-Anstrichm.*, 87: 61–64.

Homberg, E. and Bielefeld, B. 1989. Sterin Zusammensetzung und Steringehalt in 41 verschiedenen Pflanzlichen und Tierischen Fetten, *Fat Sci. Technol.*, 91: 23–27.

Itoh, T., Tamura, T., and Matsumoto, T. 1973a. Sterol composition of 19 vegetable oils, *J. Am. Oil Chem. Soc.*, 50: 122–125.

Itoh, T., Tamura, T., and Matsumoto, T. 1973b. Methylsterol composition of 19 vegetable oils, *J. Am. Oil Chem. Soc.*, 50: 300–303.

Johnsson, L. and Dutta, P. C. 2003. Side-chain oxidation products of sitosterol and campesterol by chromatographic and spectroscopic methods, *J. Am. Oil Chem. Soc.*, 80: 767–776.

Johnsson, L. and Dutta, P.C. 2006. Determination of phytosterol oxides in some food products by using an optimized transesterification method, *Food Chem.*, 97, 606–613.

Johnsson, L., Andersson R.E., and Dutta, P.C. 2003. Side-chain autoxidation of stigmasterol and analysis of a mixture of phytosterol oxidation products by chromatographic and spectroscopic methods, *J. Am. Oil Chem. Soc.*, 80: 777–783.

Kamal-Eldin, A., Appelqvist, L.-Å., Yousif, G., and Iskander, G.M. 1992. Seed lipids of *Sesamum indicum* and related wild species in Sudan. The sterols, *J. Sci. Food Agric.*, 59: 327–334.

Kamal-Eldin, A. and Appelqvist, L.-Å. 1994. Variations in the composition of sterols, tocopherols and lignans in seed oils from four sesamum species, *J. Am. Oil. Chem. Soc.*, 71: 149–156.

Kaufmann, H. P., Vanneckel, E., and Hamza, Y. 1970. Űber die Veränderung der Sterine in Fetten und Őlen Bei der Industriellen Bearbeitung Derselben I., *Fette, Seifen, Anstrichm.*, 72: 242–246.

Kaufmann, H. P. and Hamza, Y. 1970. Űber die Veränderung der Sterine in Fetten und Őlen bei der Industriellen Bearbeitung Derselben II., *Fette, Seifen, Anstrichm.*, 72: 432–433.

Kikugawa, K., Arai, M., and Kurechi, T. 1983. Participation of sesamol in stability of sesame oil, *J. Am. Oil Chem. Soc.*, 60: 1528–1533.

Kornfeldt, A. and Croon, L. 1981. 4-Demethyl, 4-monomethyl and 4,4-dimethylsterols in some vegetable oils, *Lipids*, 16: 306–314.

Kudchodkar, B.J., Horlick, L., and Sodhi, H.S. 1976. Effect of plant sterols on cholesterol metabolism in man, *Atherosclerosis*, 23: 239–248.

Lambelet, P., Grandgirard, A., Gregoire, S., Juaneda, P., Sebelio, J. L., and Bertoli, C. 2003. Formation of modified fatty acids and oxyphytosterols during refining of low erucic acid rapeseed oil, *J. Agric. Food Chem.*, 51: 4284–4290.

Lampi, A.-M., Juntunen, L., Toivo, J., and Piironen, V. 2002. Determination of thermoxidation products of plant sterols, *J. Chromatogr. B*, 777: 83–92.

Lampi, A.-M., Kemmo, S., Mäkelä, A., Heikkinen, S., and Piironen, V. 2009. Distribution of monomeric, dimeric and polymeric products of stigmasterol during thermoxidation, *Eur. J. Lipid Sci. Technol.*, 111: 1027–1034.

Lee, K., Herian, A., and Higley, N. 1985. Sterol oxidation products in french fries and in stored potato chips, *J. Food Protection*, 48: 158–161.

Lavanjinha, J., Vilira, O., Almeida, L., and Madeira, V. 1996. Inhibition of metmyoglobin/H_2O_2-dependent low density lipoprotein lipid peroxidation by naturally occurring phenolic acids, *Biochem. Pharmacol.*, 51: 395–402.

Lea, L.J., Hepburn, P.A., Wolfreys, A.M., and Badrick, P. 2004. Safety evaluation of phytosterol esters. Part 8. Lack of genotoxicity and subchronic toxicity with phytosterol oxides, *Food Chem. Toxicol.*, 42: 771–783.

Louter, A. 2004. Determination of plant sterol oxidation products in plant sterol enriched spreads, fat blends, and plant sterol concentrates, *J. AOAC Int.*, 87: 485–492.

Maguire, L., Konoplyannikov, M., Ford, A., Maguire, A.R., and O'Brien, N. 2003. Comparison of cytotoxic effects of β-sitosterol oxides and α cholesterol oxide, 7β-hydroxycholesterol, in cultured mammalian cells, *Brit. J. Nutr.*, 90: 767–775.

Malecka, M., Rudzińska, M., Pacholek, B., and Wąsowicz, E. 2003. The effect of raspberry, black currant and tomato seed extracts on oxyphytosterol formation in peanuts, *Pol. J. Food Nutr. Sci.*, 12/53, SI 1: 49–53.

Marinova, E. and Yanishlieva, N.V. 1994. Effect of lipid unsaturation on the antioxidative activity of some phenolic acids, *J. Am. Oil Chem. Soc.*, 71: 427–434.

Mennie, D., Moffat, C.F., and McGill, A.S. 1994. Identification of sterene compounds produced during the processing of edible oils, *J. High Res. Chromatogr.*, 17: 831–838.

Nagata, M., Osawa, T., Namiki, M., Fukuda, Y., and Ozaki, T. 1987. Stereochemical structures of antioxidative bisepoxylignans, sesaminol and its isomers, transformed from sesamolin, *Agric. Biol. Chem.*, 51: 1285–1289.

Nakhasi, D., Daniels, R., and Eartly, J. 2007. Phytosterol containing deep-fried foods and methods with health promoting characteristics, Patent, US WO/2007/067884.

Namiki, M. 1995. The chemistry and physiological functions of sesame, *Food Rev. Int.,* 11: 281–329.

Niewiadomski, H. and Sawicki, J. 1964. Űber die bildung und beseitigung verschiedener steroid-derivate während der raffination der spreiseőle, *Fette, Seifen, Anstrichm.*, 66: 830–935.

Nourooz-Zadeh, J. and Appelqvist, L.-Å. 1992. Isolation and quantitative determination of sterol oxides in plant-based foods: soybean oil and wheat flour, *J. Am. Oil Chem. Soc.*, 69: 288–293.

Oehrl, L., Hansen, A., Rohrer, C., Fenner, G., and Boyd, L. 2001. Oxidation of phytosterols in a test food system, *J. Am. Oil Chem. Soc.*, 78: 1073–1078.

Peng, S.-K., Hu, B., and Morin, R.J. 2005. Angiotoxicity and atherogenicity of cholesterol oxides, *J. Clin. Labor. Anal.*, 5: 144–152.

Peterson, D.W., Robbins, R., Shneour, E.A., and Myers, W.D. 1951. Effect of soybean sterols in the diet on plasma and liver cholesterol in chicks, *Proc. Soc. Exptl. Biol. Med.*, 78: 143–147.

Rogers, E.J., Rice, S.M., Nicolosi, R.J., Carpenter, D.R., McClelland, C.A., and Romanczyk, L.J. 1993. Identification and quantitation of γ-oryzanol components and simultaneous assessment of tocols in rice bran oil, *J. Am. Oil Chem. Soc.*, 70: 301–307.

Romero, A., Cuesta, C., and Sanchez-Muniz, F.J. 1995. Quantitation and distribution of polar compounds in an extra virgin olive oil used in frying and turnover of fresh oil, *Fat Sci. Technol.*, 97: 403–407.

Rudzińska, M., Korczak, J., Gramza, A., Wąsowicz, E., and Dutta, P.C. 2004. Inhibition of stigmasterol oxidation by antioxidants in purified sunflower oil, *J. AOAC Int.*, 87: 499–504.

Rudzińska, M., Uchman, W., and Wąsowicz, E. 2005. Plant sterols in food technology, *Acta Sci. Pol. Technol. Aliment.*, 4: 147–156.

Rudzińska, M., Przybylski R., and Wąsowicz, E. 2009. Products formed during thermo-oxidative degradation of phytosterols, *J. Am. Oil Chem. Soc.*, 86: 651–662.

Ryan, E., Chopra, J., McCarthy, F., Maguire, A.R., and O'Brien, N.M., 2005. Qualitative and quantitative comparison of the cytotoxic and apoptotic potential of phytosterol oxidation products with their corresponding cholesterol oxidation products, *Brit. J. Nutr.*, 94: 443–451.

Ryan, E., McCarthy, F.O., Maguire, A.R., and O'Brien, N.M, 2009. Phytosterol oxidation products: Their formation. occurrence, and biological effects, *Food Rev. Int.*, 25: 157–174.

Sah, A., Agrawal, B.K.D., and Shukla, L.J. 1983. A new approach in dewaxing and refining rice bran oil, *J. Am. Oil Chem. Soc.*, 60: 466.

Säynäjoki, S., Sundberg, S., Soupas, L., Lampi, A.-M., and Piironen, V. 2003. Determination of stigmasterol primary oxidation products by high-performance liquid chromatography, *Food Chem.*, 80: 415–421.

Scott, C.B., Butler, J., Halliwell, B., and Aruoma, O.I. 1993. Evaluation of antioxidant actions of ferulic acid and catechins, *Free Radic. Res. Commun.*, 19: 241–253.

Sims, R.J., Fioriti, J.A., and Kanuk, M.J. 1972. Sterol additives as polymerization inhibitors for frying oils, *J. Am. Oil Chem. Soc.*, 49: 298–301.

Soupas, L., Juntunen, L., Lampi, A.-M., and Piironen, V. 2004. Effects of sterol structure, temperature, and lipid medium on phytosterol oxidation, *J. Agric. Food Chem., 52:* 6485–6491.

Soupas, L., Huikko, L., Lampi, A.-M., and Piironen, V. 2005. Esterification affects phytosterol oxidation, *Eur. J. Lipid Sci. Technol.*, 107: 107–118.

Soupas, L. 2006. Oxidative Stability of Phytosterols in Food Models and Foods (Doctoral dissertation), University of Helsinki.

Soupas, L., Huikko, L., Lampi, A.-M., and Piironen, V. 2007. Pan-frying may induce phytosterol oxidation, *Food Chem.*, 101, 286–297.

Tabee, E., Azamard-Damirchi, S., Jägerstad, M., and Dutta, P.C. 2008a. Lipids and phytosterol oxidation products in commercial french fries commonly consumed in Sweden, *J. Food Compos. Anal.*, 21: 169–177.

Tabee, E., Azamard-Damirchi, S., Jägerstad, M., and Dutta, P.C. 2008b. Effects of α-tocopherol on oxidative stability and phytosterol oxidation during heating in some regular and high-oleic vegetable oils, *J. Am. Oil Chem. Soc.*, 85: 857–867.

Tai, C., Chen, Y., and Chen, B. 1999. Analysis, formation and inhibition of cholesterol oxidation: An overview (part 1), *J. Food Drug Anal.*, 7: 243–257.

Tomoyori, H., Kawata, Y., Higuchi, T, Ichi, I., Sato, H., Sato, M., Ikeda, I., and Imaizumi, K. 2004. Phytosterol oxidation products are absorbed in the intestinal lymphatics in rats but do not accelerate atherosclerosis in apolipoprotein E-deficient mice, *J. Nutr.*, 134: 1690–1696.

Verleyen, T., Cortes, E., Verhe, R., Dewettinck, K., Huyghebaert, A., and De Greyt, W. 2002. Factors determining the steradiene formation in bleaching and deodorisation, *Eur. J. Lipid Sci. Technol.*, 104: 331–339.

Vuovisto, M. and Miettinen, T.A. 1994. Absorption, metabolism and serum concentrations of cholesterol in vegetarians: Effects of cholesterol feeding, *Am. J. Clin. Nutr.*, 59: 1325–1331.

Weber, N., Bergenthal, D., Brűhl, L., and Schulte, E. 1992. Disterylether-Artefakte der Fettbleichung, *Fat Sci. Technol.*, 94: 182–192.

Weihrauch, J.L. and Gardner, J.M. 1978. Sterol content of foods of plant origin, *J. Am. Diet. Assoc.*, 73: 39–47.

Weststrate, J.A. and Meijer, G.W. 1998. Plant sterol-enriched margarines and reduction of plasma total- and LDL-cholesterol concentrations in normocholesterolaemic and mildly hypercholesterolaemic subjects, *Eur. J. Clin. Nutr.*, 52: 334–343.

White, P.J. and Armstrong, L.S. 1986. Effect of selected oat sterols on the deterioration of heated soybean oil, *J. Am. Oil Chem. Soc.*, 63: 525–529.

Winkler, J.K., Warner, K., and Glynn, M.T. 2007. Effect of deep-fat frying on phytosterol content in oils with differing fatty acid composition, *J. Am. Oil Chem. Soc.*, 84: 1023–1030.

Winkler, J.K. and Warner, K. 2008a. The effect of phytosterol concentration on oxidative stability and thermal polymerization of heated oils, *Eur. J. Lipid Sci. Technol.*, 110: 455–464.

Winkler, J.K. and Warner, K. 2008b. Effect of phytosterol structure on thermal polymerization of heated soybean oil, *Eur. J. Lipid Sci. Technol.*, 110: 1068–1077.

Yan, P.S. and White, P.J. 1990. Linalyl acetate and other compounds with related structures as antioxidants in heated soybean oil, *J. Agric. Food Chem.*, 38: 763–768.

Yan, P.S. and White, P.J. 1991. Linalyl acetate and methyl silicone effects on cholesterol and triglyceride oxidation in heated lard, *J. Am. Oil Chem. Soc.*, 68: 763–768.

Yanishlieva, N.V., Schiller, H., and Marinova, E. 1980. Autoxidation of sitosterol. II. Main products formed at ambient and high temperature treatment with oxygen, *Riv. Ital. Sost. Grasse*, 57: 572–576.

Yanishlieva, N.V. 1981. Über den Einfluss der Hauptoxidationsprodukte der Δ^5-Sterine auf die Autoxidationsstabilität der Sonnenblumenöls, *Seife-Fette-Öle-Wachse*, 107: 591–594.

Yanishlieva, N.V. and Marinova, E. 1992. Inhibited oxidation of lipids. I. Complex estimation and composition of the antioxidative properties of some natural and synthetic antioxidants, *Fat Sci. Technol.*, 94: 374–379.

Yanishlieva, N.Y. and Schiller, H. 1983. Effect of sitosterol on autoxidation rate and product composition in a model system, *J. Sci. Food Agric.*, 35: 219–224.

Yanishlieva, N.Y., Marinova, E., and Popov, A. 1982. Über den Einfluss des Sitosterins und dessen Stearats auf die Oxidationsstabilität verschiedener Lipidsysteme, *Bulgarian Acad. Sci.—Commun. Dept. Chem.*, 15: 301–309.

Yen, G.-C. 1990. Influence of seed roasting process on the changes in composition and quality of sesame oil, *J. Sci. Food Agric.*, 50: 563–570.

Zhang, X., Julien-David, D., Miesch, M., Geoffroy, P., Raul, F., Roussi, S., Aoude-Werner, D., and Marchioni, E. 2005. Identification and quantitative analysis of β-sitosterol oxides in vegetable oils by capillary gas chromatography-mass spectrometry, *Steroids*, 70: 896–906.

Zhang, X., Julien-David, D., Miesch, M., Raul, F., Geoffroy, P., Aoude-Werner, D., Ennahar, S., and Marchioni, E. 2006. Quantitative analysis of β-sitosterol oxides induced in vegetable oils by natural sunlight, artificially generated light, and irradiation, *J. Agric. Food Chem.*, 54: 5410–5415.

Zunin, P., Calgagno, C., and Evangelisti, F. 1998. Sterol oxidation in infant milk formulas and milk cereals, *J. Dairy Res.*, 65: 591–598.

Zunin, P., Salvadeo, P., Boggia, R., and Evangelisti, F. 2006. Sterol oxidation in meat- and fish-based homogenized baby foods containing vegetable oils, *J. AOAC Int.*, 89: 441–446.

10 Chemical and Biological Modulations of Frying Fats—Impact on Fried Food

Karl-Heinz Wagner and Ibrahim Elmadfa

CONTENTS

10.1 INTRODUCTION

Many people worldwide appreciate the taste and convenience of fried food products, which very often become a predominant food choice. From the health point of view, concern is raised due to the high fat content and the impact of the fried food product on the nutritional quality. For decades, partially hydrogenated frying fats (PHF) were used to prolong shelf life. This kind of frying fat was found to be the highest source of *trans* fatty acids (TFA). In the last decade, most of the PHFs were replaced by other frying fats; however, such fats are still in use in countries where

TFA intakes are not or less calculated at the population level. In this chapter, alternative methodology to partial hydrogenation, the change in the strategy of using PHFs, and the consequent reduction in TFA in foods are discussed. Other aspects covered are the interactions between frying fats and foods, and their impact on the fried end product.

10.2 FAT–FOOD INTERACTIONS

10.2.1 Oil Uptake

Various factors seem to be responsible for the fat level of the fried end product (Table 10.1) such as the frying temperature, the origin and quality of the frying oil, the surface of the product (product shape and product size), a prefrying process and, most important, the postfrying treatment of the product (Figure 10.1) (Ziaiifar et al., 2008).

Overall, three mechanisms have been proposed to describe the oil absorption process: water replacement, cooling-phase effects, and surface-active agents (Dana and Saguy, 2006).

10.2.1.1 Water Replacement

After adding the food to the heated oil, the temperature of the oil decreases and the temperature in the food increases slowly. At the same time the water penetrates to the exterior part. During the frying process heat is transferred to the food, which is quickly heated and covered with steam since the water at the surface of the product

TABLE 10.1
Mean Oil Content of Some Fried Foods

Food Item	Oil Content (g/100 g Edible Product)
Potato chips	37.5
Tortilla chips	23.4
Donuts	22.9
Fried tofu	20.2
Onion rings	18.7
Chicken, battered/breaded, light meat	18.1
French fries	17.1
Shrimp, breaded	15.2
Sardines	13.3
Fish fillet, battered/breaded	12.3
Pork leg schnitzel	9.1

Source: Candela et al., (1996). *Journal of Food Composition and Analysis*, 9 (3): 277–282; Clausen, I. and Ovesen, L. (2005). *Journal of Food Composition and Analysis,* 18 (2–3): 201–211; Dana, D. and Saguy, I.S. (2006). *Advances in Colloid and Interface Science,* 128–130: 267–272.

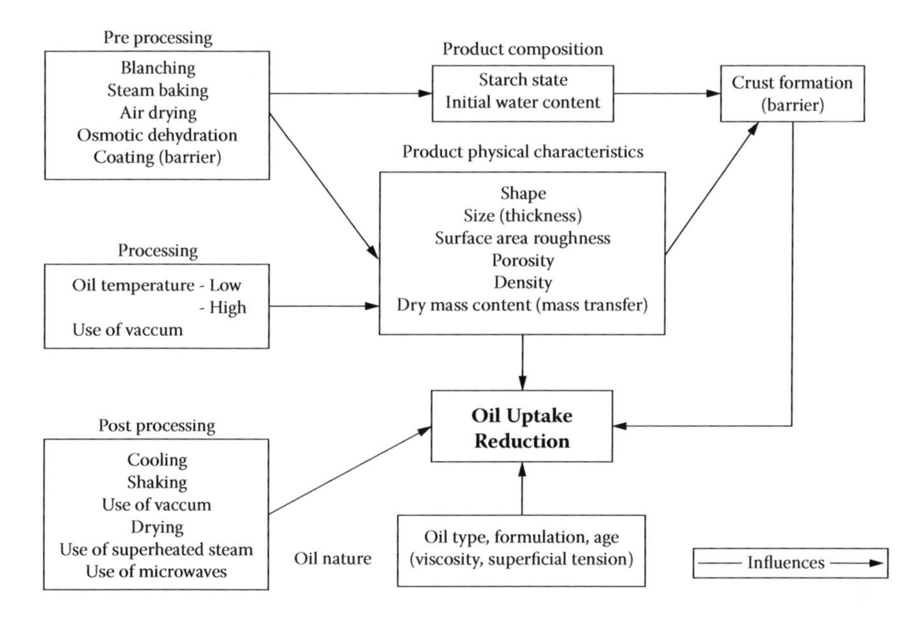

FIGURE 10.1 Factors affecting the oil uptake. (From Ziaiifar, A.M., et al., 2008. *International Journal of Food Science and Technology*, 43 (8): 1410–1423.)

is reaching the boiling point. The steam limits the penetration of the oil through the surface to the center, so the result is a dehydrated surface and a core where the temperature is lower than 100°C (Dobarganes et al., 2000). As the process progresses, oil adheres to the food, entering the large voids, product imperfections, and crevices left by the changes in structure. As the voids are quite large, there is no inner resistance due to positive water vapor pressure. This mechanism could furnish a possible explanation for the direct relationship observed between water loss and oil uptake; however, the oil uptake is rather low (Figure 10.1).

10.2.1.2 Cooling of the Fried Product

When the frying process is completed, the food is removed from the fryer and the product starts to cool down immediately. This leads to water vapor condensation and a decrease in the internal pressure. Oil, which is adhered to the surface of the product, is sucked into the product similar to a "vacuum effect."

In this phase, the oil uptake is a surface phenomenon involving equilibrium between adhesion and drainage of oil as the food is removed from the oil bath (Ufheil and Escher, 1996).

Further, the oil viscosity increases with extended frying time; for example, soybean oil viscosity rose from 53 to 208 mPa s for fresh oil and 60 h degraded oil, respectively. Olive oil viscosity increased from 50 to 100 mPa s after 16 h heating at 200°C (Benedito et al., 2002). The increased viscosity and the product's surface characteristics are probably the most critical factors leading to higher oil uptake when food is removed from the fryer, because of the decreased drainage of oil from the fried product (Figure 10.2). Two effective approaches could be utilized to reduce

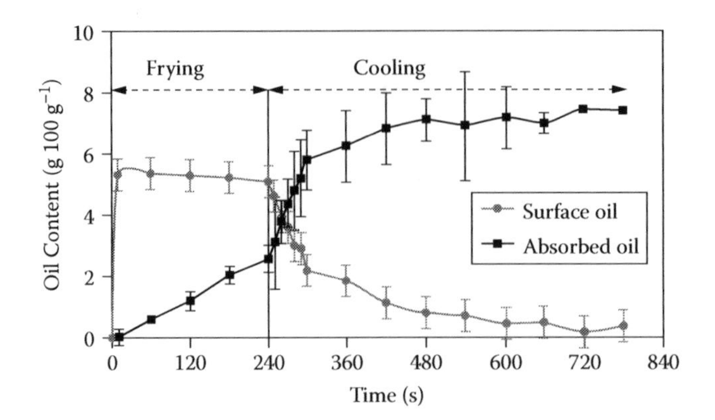

FIGURE 10.2 Oil content absorbed or remained on the surface of french fries during frying (170°C) and cooling (20°C). (From Ziaiifar et al., 2008. *International Journal of Food Science and Technology*, 43 (8):1410–1423).

oil uptake after frying. The food industry approach is to use hot air or superheated steam that blows away most of the surface oil and simultaneously maintains the product temperature preventing cooling and vacuum. The household approach is blotting the surface with an adsorbent paper.

10.2.1.3 Surface-Active Agents

Oil degrades during the frying process from a mixture of triglycerides into a mixture of hundreds of compounds. The water evaporation induces hydrolytic reactions, high frying temperatures, and the subsequent formation of diglycerides, monoglycerides, free fatty acids, and glycerol.

Mono- and diglycerides are surface-active compounds which increase the foaming tendency of the oil. The formation of surfactants enhances the contact between the food and the frying oil, which results in excessive oil absorption. Surfactant formation also affects heat transfer at the oil–food interface and reduces the surface tension between the two immiscible materials. According to *Surfactant Theory of Frying* (Blumenthal and Stier, 1991), surfactants are responsible for the surface and interior differences in the fried-food quality induced by the aging oil. As contact time between the food and the frying oil increases, more heat is transferred from the frying oil to the food, causing higher dehydration at the surface and water migration from the core to the exterior of the fried food. Thus, higher surfactant concentrations produce oil-soaked products with an overcooked exterior and undercooked interior (Blumenthal, 2001; Blumenthal and Stier, 1991). More about the interactions can be read in recent reviews by Dana and Saguy (2006) and Ziaiifar et al. (2008).

10.2.2 Important Factors Influencing Oil Uptake

10.2.2.1 Product Properties

Since oil uptake is a surface phenomenon, the dimension of the food will determine the oil taken up. Generally absorption increases when the product thickness decreases. French fries absorb less oil than chips due to their lower surface/volume ratio (see Table 10.1). Further, the surface area (functional properties of the outer layer) is important since the oil penetrates in the food through pores in the crust. If cells are broken during cutting, this is a favorable location for oil absorption.

The solid/water content is an important factor that influences the fat uptake. The higher the initial water content, the higher the oil uptake. The moisture loss that occurs during frying of thin products (i.e., chips) leads to considerable fat uptake because of the extensive void volumes created by the water escape. Also, the food density may play a critical role; the higher the density, the lower the fat uptake.

10.2.2.2 Frying Oil Properties

In general the type of oil and the frying temperature have a substantial influence on the oil uptake. Some data indicate that the oil uptake is higher when the content of polyunsaturated fatty acids (PUFA) in the oil is higher compared to monounsaturated fatty acids (MUFA) (Kita and Lisinska, 2005). Vitrac et al. (2002) showed the opposite; oil uptake was weaker with PUFA-rich oils (Vitrac et al., 2002). It seems that a very important factor within these processes is the oil viscosity. The higher the viscosity, the slower the oil migration. In this respect not only the fatty acid pattern is important but also the frying temperature. Increasing the temperature means increasing the oil viscosity, which leads to a lower oil uptake in the fried product. Frying at 190°C leads to a significantly lower oil uptake than at 150°C –170°C (Kita and Lisinska, 2005). However, since concern has risen on frying at high temperatures due to the formation of Maillard reaction products such as acrylamide, in many countries the frying temperatures were restricted to 170°C or less. The highest oil uptake was observed at lower temperatures of 120°C. Also, oil aging seems to have an effect; however, the effect is less than expected (Mellema, 2003).

The most popular processes applied before frying at the industrial scale are blanching, air drying, osmotic dehydration, steam baking, and surface treatment (coatings).

Blanching is a process where the food is plunged into hot water or steam for enzyme and microorganism inactivation. It has also been stated that blanching before frying decreases the dry mass content of the product because of the migration of water-soluble components from the product to the blanching water. As a result, water content is increased and thus final oil content (Alvarez et al., 2000; Pedreschi et al., 2005). On the other hand, surface starch gelatinization that occurs during blanching could form a firm thin layer that reduces the oil absorption. In addition, blanching activates pectinesterases that can damage the surface cell walls, thereby causing a decrease in the porosity and oil content of the product (Aguilera-Carb et al., 1999). Drying prior to frying is another way used to create a firm and dried surface material matrix around the product. This technique decreases the total water content of the product and limits oil absorption (Vitrac et al., 2002). Moreover, the shrinkage that occurs during drying reduces total surface area and consequently diminishes mass transfer. Finally, this preprocess affects in two more ways: it decreases oil uptake while improving organoleptic properties of the product by increasing crispness.

Significant effects were also reported by osmotic dehydration. Krokida et al. (2001) showed that french fries soaked in a sugar solution (40% w/w) exhibited 60%

reduction in fat content, while soaking in NaCl (20% w/w) and maltodextrine solutions (20% w/w) for the same treatment times resulted in lower reductions in oil content (35% and 15%, respectively) (Krokida et al., 2001).

Finally, steam baking, like blanching and air drying, can be used to form a tight barrier at the outer surface of the product because of severe starch gelatinization. This outer layer presents strong resistance to oil entrance into the product (Ziaiifar et al., 2008).

Coating is also a popular prefrying technique that consists of dipping a raw product in a coating suspension for a short time just before frying. This treatment reduces surface porosity and also builds a barrier against oil absorption. Indeed, coating decreases not only fat uptake but also water loss during frying. The most important properties of hydrocolloids are their film formation ability, heat stability, as well as fat and water transfer properties. In addition there is no impact on organoleptic properties. Some of the most commonly used and efficient hydrocolloids are cellulose derivatives such as hydroxypropyl cellulose (HPC), methyl cellulose (MC), and hydroxypropylmethyl cellulose (HPMC) (Albert and Mittal, 2002). Some studies have demonstrated that oil uptake could be reduced by 40% in french fries using cellulose coatings before frying although their water content was high (Garcia et al., 2002; Mellema, 2003).

10.3 NUTRITIONAL CHANGES

As discussed in the first part of the chapter, frying leads to an oil uptake and an oil penetration into the food. This means a change toward a higher fat food accompanied by changes in the organoleptic properties, the food fatty acid pattern and the content of fat soluble substances such as lipid soluble vitamins.

Researchers have investigated the impact of frying on the fatty acid pattern of the final product. Sanchez-Muniz et al (1992) determined the effect of frying oils (sunflower oil, olive oil, and lard) on the fatty acid composition of sardines, which are a very good source of n-3 fatty acids. Saturated fatty acid content increases when lard is used for frying. A 4-times increase in MUFA was noted in the fish fried in sunflower oil and 8 times in lard, while a 10-times increase was noted in the fish fried in olive oil. PUFA n-6 content rose 4 times with olive oil, 6.3 times with lard, and 19.9 times with sunflower oil. The n-3 PUFA fell 3.3 times in the case of sunflower oil and 2.2 times with olive oil; no changes were observed with lard. The content of polar methyl esters significantly increased in lard after the frying of sardines but remained unmodified in olive oil and sunflower oil. Sardine cholesterol content significantly decreased after frying (Sanchez-Muniz et al., 1992). Similar findings were observed in later studies with meat dishes, where also all fatty acids increased in chicken breast, pork steak, and pork loin (Candela et al., 1996) or pork loin chips (Ramirez and Cava, 2005). In another study meat with different fat contents were pan-fried in margarine or oil. Beefsteaks and pork leg schnitzel having about 6% and 2% initial fat gained not more than 2 g fat/100 g raw meat, even when pan-fried in a relatively large amount of margarine. Slight differences were observed, depending on the frying time, slice thickness, frying fat type (oil or margarine), quantity of frying fat, or "resting" on the pan. Breaded pork schnitzel gained up to 8 g fat/100 g raw breaded schnitzel due to the higher absorption properties of the coating. High-fat pork patties (18% fat initially) lost 2.7 g fat/100 g raw weight even when pan-fried with large amounts of margarine, while low-fat pork patties (11% fat initially) gained 0.4 g. Similar results were obtained for low- (8%) and high-fat (12%) beef patties (Clausen and Ovesen, 2005).

Significant changes in the fatty acid pattern were observed by Romero et al. (2000), but they found that the frequent addition of fresh oil throughout the frying process was able to minimize the fatty acid changes (Romero et al., 2000).

Changes in the initial fatty acid pattern can also be favorable, in particular if the initial saturated fatty acid (SFA) content is high. In medium fat steaks with moderate initial fat content (E% fat) fried in extra virgin olive oil total fat and SFA content decreased, while MUFA and PUFA content increased (Librelotto et al., 2008).

Similar observations were reported by Saghir et al. (2005), who found an increase in the *trans* fatty acids in beef filets after short pan-frying with olive oil, corn oil, or partially hydrogenated plant oil. In general, SFAs in the beef were reduced. At the same time, there was an increase in vitamin E, particularly after pan-frying with corn oil, which is a good source of γ-tocopherol. Although lipid oxidation products increased moderately after pan-frying, they were always within a low range and without hazard for the consumers (Saghir et al., 2005).

In the same investigation, similar changes were found for salmon and pork. The pan-frying process did not affect the vitamin D concentration in salmon (Elmadfa et al., 2006).

Very recently, home cooking methods were investigated for their potential to change the antioxidant activity of vegetables. Frying vegetables was either able to increase the total antioxidative capacity or had no/decreasing effect (Jimenez-Monreal et al., 2009).

10.3.1 *Trans* Fatty Acids

Evidence from the last 15 years indicates that diets containing TFAs result in increased serum low-density lipoprotein cholesterol (LDL-C) and decreased high-density lipoprotein cholesterol (HDL-C), if compared to diets containing MUFA- or PUFA- rich fats; the latter are believed to increase the ratio of total Cholesterol:HDL-C, which is associated with a risk of cardiovascular disease (CVD) (EFSA, 2004). Consumption of TFA is associated with higher risk of coronary heart disease, sudden death, and possibly diabetes mellitus (Burlingame et al., 2009; Mozaffarian et al., 2006).

These links are greater than would be predicted by the effects of TFA on serum lipoproteins alone, suggesting that TFA intake may also influence other, nonlipid risk factors.

TFA intake has also been suggested to influence inflammation, early development, and fetal growth, cancer, and allergies, although data are not always consistent (EFSA, 2004; Innis, 2006; Wijga et al., 2006). Due to the public and scientific concern, TFA content of foods has been reduced worldwide (Craig-Schmidt, 2006). In 1976, the average intake in Europe was 6 g/d (Stender et al., 2006a); in the TRANSFAIR study, the intake varied between 1.2 and 6.7 g/d, with lower intake in the south (Hulshof et al., 1999).

TFAs are formed by three different ways: hydrogenation of PUFA-rich liquid oils to produce hardstocks, biohydrogenation in rumens, and deep-frying. Partial hydrogenation of PUFA-rich oils causes the highest TFA concentrations in food. The hydrogenation process has been known for a century; its application in industry started in early 1900 when fats were prepared from sperm whale oil (Martin et al., 2007). Since then, partially hydrogenated and TFA, rich fat have been used for margarines, spreads, shortenings, and all kind of ready meals. It was also used for frying since it improved the physical, chemical, and organoleptic characteristics of the food; and particularly for frying fats, it increased shelf life. Traditionally, frying oils were solid and rich in high SFAs, to minimize the oxidation process. Shortenings for this application were first introduced in the 1950s. Frying oils used by restaurants, fast-food companies, and food service operations ranged in TFA content from 0% to about 35% of total FA.

Since the first alarming reports of the negative health effects in the 1990s (Hu et al., 1997; Willett et al., 1993), the frying fats were slowly replaced by nonhydrogenated fats or alternatively produced frying oils. However, due to economic reasons, the exchange was very slow and often forced by food laws.

Three types of hydrogenated fats have been produced since 1900. The first was a solid fat, similar to lard or beef tallow that was used as a frying and baking fat. This fat was used until 2004. It contained 45% TFA and 0% essential fatty acids (linoleic and linolenic acid), according to its producer. A second type of hydrogenated fat was used from 1968 to 2004. It contained 20%–27% *trans* fatty acid with approximately 24% linoleic acid (Kummerow, 1993). The third type of hydrogenated fat now being used extensively in United States contains 39.7% isomeric *cis* and *trans* fat, 16.6% linoleic acid, and 0.7% linolenic acid (Kummerow et al., 2007). There were 16 billion pounds of soybean oil produced in the United States in 2006, of which 8 billion pounds were hydrogenated.

In the TRANSFAIR study, which was performed in the 1990s, the TFA content in fats for frying ranged from less than 1% to 50% of total fatty acids. The Nordic countries had much higher TFA concentrations in their foods, the South European countries had lower TFAs (Hulshof et al., 1999).

There was a need within the national food laws to cut down the TFA food content. Denmark introduced legislation effective at the beginning of 2004 restricting the use of TFA (only industrially produced and not the TFAs which are produced by biohydrogenation) to a maximum of 2% of the fat in the product. Several countries followed this example or are discussing to do so, such as France, Switzerland, Austria, etc. (Wagner et al., 2008). In July 2003, the U.S. Food and Drug Administration issued regulations requiring the labeling of TFAs on packaged foods on or before January 1, 2006. Some more countries followed this labeling in 2007, such as Argentina, Brazil, Paraguay, and Uruguay (L'Abbe et al., 2009). In addition, many food manufacturers who have used partially hydrogenated oils in their products have developed or are considering ways to reduce or eliminate TFAs from these products.

In December 2006, the Board of Health required that artificial *trans* fat be phased out of restaurant food in New York. By November 2008, the restriction was in full effect in all New York City restaurants and estimated restaurant use of artificial *trans* fat for frying, baking, or cooking or in spreads had decreased from 50% to less than 2%. Within

2 years, dozens of national chains had removed artificial *trans* fat, and 13 jurisdictions, including California, had adopted similar laws (Angell et al., 2009; Mello, 2009).

Investigations showed that replacement of fats rich in TFA has resulted in products with more healthful fatty acid profiles. For example, in major restaurant chains in New York, total saturated fat plus TFAs in french fries decreased by more than 50% (Angell et al., 2009). Similar is the situation in Denmark and Central Europe, where the TFA content in general, but particularly in fried products, was reduced. Very positive in this respect is the fact that TFAs were not only replaced by SFAs but also by MUFAs and PUFAs. This nutritional benefit is even greater than the benefit of a one-to-one substitution of TFA by SFA (Stender et al., 2009).

However, data also indicate that especially in countries where there is no discussion of TFAs on a public level, fats and frying oils are still partially hydrogenated and contain high amounts of TFAs. But also in other countries the TFAs topic is still present (Figure 10.3).

French fries and chicken portions of two fast-food restaurants were collected worldwide and analyzed for their TFA content. A high-TFA menu with more than 10 g TFA was served in many countries such as Hungary, Czech Republic, Poland, Bulgaria, and the United States (Stender et al., 2006b).

Although TFAs are also formed at deep frying conditions, the content in fried food is negligible compared to the hydrogenation process. TFA production becomes significant at 250°C but such high temperatures are not used in commercial or household frying (Liu et al., 2008; Martin et al., 2007).

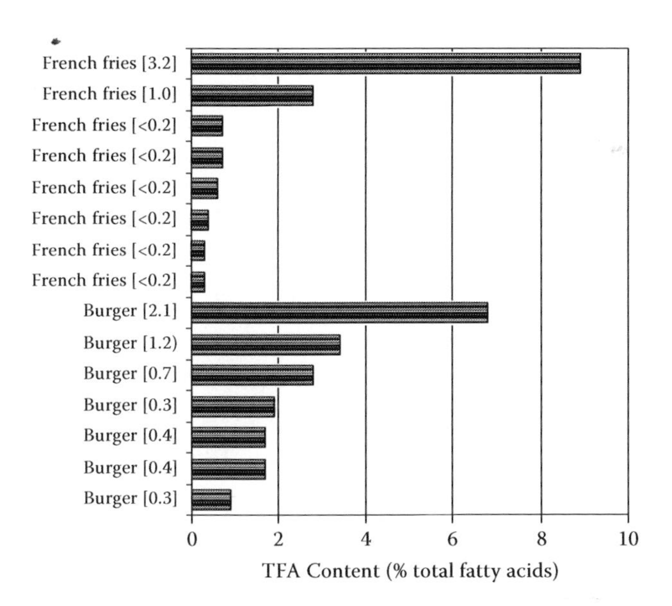

FIGURE 10.3 Comparison of TFA concentrations of french fries and burgers. Values in parentheses are g of TFA in a 200 g portion. (Reprinted from Food Chemistry, 108 (3), Wagner, K.H., Plasser, E., Proell, C. and Kanzler, S., *Comprehensive studies on the trans fatty acid content of Austrian foods: Convenience products, fast food and fats*: 1054–1060 (2008) with permission from Elsevier.)

10.3.2 Strategies to Replace or Reduce *Trans* Fatty Acids in Frying Fats

Food manufacturers are using or developing basically four technologies as options to reduce or eliminate *trans* fatty acids in their products. These options include (1) modification of the hydrogenation process, (2) interesterification, (3) fractions from natural oils, high in solids, and (4) trait-enhanced oils.

10.3.2.1 Hydrogenation and Interesterification Processes

Solid fats usually can be obtained in two different ways: partial hydrogenation leading to TFA-rich fats or full hydrogenation forming mainly SFA rich fats. Modifying the conditions of hydrogenation (e.g., pressure, temperature, and catalyst) affects the fatty acid composition of the resulting oil, including the amount of TFA formed, and properties such as melting point and solid fat content. It is possible to make products of equivalent performance with low *trans* fats by increasing the degree of hydrogenation, which reduces the level of TFA but increases the level of saturated fatty acids (Table 10.2). Such fats will differ analytically from fats hydrogenated to a lesser degree because SFA will contribute most to the solid content (Hunter, 2005).

The interesterification process involves the rearrangement of the fatty acids in the glycerol backbone of the fat in the presence of a chemical catalyst, such as sodium methoxide, or an enzyme. Interesterification modifies the melting and crystallization behavior of the fat and the resulting fat is trans-free. Interesterified products include cocoa butter substitutes and lard. In chemical interesterification, a nonhydrogenated liquid plant oil with zero TFA and a fully hydrogenated vegetable oil (a hardstock) that has a low iodine value and TFA are mixed and heated. A typical blend may contain approximately 85% nonhydrogenated oil and approximately 15% hardstock. A newer method uses a specific lipase derived from microorganisms and immobilized on a granulated silica matrix.

An interesterified fat has almost no TFA but higher levels of stearic, linoleic, and linolenic acids.

TABLE 10.2
Changes in Unsaturation (Iodine Value) and Percentages of Saturated (SFA) and *Trans* Fatty Acids (TFAs) with Progressive Hydrogenation of Soybean Oil

Iodine Value	% SFA	% TFA
130	14	0
110	14	6
68	23	26
55	33	45
15	82	1.5
5	92	0

10.3.2.2 Fractionation

The controlled crystallization of a triacylglycerol can result in the separation of a solid phase (stearin) from a liquid phase (olein), and these in turn can be further fractionated. If fat is melted and cooled slowly to below its melting point, the triglycerides with a higher melting point than the tempering temperature will eventually form crystalline material, which can be relatively easily filtered off from the liquid part. Palm oil is an excellent natural candidate for fractionation and fat blending; it is not only a significant portion of high-melting triglycerides, but tends to crystallize in the favored β' configuration (Lida and Md. Ali, 1998). However, to maximize the potential of palm oil as a blending source, it is occasionally necessary to fractionate it (Jeyarani and Reddy, 2003). This can be done in one of two ways: either dry fractionation or solvent fractionation. Under dry conditions the palm oil is heated and then cooled to a given temperature under agitation such that the partially crystallized mass can be filtered off under vacuum to give the palm stearin. The remaining liquid fraction gives the palm olein. Such fractions can be blended with liquid oils, such as soybean and canola oils. The resulting fats are almost TFA free.

10.3.2.3 Trait-Enhanced Oils

Trait-enhanced oils can be categorized into three different types of fats: high oleic acid oils, such as high oleic sunflower or canola oils; midrange oleic acid oils, such as mid-oleic sunflower and soybean oils; and low linolenic acid oils, such as low-linolenic canola and soybean oils. These types of oils are derived through traditional plant breeding or by biotechnological methods. The high-range and midrange oleic acid oils are less susceptible to lipid oxidation than traditional oils high in linoleic acid, such as soybean, corn, and sunflower oils. Low linolenic acid oils have significantly reduced levels of α-linolenic acid. In France, fats for frying purposes are limited to a α-linolenic acid content of 3%. All of these trait-enhanced oils have a high oxidative stability (i.e., a relatively high content of oleic acid and comparatively low levels of linoleic and α-linolenic acids), making them suitable for frying, spraying, beverages, and some bakery applications.

At present, the plant varieties that produce high oleic and low linolenic, as well as some mid-oleic oils, are not widely cultivated, and are more expensive than the corresponding polyunsaturated oils. Moreover, the high oleic oils are liquid at room temperature and thus would not be particularly useful in spread and baking applications that require a relatively high solid content. Nevertheless, there is considerable interest in the food industry in developing these oils (Wassell and Young, 2007).

In a recent study it was shown that cookies fried in high oleic sunflower oil, which was prepared by interesterification, were able to show the same results of sensory analysis as a commercial shortening (Ahmadi and Marangoni, 2009).

10.4 CONCLUSION

The interactions between the frying fat and the fried food are of great relevance for the nutritional quality of the final product. In particular, the amount of fat taken up can be positively influenced at the industrial and also at the household level by different pre- and post-frying conditions. The fat uptake can also lead to a completely different fatty acid pattern of the product. The change is always toward the predominant fatty acids in the frying fat, which can be beneficial when replacing SFAs by MUFAs or PUFAs, but can also be a disadvantage in the case of fish when the initial content of n-3 fatty acids is significantly reduced. This must also be considered for the nutritional calculation so as not to mislead the nutrient composition of the fried product.

Developments in producing frying fats that are low in TFAs are very positive. In past decades, many efforts have been made, sometimes under legal pressure, to reduce the TFAs of the frying fats. However, improvements still have to be made in order to minimize TFA content in frying fats worldwide.

REFERENCES

Aguilera-Carb, A., Montanez, J.C., Anzaldua-Morales, A., De La Luz Reyes, M., Contreras-Esquivel, J., and Aguilar, C.N. (1999). Improvement of color and limpness of fried potatoes by in situ pectinesterase activation. *European Food Research and Technology*, 210(1):49–52.

Ahmadi, L. and Marangoni, A.G. (2009). Functionality and physical properties of interesterified high oleic shortening structured with stearic acid. *Food Chemistry*, 117(4):68–673.

Albert, S. and Mittal, G.S. (2002). Comparative evaluation of edible coatings to reduce fat uptake in a deep-fried cereal product. *Food Research International*, 35(5):445–458.

Alvarez, M.D., Morillo, M.J., and Canet, W. (2000). Characterization of the frying process of fresh and blanched potato strips using response surface methodology. *European Food Research and Technology*, 211(5):326–335.

Angell, S.Y., Silver, L.D., Goldstein, G.P., Johnson, C.M., Deitcher, D.R., Frieden, T.R., and Bassett, M.T. (2009). Cholesterol control beyond the clinic: New York City's trans fat restriction. *Annals of Internal Medicine*, 151(2):129–134.

Benedito, J., Mulet, A., Velasco, J., and Dobarganes, M.C. (2002). Ultrasonic assessment of oil quality during frying. *Journal of Agricultural and Food Chemistry*, 50(16):4531–4536.

Blumenthal, M.M. (2001). A new look at frying science. *Cereal Foods World*, 46(8):352–354.

Blumenthal, M.M. and Stier, R.F. (1991). Optimization of deep-fat frying operations. *Trends in Food Science and Technology*, 2(C):144–148.

Burlingame, B., Nishida, C., Uauy, R., and Weisell, R. (2009). FAO/WHO 2009. Fats and fatty acids in human nutrition. *Annals of Nutrition and Metabolism*, 55:1–3.

Candela, M., Astiasaran, I., and Bello, J. (1996). Effect of frying on the fatty acid profile of some meat dishes. *Journal of Food Composition and Analysis*, 9(3):277–282.

Clausen, I. and Ovesen, L. (2005). Changes in fat content of pork and beef after pan-frying under different conditions. *Journal of Food Composition and Analysis*, 18(2–3):201–211.

Craig-Schmidt, M.C. (2006). World-wide consumption of trans fatty acids. *Atherosclerosis Supplements*, 7(2):1–4.

Dana, D. and Saguy, I.S. (2006). Review: Mechanism of oil uptake during deep-fat frying and the surfactant effect-theory and myth. *Advances in Colloid and Interface Science*, 128–130:267–272.

Dobarganes, C., Marquez-Ruiz, G., and Velasco, J. (2000). Interactions between fat and food during deep-frying. *European Journal of Lipid Science and Technology*, 102(8–9):521–528.

EFSA (2004). Opinion of the scientific panel on dietetic products, nutrition and allergies on a request from the commission related to the presence of trans fatty acids in foods and the effect on human health of the consumption of trans fatty acids. *The EFSA Journal*, 81:1–49.

Elmadfa, I., Al-Saghir, S., Kanzler, S., Frisch, G., Majchrzak, D., and Wagner, K.H. (2006). Selected quality parameters of salmon and meat when fried with or without added fat. *International Journal for Vitamin and Nutrition Research*, 76(4):238–246.

Garcia, M.A., Ferrero, C., Bertola, N., Martino, M., and Zaritzky, N. (2002). Edible coatings from cellulose derivatives to reduce oil uptake in fried products. *Innovative Food Science and Emerging Technologies*, 3(4):391–397.

Hu, F.B., Stampfer, M.J., Manson, J.E., Rimm, E., Colditz, G.A., Rosner, B.A., Hennekens, C.H., and Willett, W.C. (1997). Dietary fat intake and the risk of coronary heart disease in women. *New England Journal of Medicine*, 337(21):1491–1499.

Hulshof, K.F.A.M., Van Erp-Baart, M.A., Anttolainen, M., Becker, W., Church, S.M., Couet, C., Hermann-Kunz, E., Kesteloot, H., Leth, T., Martins, I., Moreiras, O., Moschandreas, J., Pizzoferrato, L., Rimestad, A.H., Thorgeirsdottir, H., Van Amelsvoort, J.M.M., Aro, A., Kafatos, A.G., Lanzmann-Petithory, D., and Van Poppel, G. (1999). Intake of fatty acids in Western Europe with emphasis on trans fatty acids: The TRANSFAIR study. *European Journal of Clinical Nutrition*, 53(2):143–157.

Hunter, J.E. (2005). Dietary levels of trans-fatty acids: Basis for health concerns and industry efforts to limit use. *Nutrition Research*, 25(5):499–513.

Innis, S.M. (2006). Trans fatty intakes during pregnancy, infancy and early childhood. *Atherosclerosis Supplements*, 7(2):17–20.

Jeyarani, T. and Reddy, S.Y. (2003). Preparation of plastic fats with zero trans FA from palm oil. *Journal of the American Oil Chemists' Society*, 80(11):1107–1113.

Jimenez-Monreal, A.M., Garcia-Diz, L., Martinez-Tome, M., Mariscal, M., and Murcia, M.A. (2009). Influence of cooking methods on antioxidant activity of vegetables. *Journal of Food Science*, 74(3):H97–H103.

Kita, A. and Lisinska, G. (2005). The influence of oil type and frying temperatures on the texture and oil content of french fries. *Journal of the Science of Food and Agriculture*, 85(15):2600–2604.

Krokida, M.K., Oreopoulou, V., Maroulis, Z.B., and Marinos-Kouris, D. (2001). Effect of osmotic dehydration pretreatment on quality of french fries. *Journal of Food Engineering*, 49(4):339–345.

Kummerow, F.A. (1993). Viewpoint on the report of the national cholesterol education program expert panel on detection, evaluation and treatment of high blood cholesterol in adults. *Journal of the American College of Nutrition*, 12(1):2–13.

Kummerow, F.A., Mahfouz, M.M., and Zhou, Q. (2007). Trans fatty acids in partially hydrogenated soybean oil inhibit prostacyclin release by endothelial cells in presence of high level of linoleic acid. *Prostaglandins and Other Lipid Mediators*, 84(3–4):138–153.

L'Abbe, M.R., Stender, S., Skeaff, C.M., Ghafoorunissa, R., and Tavella, M. (2009). Approaches to removing trans fats from the food supply in industrialized and developing countries. *European Journal of Clinical Nutrition*, 63(Suppl. 2):S50–S67.

Librelotto, J., Bastida, S., Serrano, A., Cofrades, S., Jimenez-Colmenero, F., and Sanchez-Muniz, F.J. (2008). Changes in fatty acids and polar material of restructured low-fat or walnut-added steaks pan-fried in olive oil. *Meat Science*, 80(2):431–441.

Lida, H.M.D.N. and Ali, A.R. (1998). Physico-chemical characteristics of palm-based oil blends for the production of reduced fat spreads. *Journal of the American Oil Chemists' Society*, 75(11):1625–1631.

Liu, W.H., Lu, Y.F., Inbaraj, B.S., and Chen, B.H. (2008). Formation of trans fatty acids in chicken legs during frying. *International Journal of Food Sciences and Nutrition*, 59(5):368–382.

Martin, C.A., Milinsk, M.C., Visentainer, J.V., Matsushita, M., and De-Souza, N.E. (2007). Trans fatty acid-forming processes in foods: A review. *Anais da Academia Brasileira de Ciencias*, 79(2):343–350.

Mellema, M. (2003). Mechanism and reduction of fat uptake in deep-fat fried foods. *Trends in Food Science and Technology*, 14(9):364–373.

Mello, M.M. (2009). New York City's war on fat. *New England Journal of Medicine*, 360(19):2015–2020.

Mozaffarian, D., Katan, M.B., Ascherio, A., Stampfer, M.J., and Willett, W.C. (2006). Trans fatty acids and cardiovascular disease. *New England Journal of Medicine*, 354(15):1601–1613.

Pedreschi, F., Hernandez, P., Figueroa, C., and Moyano, P. (2005). Modeling water loss during frying of potato slices. *International Journal of Food Properties*, 8(2):289–299.

Ramirez, M.R. and Cava, R. (2005). Changes in colour, lipid oxidation and fatty acid composition of pork loin chops as affected by the type of culinary frying fat. *LWT—Food Science and Technology*, 38(7):726–734.

Romero, A., Snchez-Muniz, F.J., and Cuesta, C. (2000). Deep fat frying of frozen foods in sunflower oil. Fatty acid composition in fryer oil and frozen prefried potatoes. *Journal of the Science of Food and Agriculture*, 80(14):2135–2141.

Saghir, S., Wagner, K.H., and Elmadfa, I. (2005). Lipid oxidation of beef fillets during braising with different cooking oils. *Meat Science*, 71(3):440–445.

Sanchez-Muniz, F.J., Viejo, J.M., and Medina, R. (1992). Deep-frying of sardines in different culinary fats. Changes in the fatty acid composition of sardines and frying fats. *Journal of Agricultural and Food Chemistry*, 40(11):2252–2256.

Stender, S., Astrup, A., and Dyerberg, J. (2009). What went in when trans went out?. *New England Journal of Medicine*, 361(3):314–316.

Stender, S., Dyerberg, J., and Astrup, A. (2006a). High levels of industrially produced trans fat in popular fast foods. *New England Journal of Medicine*, 354(15):1650–1651.

Stender, S., Dyerberg, J., Bysted, A., Leth, T., and Astrup, A. (2006b). A trans world journey. *Atherosclerosis Supplements*, 7(2):47–52.

Ufheil, G. and Escher, F. (1996). Dynamics of oil uptake during deep-fat frying of potato slices. *LWT—Food Science and Technology*, 29(7):640–644.

Vitrac, O., Dufour, D., Trystram, G., and Raoult-Wack, A.L. (2002). Characterization of heat and mass transfer during deep-fat frying and its effect on cassava chip quality. *Journal of Food Engineering*, 53(2):161–176.

Wagner, K.H., Plasser, E., Proell, C., and Kanzler, S. (2008). Comprehensive studies on the trans fatty acid content of Austrian foods: Convenience products, fast food and fats. *Food Chemistry*, 108(3):1054–1060.

Wassell, P. and Young, N.W.G. (2007). Food applications of trans fatty acid substitutes. *International Journal of Food Science and Technology*, 42(5):503–517.

Wijga, A.H., Van Houwelingen, A.C., Kerkhof, M., Tabak, C., De Jongste, J.C., Gerritsen, J., Boshuizen, H., Brunekreef, B., and Smit, H.A. (2006). Breast milk fatty acids and allergic disease in preschool children: The Prevention and Incidence of Asthma and Mite Allergy birth cohort study. *Journal of Allergy and Clinical Immunology*, 117(2):440–447.

Willett, W.C., Stampfer, M.J., Manson, J.E., Colditz, G.A., Speizer, F.E., Rosner, B.A., Sampson, L.A., and Hennekens, C.H. (1993). Intake of trans fatty acids and risk of coronary heart disease among women. *Lancet*, 341(8845):581–585.

Ziaiifar, A.M., Achir, N., Courtois, F., Trezzani, I., and Trystram, G. (2008). Review of mechanisms, conditions, and factors involved in the oil uptake phenomenon during the deep-fat frying process. *International Journal of Food Science and Technology*, 43(8):1410–1423.

11 Food Hazards Associated with Frying

George Boskou and Nikolaos K. Andrikopoulos

CONTENTS

11.1 INTRODUCTION

Frying in fat or oil is one of the most commonly used methods for the preparation of food. Fried food is very popular all around the world and comprises a wide variety of different products. Man has enjoyed fried food for thousands of years; ancient Egyptians used it as early as 1600. Food frying is a process of controlled dehydration and browning, using hot oil as a heat transfer medium, where the oil plays a significant role in the food flavor, texture, and stability. Similar to cooking with hot water, heat transfer is managed exclusively by the hot liquid medium. Frying processes are executed at temperatures between 140°C and 200°C, far higher than that of boiling water, close to that of roasting, and less than that for grilling. It is easily understood that under these extreme conditions there are physical and chemical changes that may lead to both desirable and undesirable effects.

The complex reactions taking place under the conditions of the frying process have been extensively studied in frying oils. Food and oil during frying are exposed to elevated temperatures in the presence of water and oxygen. The degree of food or oil deterioration depends on the nature of the frying oil and many other variables such as the construction of the fryer, the kind of food, etc. Foods subjected to frying undergo intense drying, generally accompanied by structural changes (formation of pores and crust), as well as color and flavor changes; loss of vitamins, fats, or solutes; starch and/or protein modifications; and microbiological inactivation. Moreover, if food products are fried in degraded oil, which is a source of harmful compounds, it will absorb large quantities of such compounds, which adversely affect the safety of the fried food.

Regulations were drawn up in each country to indicate when a frying oil should be discarded; the level of total polar compounds (TPC) and polymerized triglycerides (PTG) are thus far considered the most reliable parameters. Limits of 24%–26% TPC and 12%–13% PTG are accepted in most of the European countries, following the recommendations of the German Society for Fat Research (DGF). Safety and quality of meals is a primary objective of foodservice systems; therefore control of quality is a major management system. Hazard Analysis Critical and Control Points (HACCP) is a food process control system developed in the early 1970s to ensure the safety of foods for the U.S. space program, maintaining food free of pathogens in order not to cause illness to the astronauts who were traveling in space. Since then it has been used throughout the food industry worldwide, and there is an increasing tendency of implementing this concept in all small restaurants and fast-food shops. In this chapter, we summarize the hazards associated with frying, and the preventive measures, critical limits, and corrective actions to be followed to avoid or eliminate deviations during the food processing.

Oil heating increases the speed of autoxidation. At a high temperature, all autoxidation reactions speed up, and the amount of alteration compounds formed depends on the heating time applied to the oil or fat. At temperatures close to 200°C, decomposition and other reactions of hydroperoxides are faster than their formation. In this case, the major compounds are dimers and polymers of triacylglycerols.

11.2 HAZARD DESCRIPTION

Formation of new compounds during the frying process has been the subject of intensive investigations that have been reviewed (see Chapter 6). Degradation products include compounds associated with the autoxidation process, with polymerization (because of the high temperatures applied), and with hydrolysis, due to the water in the food subjected to frying.

Due to the high temperature, the fat degrades rapidly during deep-frying. The rate of degradation depends on several parameters such as the type of fat, the fried food, and the frying conditions (Schwarz, 2000). The chemistry of vegetable oils at frying temperatures is more complex than just thermal oxidation or autoxidation. Initially, the oil undergoes autoxidation, followed by thermal oxidation and other reaction such as hydrolysis, polymerization, and dehydration. The latter takes place at temperatures above 150°C (Gertz et al., 2000). Such unavoidable chemical reactions cause the formation of both volatile and nonvolatile decomposition products (Chang et al., 1978). The volatile products are constantly removed by volatilization, which is aided by the evolving steam, but they are also formed by continuing oxidation, hydrolysis, thermal reactions, and pyrolysis. The nonvolatile products are formed mainly through the oxidation reactions, but they are also formed via thermal or pyrolytic pathways. Nonvolatile by-products are absorbed by the fried food either by migration or by oil uptake. The polymerized triglycerides are also deposited on the wall of the frying pan or the grid of the frying canister or the filters for frying oils (Fritsch, 1981).

In general, the food hazards that may occur at frying can be related to the frying medium, the frying conditions (time and temperature), and the nature of the food itself. In Table 11.1 there is a brief description of well-known frying hazards in alphabetical order (see also Chapter 4).

11.3 HAZARD CHARACTERIZATION

11.3.1 ACRYLAMIDE

Acrylamide is probably the most-cited food hazard in the scientific literature, with approximately 1000 citations. Most of these citations came after a press release of the Swedish National Food Authority (SNFA) in April of 2002, announcing that acrylamide, a toxic and potential cancer-causing chemical, is formed in many types of food prepared/cooked at high temperatures. According to Reynolds (2002), a railway accident in 1997 with a leak of acrylamide in a tunnel prompted the association of the neurotoxicity of the workers exposed to the chemical with that in cattle herds not exposed to the chemical but fed with heat-processed forage. This association triggered an investigation for the presence of acrylamide in cooked foods and feeds. Tareke et al. (2000) made the first hypothesis that acrylamide is a food hazard of cooking in their article titled "Acrylamide: A Cooking Carcinogen?" The same scientific group confirmed their hypothesis 2 years later, providing scientific support for the press release of the SNFA (Tareke et al., 2002).

TABLE 11.1
Hazards Associated with Frying

Hazard	Description
Acrylamide *(origin: food and oil)*	Is mostly associated with the nature of the food since it is produced by Maillard reactions. At extended heating acrylamide may be formed by glycerol condensation to acrolein.
Chloropropanols *(origin: food and oil)*	There is no clear explanation for the origin of these substances. However, it is apparent that they are related to either hydrolyzed vegetable proteins in food or the heat-induced formation from the triglycerides under acid conditions. Typical representative of this group is 3-MCPD (3-monochloropropane-1,2-diol or 3-chloro-1,2-propanediol)
Decadienal *(origin: oil)*	Is a medium-polarity volatile substance that is produced by the decomposition of unsaturated triglycerides. Together with decadienal, other aldehydes with 7 to 12 carbon atoms are found. They are responsible for the heavy frying odor and the air contamination of the kitchen.
Heterocyclic amines *(origin: food and oil)*	Substances produced when high protein food is cooked at high temperatures, particularly poultry or fish. A typical representative of this group is PhIP (phenylimidazo-pyridine).
Polar compounds *(origin: oil)*	Are the major decomposition products of frying oil. Various chemical reactions including oxidation, hydrolysis, and polymerization give rise to the formation of polar compounds such as free fatty acids, aldehydes, di- and monoglycerides, and polymerized triglycerides. They are responsible for the major chemical and physicochemical alteration in frying oil and therefore are used as quality parameters.
Polyaromatic hydrocarbons *(origin: oil and food)*	Are due to extended decomposition of the frying oil or the fat of the fried food. Usually, they are produced in shallow frying processes, like Asian cooking, and are also responsible for the chemical pollution in the air of the kitchen.
***Trans* isomers** *(origin: oil)*	Originate from hydrogenated frying oils or fats. Several studies show an increase in *trans* fatty acid composition during frying. The *cis*-to-*trans* isomerization process of unsaturated fatty acids is favored by high frying temperatures.
Others *(origin: food and oil)*	A limited number of studies show that residues of pesticides or veterinary drugs in the food may be stable during frying, with restricted increase or decrease in the initial concentration.

The adverse health effects of acrylamide were already known. According to the International Agency for Research on Cancer, acrylamide has been classified since 1994 (class 2A) as probably carcinogenic to humans. In 2002, FAO/WHO issued a consultation in which they suggested that on the basis of current knowledge, the human cancer risk could not be calculated directly (FAO/WHO 2002b). It was not possible then to say what level of exposure is safe and how to avoid formation of acrylamide in food (Sharp, 2003). Later work in the United Kingdom, Switzerland, and Canada (Mottram et al., 2002; Stadler et al., 2002; Becalski et al., 2003) provided an answer to that, linking the formation of acrylamide with the Maillard reactions described beginning 1912 by Louis Camille (Sharp, 2003). Maillard reactions contribute to the color and flavor of cooked foods, but some reaction between asparagine, an amino acid found in potatoes and cereals, and a reducing sugar could yield acrylamide (Mottram et al., 2002). Researchers at the Nestlé Research Centre in Switzerland focused on *N*-glycosides formed at early stages of Maillard reactions, and the use of isotopes confirmed asparagine as a supplier of carbon and nitrogen in acrylamide molecules (Stadler et al., 2002). Except for Maillard reactions, there are other pathways such as the glycerol condensation to acrolein and the thermolytic release of acrylamide from gluten in wheat bread rolls (Lantz, et al., 2006). These mechanisms are only marginal contributors to the overall acrylamide concentration in foods. Some of the most commonly proposed mechanisms of formation are (Figures 11.1 and 11.2):

- Direct formation from amino acids with reducing sugars (Maillard reactions)
- Formation via acrolein or acrylic acid, which may be derived from the degradation of lipids, carbohydrates, or free amino acids
- Formation via the dehydration/decarboxylation of certain common organic acids, including malic acid, lactic acid, and citric acid

In the first two cases, the source of nitrogen in the acrylamide molecule is possibly ammonia released in deamination processes. It is quite obvious that acrylamide is a common reaction product generated in a wide range of cooking processes, and therefore it has been present in human foods probably since humans first discovered the cooking of food with heat. Our knowledge on the adverse health effects has increased in the last two decades, and our knowledge of its levels in food, together with the application of preventive measures, has increased in the last decade.

It has been confirmed that a wide range of cooked foods contain acrylamide at levels of a few ppb (μg/kg) up to 1000 ppb in some cases (Confederation of the Food and Drink Industries of EU CIAA 2009). This range of foods includes staple foods such as bread, fried potatoes, and coffee as well as specialty products such as potato crisps, biscuits, crisp bread, and a range of other heat-processed products. Exposure at high levels causes damage to the nervous system (Tilson, 1981; Lopachin and Lehning, 1994). Acrylamide is also considered a reproductive toxin, with mutagenic and carcinogenic properties *in vitro* and *in vivo* (Dearfield et al., 1988; Costa et al., 1992). In accordance with a No Observable Adverse Effect Level (NOAEL) for acrylamide of 0.1 mg/kg bodyweight, it is estimated that people eating fried potato and crisps are at a very low risk of cancer from this source (Shaw and Thomson, 2003). Glycidamide is the genotoxic metabolite of acrylamide, formed by the epoxidation of acrylamide with fatty acid hydroperoxides (CIAA, 2009). Fatty acid hydroperoxides can mediate the formation of glycidamide from acrylamide, in very small amounts and only in some fried foods (1.51 μg/kg in potato chips and 0.002–0.29 μg/kg in precooked french fries) (CIAA, 2009, Granvogl et al., 2008). Friedman and Levin (2008) in their review on the methods for the reduction of dietary content and toxicity of acrylamide listed about 30 mitigation factors concerning the safety of dietary acrylamide. According to the CIAA Acrylamide Toolbox (2009), the 14 parameters given in Table 11.2 have been identified to play a role in the formation of acrylamide.

The above-mentioned parameters are thoroughly described in the CIAA Acrylamide Toolbox (2009) and some of them are assessed further in the section

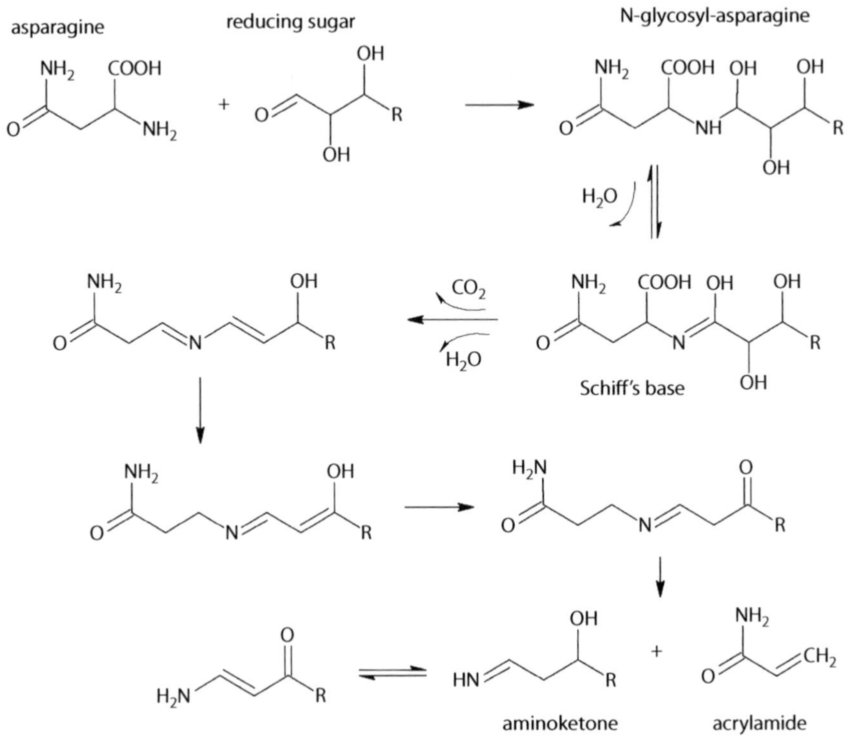

FIGURE 11.1 Formation of acrylamide with Maillard reactions from asparagine and reducing sugars.

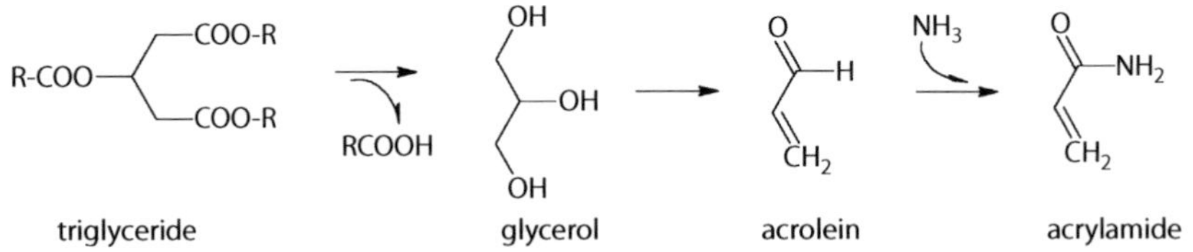

FIGURE 11.2 Formation of acrylamide via the acrolein pathway from lipids.

on culinary processes and preventive measures. The National Heart Foundation of New Zealand suggested five general measures for the control of acrylamide in fried potatoes as follows (adapted from Drummond, 2005):

1. Selection of potatoes and potato varieties with *low levels of reducing sugars*. Reducing sugar levels and asparagine in potatoes are highly variable and related to cultivar and agricultural practices (fertilizer application, pesticide/herbicide application, time of harvest, etc.). Selection of the appropriate potato cultivars and control of reducing sugar levels are found

TABLE 11.2
Parameters That Play a Role in the Formation of Acrylamide in Food Processing

Parameter	Category
Sugars Asparagine	Agronomical
Raising agents Other minor ingredients (e.g., glycine and bivalent cations) pH Dilution Rework	Recipe
Fermentation Thermal input and moisture control Pre-treatment (e.g., washing, blanching, bivalent cations) Asparaginase	Processing
Color endpoint Texture/flavor Consumer guidance	Final Preparation

Source: Adapted from CIAA Acrylamide Toolbox, rev. 12, Confederation of the Food and Drink Industries of the EU, Brussels.

to be significant factors in minimizing acrylamide formation in deep-fried potato products. Postharvest handling is also important in the minimization of reducing sugar levels and can be supported by ensuring that storage temperature of potatoes prior to processing is not less than 6°C to avoid the initiation of starch conversion to reducing sugars. Dips in sugar or hydrolyzed starch are to be avoided as well to minimize the uptake of reducing sugars.

2. *Blanching* and/or *soaking* can minimize the surface levels of reducing sugars; however, this is not always effective. Blanching has been shown to be effective in the reduction of asparagine content as well.
3. *Lowering the pH* by the addition of citric acid to a soaking solution is successful; however, care is required to ensure correct conditions to avoid souring and rapid development of rancidity in the frying oil. In some cases soaking in a glycine solution (3%) was effective to reduce the final acrylamide formation.
4. *Cooking temperature* is an influential parameter in controlling acrylamide formation, particularly on the food–oil interface. Several researchers have found this to be more important for products with a low food/oil ratio than

for those with a high ratio. It is more likely that the final frying temperature of the food, and not the interim temperature, is related to the final acrylamide levels in the fried food. Reducing the frying temperature significantly reduces the acrylamide content of the fried product, since acrylamide formation occurs at normal frying temperature (160°C–180°C). The rate of acrylamide formation is significantly increased at temperatures above 175°C. Temperatures as low as 150°C are suggested; however, some quality factors will be adversely affected at these low frying temperatures.

5. The *type of frying oil*, or the use of silicone oil as antifoaming, has a very small effect on acrylamide formation. Repeated reuse and extended heating time of the oil is closely associated with an increase in acrylamide formation. The use of magnesium silicate, typically used for the adsorption of polar compounds, could be a successful oil adsorbent treatment for acrylamide as well.

It should to be emphasized, though, that in most cases, not a single solution, but a carefully selected combination of measures has to be applied to reduce acrylamide in food. The selection depends on the design and the flexibility of the existing culinary outlet. There are, of course, drawbacks, since mitigation of acrylamide formation through changes in product composition and/or process conditions may have an impact on the overall quality and nutritional value or even the safety of the food (upon other aspects of food safety). Romani et al. (2008) reported the appearance of acrylamide after a 4 min frying of french fries at 180°C. Frying potatoes at lower temperatures can reduce acrylamide formation, but longer frying time can consequently increase the fat uptake. Excessive blanching of potatoes results in loss of minerals and vitamins. Use of refined flour for coating reduces acrylamide formation, but nutritionally this is less desirable compared with whole-bran flour. Replacing ammonium bicarbonate with sodium bicarbonate helps control acrylamide formation, but if applied systematically will increase sodium levels (Seal et al., 2008). In a potato powder model system, sodium acid pyrophosphate, citric, and acetic and L-lactic acids significantly reduced the final acrylamide content, by the lowering of the pH. Addition of glycine, L-lysine, and L-cysteine also lowered acrylamide, at normal pH level, while L-glutamine increased the formation of acrylamide (Mestdagh et al., 2008). There is also a potential loss of beneficial compounds during frying or pre-frying processes that may have protective health effects, e.g., antioxidants with *in vitro* antioxidant capacity in heated foods (Summa et al., 2006; Andrikopoulos et al., 2002a; Andrikopoulos et al., 2003; Kalantzakis et al., 2006).

According to meta research of Seal et al. (2008), the products that contribute mostly to acrylamide formation are baked or fried potatoes (~28%) and then crispy products (~16%). The mean dietary intake of acrylamide is below 1 μg/kg of body weight per day in adults (0.28–0.71 μg/kg of bodyweight/day) and about 1.5 μg/kg among children and adolescents, depending on the studies and age categories considered (Dybing et al., 2005). This is in accordance with the NOAEL for genotoxicity that is 2 mg/kg bodyweight per day and the NOAEL for neurotoxicity that is 10 mg/kg bodyweight per day (FAO/WHO, 2005). Among foods with higher concentrations of acrylamide are potato crisps (average 0.754 mg/kg) and fried potatoes (average

0.334 mg/kg). With moderate acrylamide concentration are flour-based baked or fried foods and croquettes (0.123 and 0.110 mg/kg, respectively). However, in some samples of crisps and fried potatoes; levels as high as 4.1 and 5.3 mg/kg, respectively, were found (FAO/WHO, 2005). In these cases, a single portion of 150 g could reach 30–50% of the NOAEL for genotoxicity.

11.3.2 Chloropropanols

The research group of Prof. Jan Velısek in Prague (Velisek et al., 1978, 1979, 1980) was the first to demonstrate that chloropropanols could be formed in hydrolyzed vegetable proteins (HVP) produced by hydrochloric acid hydrolysis of proteinaceous by-products from edible oil extraction such as soybean meal, rapeseed meal, and maize gluten (Stadler and Goldmann, 2008). Hydrochloric acid, associated with proteinaceous materials, can react with residual glycerol and lipids to yield a range of chloropropanols. The most abundant chloropropanol is 3-chloropropane-1,2-diol (3-MCPD), together with lesser amounts of 2-chloropropane-1,3-diol (2-MCPD), 1,3-dichloropropanol (1,3-DCP), 2,3-dichloropropanol (2,3-DCP), and 3-chloropropan-1-ol. The relative proportions of the major chloropropanols (3-MCPD, 2-MCPD, 1,3-DCP, and 2,3-DCP) occurring in HVPs were approximately in the ratio of 1000:100:10:1 (Stadler and Goldmann, 2008) (Figure 11.3).

Studies on domestic cooking have shown elevated levels of 3-MCPD formation in toasted bread, grilled cheeses, and fried batters (Crews et al., 2001). In 1994 the EU Scientific Committee for Food concluded that 3-MCPD should be regarded as a genotoxic carcinogen and in 1997 suggested that residues in food are undetectable by the most sensitive analytical method (EC, 1997). Another evaluation of 3-MCPD by the Joint FAO/WHO Expert Committee on Food Additives (JECFA) in 2001, upon current mutagenicity and carcinogenicity data, provided reassurance that the mutagenic activity seen *in vitro* was not expressed *in vivo* (Schlatter et al., 2001). The EU Scientific Committee on Food recommended a provisional maximum tolerable daily intake of 2 mg/kg bodyweight for 3-MCPD, considering it a nongenotoxic carcinogen (Scientific Committee on Food SCF, 2001). Further studies came upon the formation of 3-MCPD in heat-processed foods, particularly fried in refined vegetable oils (Hamlet et al., 2002; Hamlet and Sadd, 2004). According to Velisek et al. (1991), fatty acid esters of 1,3-DCP and 3-MCPD (mono- and diesters) are formed in acid-HVP, found only in traces, since the majority are removed during the filtration

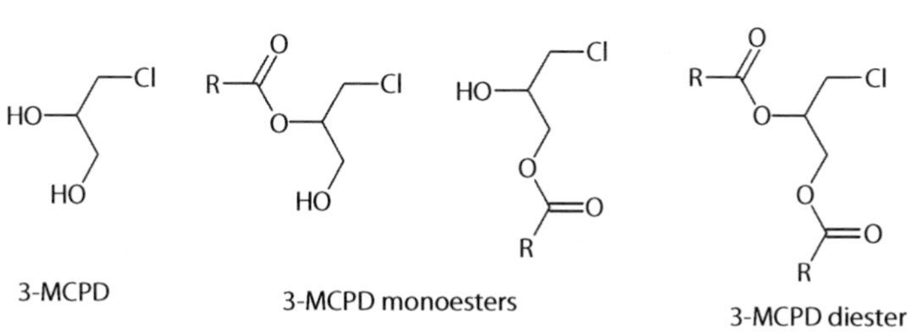

FIGURE 11.3 Chemical structures of 3 MCPD, mono- and diesters.

of the hydrolysates. Such chloroesters are intermediate products in the formation of DCPs and MCPDs from lipids, and the relative amount of each chloropropanol in acid-HVP depends on the composition of the lipid in the raw materials (Collier et al., 1991). Later work on the formation of chloroesters related their occurrence to thermal treatment of foods, including fried products among others (Hamlet et al., 2004; Stadler et al., 2007).

As mentioned before, the occurrence of chloroesters in food was first reported in 1978 by Velisek et al. Several heat-processed foods as well as vegetable oils were examined for the presence of chloroesters (Hamlet and Sadd, 2004; Svejkovska et al., 2004; Zelinkova et al., 2006). Due to their structural similarity to mono- and diacylglycerols, MCPD-monoesters and MCPD-diesters represent potential substrates for lipases and can thus be converted into MCPD in the gastrointestinal tract (Stadler and Goldmann, 2008). Based on toxicological assessments, a maximum tolerable daily intake of 2 mg/kg of bodyweight has been proposed (FAO/WHO, 2002a). The European Commission has set a regulatory limit of 20 μg/kg for 3-MCPD in HVP and soy sauce (European Regulation 1881/2006) but the limit for cooked products is expected to be zero or less than the detectable limit.

In vegetable fat refining processes, deodorization seems to be a critical step for the formation of 3-MCPD esters. Weisschar (2008) reported values in refined oils ranging from 0.2 mg/kg to 20 mg/kg. The main factor affecting the formation of 3-MCPD esters is the presence of chloride ions (from water and other materials used during the refining), glycerol, tri-, di-, and monoacylglycerols, as well as temperature and time. Increasing amounts of mono- and diacylglycerols in the oil show a linear correlation with the formation of 3-MCPD esters (Larsen, 2009). This supports the assumption that higher amounts of mono- and diacylglyceroles are responsible for the higher amounts of 3-MCPD esters found in palm oils and palm oil products. The highest concentrations of 3-MCPD esters are found in unused frying fat. In used frying fat, 3-MCPD levels decrease during frying time. During the deep-frying process nearly no additional 3-MCPD is formed. Doledzal et al. (2008) reported levels of 3-MCPD that range approximately from 0.1 mg/kg to 0.6 mg/kg for potato chips fried at 170°C. The level of 3-MCPD esters in french fries and other fried foodstuffs only depends on its concentration in the used frying fat (Larsen, 2009). Some commercially available refined vegetable oils and fats were conventionally classified into three groups, according to the level of 3-MCPD found to be ester-bound (Larsen, 2009):

- Low level (0.5–1.5 mg/kg): rapeseed, soybean, coconut, sunflower oil
- Medium level (1.5–4 mg/kg): safflower, groundnut, corn, olive, cottonseed, rice bran oil
- High level (>4 mg/kg): hydrogenated fats, palm oil and palm oil fractions, solid frying fats

Although there is a lack of data about 3-MCPD esters for many foodstuffs, it is obvious that thermally processed foods and refined fats and oils (as such or as a component of other foodstuffs) are the most significant sources of 3-MCPD esters for consumers. In particular, refined oils in different kinds of foodstuffs are responsible for a significant part of the exposure. In contrast to acrylamide that is formed during the frying process, formation of 3-MCPD does not take place due to frying. Its presence is a result of the carry-over from the oil used for frying to the fried food. The reasons for the formation of chloropropanols and chloropopanolesters are not yet fully elucidated. Further work is needed as well on the toxicity of these compounds and the exposure of consumers to these hazards. With the exception of the precautions with food products that contain hydrolyzed vegetable proteins of soy, beer yeast, corn or starch gluten, grounded pulses, etc., there are no particular preventive measures suggested. Major concern is focused on Asian food where soy sauce is widely used before and after stir-frying in a wok (Stadler and Goldmann, 2008). Another concern is raised concerning dough-based products, which are leavened with yeast, such as donuts and dumplings (Hamlet and Sadd, 2005), particularly when these contain sugar and an acidifying ingredient, such as citric acid or tartaric acid (to a final pH close to 4 or lower). The promotion of chloropropanol formation in fried food of leavened dough seems to be associated both with Maillard reactions and the presence of glucose and glycerol (Hamlet and Sadd, 2005).

11.3.3 Decadienal

Flavor compounds in fried foods are mainly volatile compounds from fatty acids like dienals, alkenals, lactones, hydrocarbons, and various cyclic compounds (Pokorny, 1989). Most of them contribute to the typical flavor of frying and are produced from the oxidation of linolenic or linoleic acid (Buttery, 1989). Different oils produce different volatile substances during frying due to the differences in nature and quantity of fatty acids (Prevot et al., 1988). Wu and Chen (1992) reported 2-heptenal, 2-octenal, 1-octen-3-ol, 2,4-heptadienal, and 2,4-decadienal as major volatile compounds in soybean oil at 200°C. Takeoka et al. (1996) reported 140 volatile constituents of frying oils, with volatile aldehydes being the major group among them. Carbonyl compounds formed during deep-fat frying can react with amino acids, amines, and proteins and produce pyrazines with Maillard reactions (Negroni et al., 2001). Under excessive heating, volatile aldehydes blocked within the oil or the food can be further condensed to produce furans (Takeoka et al., 1996; Stadler and Goldmann, 2008). Volatile aldehydes like 2,4-decadienal are produced by the peroxidation of linoleic and arachidonic acid in polyunsaturated vegetable seed oils (Esterbauer et al., 1990; Spiteller et al., 2001) (Figure 11.4).

In 2003, Doleschall et al. detected certain aldehydes in frying vegetable oils by using headspace SPME (solid phase micro-extraction). In particular, they detected hexanal, *t*-2-hexenal, *t*-2-heptenal, *t*-2-octenal, nonanal, *t,t*-2,4-nonadienal, *t*-2-nonenal, *t*-2-decenal, *t,c*- and *t,t*-2,4-decadienal during deep-frying in sunflower oil. In the case of sunflower oil, the most important volatile compounds were hexanal and *t,t*-2,4-decadienal, which appears first from the secondary oxidation products. A year before, Andrikopoulos et al. (2002b) reported some medium-polarity materials in vegetable frying oils that were detected with liquid chromatography. The group of Andrikopoulos et al. (2004) later identified these medium-polarity materials as being mainly *t,t*-2,4-decadianal appearing in several samples from frying oils in restaurants, particularly in sunflower oil after frying potatoes. According to Boskou et al. (2006), decadienal formation is related mostly to the type of oil used for frying

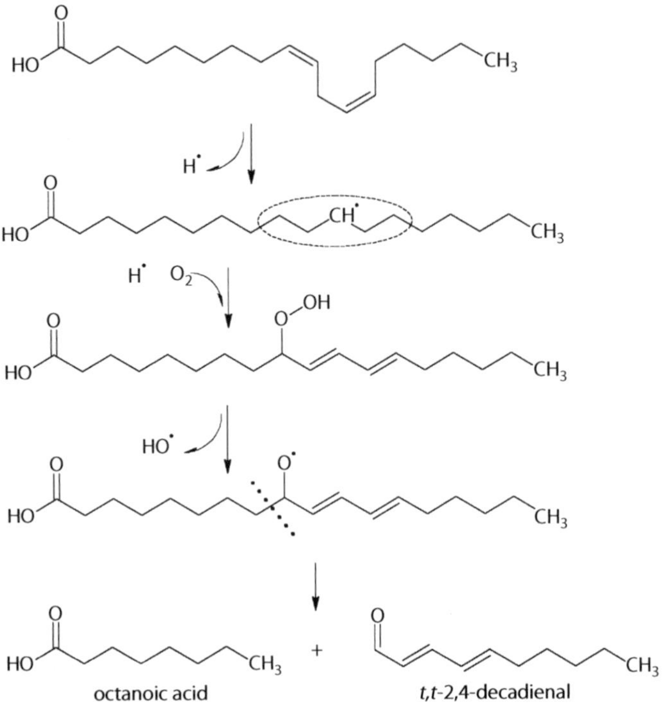

FIGURE 11.4 Oxidation of linoleic acid leading to the formation of *t,t*-2,4-decadienal and octanoic acid.

and to the frying method rather than to the thermal deterioration of the oil. In deep-frying, higher quantities of 2,4-decadienal were observed both in oil and french fries, and sunflower oil was found to produce the higher quantities. Based on a quadratic response surface model, upon raw data from frying fats, Andrikopoulos et al. (2004) supported the hypothesis that 2,4-decadienal formation correlates more with the formation of polymerized triglycerides than to the formation of polar compounds.

The ability of *t,t*-2,4-decadienal to oxidize *in vitro* the LDL is reported by Kaliora et al. in 2003. Decadienal is also detected in cooking fumes resulting from heating edible oils such as rapeseed oil, soybean oil, and peanut oil. It is considered the major mutagenic and cytotoxic compound in oil fumes (Zhu et al., 2001). According to the findings of Wu and Yen (2004), 2,4-decadienal induces DNA damage and so is related to the development of lung cancer in cooks (nonsmokers). Other studies indicate an association of this aldehyde with genotoxic effects due to reaction with nucleic acid bases (Loureiro et al., 2000). The study of Girona et al. (2001) indicated that low levels of *t,t*-2,4-decadienal may modulate inflammatory action by inhibiting TNF-α mRNA expression and that the biological activity of 2,4-decadienal may be involved in the development of atherosclerosis. The cytotoxicity and the effect on glutathione levels in the human erythroleukemia cell lines were investigated, showing that 2,4-decadienal inhibits cell growth and affects cell viability (Nappez et al., 1996). Cabre et al. (2003) investigated the cytotoxic effects of 2,4-decadienal and hexanal in vascular smooth muscle cells through the release of lactate dehydrogenase and the changes in cell morphology. Results showed that hexanal at concentrations up to 50 mM for 24 h is not cytotoxic, while on the other hand, 2,4-decadienal induced a 48% lactate dehydrogenase leakage. The same scientific team also demonstrated that 2,4-decadienal increases the expression of tissue factor, which is expressed in atherosclerotic plaques and co-localizes with oxidized lipids, initiating the thrombogenic process (Cabre et al., 2004). Reaction *in vitro* of 2,4-decadienal with 2′-deoxyguanosine in the presence of peroxide has a DNA-damaging potential (Loureiro et al., 2004).

Based on the findings of Kaliora et al. (2003), the maximum decadienal quantity (1650 mg) in a portion of french fries (150 g), deep-fried in sunflower oil, can induce 97% LDL oxidation as compared to the oxidation produced by copper sulfate (5 mM). The maximum 2,4-decadienal quantity (450 mg) in french fries fried in olive oil would cause only 30% oxidation of LDL. Considering the oil type, differences in the produced quantity of 2,4-decadienal in french fries are related to the degree of unsaturation of the oils or the presence of antioxidants in olive oil. The most unsaturated oils (sunflower oil, cottonseed oil, vegetable shortenings) are expected to produce higher quantities of decadienal as compared to olive oil and palm oil. Frying potatoes in olive oil and palm oil presented 70% less content of the aldehyde in olive oil and 43% less in palm oil during deep-frying (Boskou et al., 2006). Respectively, it was 64% less for both oils during pan-frying, as compared to the 2,4-decadienal content of french fries prepared in sunflower oil. It is estimated that the 2,4-decadienal intake is 150–300 mg per portion french fries (150 g), if pan-frying process is performed in olive oil, within three consecutive frying sessions (Boskou et al., 2006). In deep-fat frying, the decadienal intake can be up to 1500 mg when fried in sunflower oil. In deep-frying, the maximum content of decadienal in oil varies after several frying processes. This can be explained by the fact that the oil is saturated with the volatile aldehyde and a kind of distillation is occurring in the fumes. The decline can be explained by the fact that 2,4-decadienal decomposed either to the 2,3-epoxy or the 4,5-epoxy derivative which further decomposed to 2-octenal and acetaldehyde or to 2-octene and glyoxal (Andrikopoulos et al., 2003).

11.3.4 Heterocyclic Amines

Heterocyclic amines (HA) are substances with a high mutagenic and carcinogenic potential (Sugimura, 1997). They occur in heated meat and fish in significant amounts when the cooking temperature exceeds 150°C. The precursor is creatine converted into creatinine at high temperatures, producing imidazoquinolines and imidazoquinoxaline. Presence of free amino acids like threonine, serine, phenylalanine, leucine, and tyrosine contribute also to the formation of HA (Övervik et al., 1989). This means that during frying, grilling, baking, or roasting of high-protein food, the formation of HA increases substantially. Pearson et al. (1992) proposed two different pathways for free radical formation of heterocyclic amines. One involves bimolecular ring formation from the enaminol form of the glycoaldehyde alkylimine, followed by oxidative formation of the free radical. The other pathway involves formation of N,N1-diakyllpyrazinium ions from glyoxal monoalkylimine followed by reduction to produce the free radicals. These mechanisms help to explain the formation in meat and, in particular, in fried fish and in fried ground beef. Two main groups of HA are identified (see also Chapter 4):

- The polar group:
    - *Quinolines* (e.g., 2-amino-3-methylimidazo[4,5-*f*]quinoline [IQ])
    - *Quinoxalines* (e.g., 2-amino-3-methylimidazo[4,5-*f*]quinoxaline [IQx])
    - *Pyridines,* (e.g., 2-amino-1-methyl-6-phenylimidazo[4,5-*b*]pyridine [PhIP])
- The nonpolar group:
    - *Pyridoindoles* (e.g., 3-amino-1,4-dimethyl-5H-pyrido[4,3-b]indole [Trp-P-1])
    - *Dipyridoimidazoles* (e.g., 2-amino-6-methyl-dipyrido[1,2-a:3′,2′-d]imidazole [Glu-P-1])

There are also furopyridines, benzoxazines, and several other structures of carbazoles, fluoranthenes, and phenylpyridines. Murkovic (2004) names about 25 heterocyclic amines found in foods. The precursors of the polar HA are amino acids, carbohydrates, and creatinine. In particular, creatinine is necessary for the formation of the HA, which are created at temperatures between 150°C and 220°C. If no creatinine is present, the HA of the IQ and IQx type are not formed. The polar HA (IQ, MeIQ, MeIQx, PhIP) are formed at normal cooking temperatures and are identified in fried meat and fish as well. At higher temperatures, the nonpolar HAs are formed as products of amino acid pyrolysis (Bordas et al., 2004; Murkovic, 2004). Although HA are omnipresent, it is only the uptake of heated meat and fish products that contributes significantly to the exposure, since the concentrations in other foods or in the environment are extremely low. The analysis in meat and fish showed concentrations in the range of 0–10 μg/kg. The amount of HA formed depends on the cooking method and type of meat. It seems that frying or grilling of poultry produces higher amounts of HA, particularly PhIP, at concentrations up to 600 μg/kg (Murkovic, 2004). A Spanish study in home-cooked meat samples (Busquets et al., 2004) found IQ and MeIQ being below the limit of quantification (0.5 μg/kg), MeIQx occurred at higher concentrations (0.7–2.9 μg/kg), 4,8-DiMeIQx from 0 to 1.8 μg/kg, and PhIP from 0.6 to 47 μg/kg. The highest concentration of PhIP was found in fried chicken breast filets.

The content of HA increases with increasing temperature. According to Nielsen, et al. (1984), the increase of the pan temperature of 50°C during frying of meat can double the HA content. Although the content of the HA can be reduced substantially with lower frying temperatures, there is definitely no temperature range in which no HA are formed. With a low frying temperature of about 150°C, only low amounts of IQ, MeIQ, MeIQx, 4,8-DiMeIQx, and PhIP are formed (Murkovic et al., 1997). Analysis of 14 cooked meat dishes and the pan residues showed that up to a temperature of 150°C the total content of HA was below 1 μg/kg. A temperature up to 175°C led to a content of about 2 μg/kg (Skog et al., 1997).

Besides temperature and heating time the composition of the meat plays a crucial role in the formation of HA. In particular, the availability of the precursors (amino acids, creatine, carbohydrates) influences the formation of the HA. The type of fat (Johansson et al., 1995) as well as the oxidative status of the frying fat shows an influence on the formation. In several publications it was demonstrated that the presence of antioxidants reduces the content of HA in meat (Persson et al., 2003; Vitaglione and Fogliano, 2004). According to Persson et al. (2003), frying in olive oil significantly reduced the presence of HA in meat. Herbs with antioxidant activity (rosemary, thyme, savory, and garlic) reduce the content of HA in fried meat (Murkovic et al., 1998). A honey-based marinade for beefsteak and chicken breast to be fried resulted in a 16–45% reduction in MeIQx, DiMeIQx, and PhIP, depending on the amount of honey in the marinade (Shin and Ustunol, 2004). Vitaglione and Fogliano (2003) studied the application of marinades with sources of antioxidants, like wine and olive oil, in the reduction of HA in fried beef burgers. Oguri et al. (1998) showed that when catechins from green tea as well as specific phenols (epigallocatechingallate, luteolin, quercetin, caffeic acid) are added to model food systems, the formation of HA is reduced. Lan et al. (2004) used antioxidant additives, such as ascorbic acid and BHT, in marinades of soy sauce and sugar but they did not observe any effect on the formation of HA. The review by Vitaglione and Fogliano (2003) concluded that there are some contradictory results concerning the influence of antioxidants, and that the whole mechanism is not as simple as common radical scavenging.

According to O'Brien et al. (2006), the benchmark dose to yield 10% of the toxicity (BDML10) for PhIP is 1.25 mg/kg bodyweight per day. The human exposure to PhIP is estimated to be much less (4.8–7.6 ng/kg bodyweight per day upon raw data analysis). Association of heterocyclic amines to fried meat carcinogenicity dates back to the year 1939 when a Swedish chemist observed that extracts of fried horsemeat applied to the skin of mice induced cancer (Widmark, 1939). But it was actually the review by Sugimura (1995) that clearly indicated heterocyclic amines as possible human carcinogens in cooked foods. Keating and Bogen (2004) used a method to estimate daily HA intake in the U.S. population by combining laboratory data on meat type, cooking method, and cooking degree with national consumption data. Pan-frying and chicken, as the basic cooking method and meat type, were found to contribute mostly to the total estimated HA exposure. The PhIP was found to comprise 70% of the U.S. mean dietary intake of total HA, estimated for an adult in the range of 11.0–19.9 ng/kg bodyweight per day. Murcovic (2004) proposed several measures to reduce HA formation in fried food (though impossible to prevent); some of these are (1) intensive frying of meat and fish should be avoided, (2) burned parts should be removed and not eaten, (3) marinades with herbs to be applied before frying, and (4) the consumption of fried meat and fish should be reduced.

11.3.5 Polar Compounds

These groups of substances are usually studied together since the mechanisms of production and the significance in frying are associated. Various chemical reactions including oxidation, hydrolysis, and polymerization give rise to the formation of volatile and nonvolatile compounds such as free fatty acids, aldehydes, di- and monoglycerides, and polymerized triglycerides (Gertz, 2000; Choe and Min, 2007). The three main causes for the formation of frying by-products are moisture, oxygen, and temperature, which result in hydrolysis, oxidation, and thermal deterioration, respectively (Saguy and Dana, 2003; Velasco et al., 2008). In particular, it was found that

- Moisture in the frying oil causes hydrolysis to free fatty acids and diacylglycerols.
- Oxygen in the frying oil causes oxidation to oxidized monomeric triacylglycerols, oxidized dimeric and oligomeric triacylglycerols, and volatile compounds (aldehydes, ketones, alcohols, hydrocarbons, etc.)
- The high temperature of the frying causes thermal deterioration of the oil to cyclic monomeric triacylglycerols, isomeric monomeric triacylglycerols, dimeric, and oligomeric triacylglycerols.

Vapor and oxygen initiate the chemical reactions during heating at frying operations. Water molecules, as a weak nucleophile, attack the ester bond of triglycerides and produce di- and mono-acylglycerols, glycerol, and free fatty acids (Chung et al., 2004). Heat-induced radical reactions proliferate when triglycerides react with oxygen and produce alkyl-hydroperoxides (ROOH) or dialkyl-peroxides (ROOR). Dimers or polymers are either aliphatic or cyclic, and their formation depends on different radical reactions and the kind of fatty acid esters of the triglycerides (Choe and Min, 2007). Formation of dimers and polymers advances as the number of fryings and the frying temperature increase (Cuesta et al., 1993; Takeoka et al., 1997). Polyunsaturated oils are more easily polymerized during frying than the monounsaturated oils (Takeoka et al., 1997; Bastida and Sanchez-Muniz, 2001). The production of polar compounds, and particularly of polymerized triglycerides, was higher in high linoleic sunflower oil, followed by high-oleic sunflower oil and high palmitic sunflower oil, when thermoxidized in laboratory conditions (Marmesat et al., 2008).

Cyclic compounds are produced in a lower rate compared to the nonvolatile polar compounds, dimers, and polymers (Takeoka et al., 1997; Dobarganes et al., 2000). Cyclic polymers are produced between or within the triglycerides, either with radical reactions or Diels–Alder reactions. The formation of cyclic compounds in frying oil increases with the degree of unsaturation and the frying temperature, particularly when linoleic acid is more than 20% of the fatty acid content (Tompkins and Perkins, 2000). Romero et al. (2006) suggested that the formation of cyclic fatty acid monomers is more intensive in common sunflower oil than in high-oleic sunflower oil (Figure 11.5).

Among the oxidation products are also monoepoxy fatty acids, with an oxirane ring, and oxo (keto) fatty acids such as *trans*-9,10-epoxystearate (*trans*-9,10-ES), *cis*-9,10-epoxystearate (*cis*-9,10-ES), *cis*-9,10-, *trans*-9,10-, *cis*-12,13- and *trans*-12-,13-epoxyoleate (EO) (Marmesat et al., 2008). They have been determined in sunflower and olive oil, and in used frying oils from restaurants and fast-food outlets (Velasco et al., 2002, 2004) and are considered to be a major group among oxidized fatty acid monomers and polar fatty acids in frying oils. Thermoxidation in monounsaturated oils promotes the formation of monoepoxystearates and in polyunsaturated oils promotes the formation of monoepoxyoleates (Kalogeropoulos et al., 2007; Marmesat et al., 2008). According to Kalogeropoulos et al. (2007), oxidized fatty acids together with PTG formation follows a more or less linear correlation

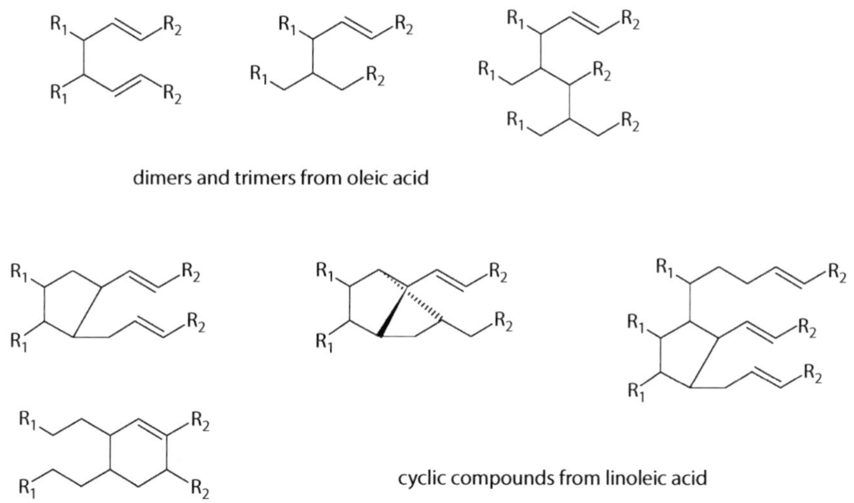

dimers and trimers from oleic acid

cyclic compounds from linoleic acid

FIGURE 11.5 Common dimeric and trimeric compounds from oleic acid and cyclic compounds from linoleic acid produced at frying.

with frying time. Pan-frying in olive oil or palm oil results in a higher formation of epoxy fatty acids than in deep-fat frying. Thermoxidized lipids are associated with atherosclerosis, liver damage, and intestinal tumors, but the dietary intake and the relationship with prooxidants and antioxidants are under investigation (Dobarganes and Márquez-Ruiz, 2003). Cholesterol is also a molecule susceptible to oxidation leading to the formation of oxysterols, which are found, among other foods, in potatoes fried in polyunsaturated oils (Valenzuela et al., 2004).

Polar compounds have been blamed for growth retardation, increased liver and kidney weights, and disorders of the enzyme system in feeding experiments with live animals. However, most of the frying techniques applied were extreme and unrealistic, and the diets introduced were unbalanced (Billek, 2000). Severely damaged fats or an unbalanced fat content, leading to increased daily consumption of polar compounds for years, are not usual in human diets. Among the polar compounds, polymerized triglycerides, cyclic fatty acid monomers, and oxidized triglycerides are considered liable for toxicity. Polymerized triglycerides may have a retarding effect on the digestion of the intact triglycerides (Márquez-Ruiz et al., 1998). Cyclic compounds stimulate the detoxification mechanism, influence the degradation of normal fatty acids, and inhibit the tricarboxylic acid cycle (Lamboni and Perkins, 1996). Oxidized triglycerides produced during the frying process demonstrate toxicity (Esterbauer, 1993) and promote the *in vivo* synthesis of low-density lipoprotein (Riemersma et al., 1994).

The estimation of total polar materials (TPM) and polymerized triglycerides (PTG) are widely accepted as parameters to decide whether or not the used oil should be replaced (Bansal et al., 2010). The TPM value is considered a better indicator since it refers to all the degraded products from initial triglycerides present in the oil. The PTG value refers to polymerized triglycerides that are the largest single class of total polar compounds. Techniques for the quantitation of oxidation compounds and oligomers are extensively discussed in Chapter 6.

In most European countries, the limits for rejection and replacement of a cooking oil range between 24% and 26% for total polar materials (TPM) and 12%–13% for polymerized triglycerides (PTG) (DGF, 2000; Firestone, 2004). Limits of 20% TPM and 10% PTG have been recommended for replenishment of the oil or fat (Poumeyrol, 1986; Gertz, 2000). The TPC/PTG ratio is almost equal to 2/1 at the aforementioned replenishment and rejection limits (Andrikopoulos et al., 2003).

Filtering of oil with certain adsorbents can lower the free fatty acid content and improve the quality of frying oil. Sunflower oil, used for deep-fat frying of potatoes at 170°C, was filtered with a mixture of 2% pekmez earth (lime soil), 3% bentonite, and 3% magnesium silicate; it had decreased contents of free fatty acids (Maskan and Bagci, 2003), but increased content of aldehyde compounds. Filtering canola oil with a mixture of Hubersorb 600 (calcium silicate), magnesol (magnesium silicate), and rhyolite (silicon oxides) decreased free fatty acids and polar compounds formation and improved the quality of the frying oil (Bheemreddy et al., 2002). The treatment of vegetable shortening fat with bleaching clay, charcoal, celite (diatomic earth), or MgO improved the oil quality for french fries (Mancini-Filho et al., 1986). Bhattacharya et al. (2008) studied the regeneration of frying oil with different absorbents or absorbent combinations, based on criteria such as free fatty acids (FFA), total polar materials (TPM), *p*-anisidine value (p-AV), conjugated dienes, viscosity, and color. Binary combination of 7.5% magnesol and 2.5% silica gel consistently reduced all the parameters and was comparatively better than single adsorbent treatments. Treatment with a quaternary combination of 3.75% magnesol, 1.25% silica gel, 3.75% activated carbon, and 1.25% aluminum hydroxide proved to be best in reducing decomposition products in frying palm and cottonseed oil.

Addition of ascorbyl palmitate to fresh oil decreased free fatty acid and dimer formation, but increased the dielectric constant (Mancini-Filho et al., 1986; Gordon and Kourimska, 1995). The application of edible films to foods before frying decreases the degradation of frying oil during deep-fat frying. Holownia et al. (2000) reported that coatings of hydroxypropylmethylcellulose on fried chicken strips decreased free fatty acid formation in peanut oil during deep-fat frying. Soaking potatoes in sodium pyrophosphate buffers inhibited the darkening in french fries and decreased the free fatty acid formation in hydrogenated canola oil during 12 to 72 h frying (Mazza and Qi, 1992). Fatty acid esters of sterols improved the oxidative stability of oil during deep-fat frying (Blekas and Boskou, 1986). Silicone forms a protective layer at the air–oil interface and inhibits convection currents in the frying oil (Sakata et al., 1985; Kusaka et al., 1985). Rosemary and sage extracts reduce oil deterioration during a 30 hours of deep-fat frying of potato chips (Che Man and Jaswir, 2000). Kim and Choe (2004) reported that the hexane extracts of burdock decreased the formation of conjugated dienoic acids and aldehydes in lard at 160°C. Enrichment of antioxidants with olive leaf extract (Salta et al., 2007) increased by 30% the Rancimat induction period (formation of polar compounds) in sunflower oil and vegetable shortening fat and by 20% in olive oil and palm oil. Similar results with olive leaf extract-enriched oils presented by Chiou et al. (2009) where in addition to the increase of the induction period, there was significant protection of tocopherols, phytosterols, and squalene.

Estimation of free fatty acid percentage (FFA%) is a widely accepted indicator of oil quality, because this percentage increases with the lifetime of the frying oil, and it is an easy test to conduct. However, there is no direct relationship between the quality of a used frying fat and its acid value (Gertz et al., 2000). Gravimetric determination of the total polar material percentage (TPM%) with glass column chromatography is a standard method for the estimation of quality in frying oils (DGF, 2000). Croon et al. (1986) compared the Food Oil Sensor (FOS), Oxifrit test, and Fritest with the FFA% and the estimation of total polar compounds. The FOS monitors the dielectric constant in frying oil, the Oxifrit test monitors oxidation products, and the Fritest is a quick acid value test. As expected, all tests had better correlation with the estimation of TPM% than with the FFA% in oils used for frying. Particularly, the linear correlation between FOS values and TPM% was estimated to be $R^2 = 0.97$ in studies of Wegmüller (1998) on 68 samples of frying oils. Similar results are demonstrated by Gertz (2000) where the FOS had very good correlation with both TPM% and PTG%, particularly when TPM is 24% and PTG is 12%. Another type of food oil sensor is the Fri-Check, based on the viscosity of the frying oil, and the results of this test are compared to standard methods such as acid value, polar compounds, polymers, and FOS values (Gertz, 2000). Similarly, with FOS the Fri-Check values had very good correlation with both TPM% and PTG%, particularly when TPM is 24% and PTG is 12%. It suggested that the analysis of suspicious frying fats and oils should utilize two different tests (TPM and PTG) to confirm abuse (Gertz, 2000). Based on the criterion of TPM 24% and PTG 12%, inexpensive frying oil sensors have been developed with "yes/no" decision-making functions (Schwarz, 2000). The review of Bansal et al. (2010) contains an evaluation of commercially available rapid test kits for frying oil quality, based on FFA% and TPM%. Five test kits were based on colorimetric reactions (FASafe, 3M Low Range Shortening Monitor, Fritest, Oxifrit-Test, and TPM Veri-Fry) and three kits were based on the dielectric constant (Food Oil Monitor 310, Testo 265, and CapSens 5000). The test kits based on physicochemical properties provide more objective and valuable results compared to those based on colorimetric reactions.

11.3.6 Polyaromatic Hydrocarbons

Polyaromatic hydrocarbons are produced when the decomposition of the frying oil or the fat of the fried food is extended. Usually they are produced in shallow-frying processes, as in Asian cooking, and are also responsible for air pollution in the kitchen. The high incidence of lung cancer observed among Chinese women has been associated with exposure to fumes from cooking oil (Purcaro et al., 2006). Polycyclic aromatic hydrocarbons (PAH) are classified as potentially mutagenic substances emitted from cooking oils heated at high temperatures. Chen and Chen (2003) observed that, when chicken legs were fried at 163°C for up to 4, PAH formation in fumes was higher when the frying was performed in soybean oil, followed by frying in canola oil and frying in sunflower oil. Li et al. (2003) estimated that the PAH emissions in Taiwan restaurants account for 76%–90% of the total emissions. In particular, the emissions were higher in Chinese restaurants, followed by Western type restaurants, fast-food outlets, and Japanese restaurants. The total emission of

PAH by gas cooking is highest during deep-frying, followed by pan frying and stir-frying (See and Balasubramanian, 2008).

Not only in the environment of the kitchen but also in cooked food, where PAH are primarily formed, considerable hazards are found. Polyaromatic hydrocarbons can be taken up into the food from the environment as well, since they are common pollutants from combustion operations, but they are also formed by the thermal treatment of food. When fat or fatty food (especially meat and fish) is cooked at high temperatures, PAH are formed as a result of incomplete thermal decomposition (pyrolysis) of the organic material. Some of the major compounds within the PAH are fluoranthene, benzo(*a*)anthracene, benzo(*k*)fluoranthene, benzo(*a*)pyrene, and benzo(*g,h,i*)perylene. Though grilling is a major source of PAH in foods a considerable amount is produced in frying operations. According to Janoszka et al. (2004) the total amount in pan-fried pork chops was 15.7 μg/kg, in very well-done grilled pork necks was 16.1 μg/kg, and in well-done roast turkey breasts was 12.7 μg/kg. The major contributor of PAH in pan-fried pork chops was benzo(*a*)anthracene, accounting for about 70% of the total PAH. These figures could be translated to a daily human exposure up to 1.6 μg per capita, calculated on the basis of 100 g daily consumption of meat (Janoszka et al., 2004). On the other hand, Purcaro et al. (2006) used peanut oil to fry french fries and fish at temperatures between 160 and 185°C and no appreciable differences in the PAH load was observed in the same oil before and after frying. White et al. (2008) evaluated the content of PAH in 333 different food samples from experimental applications of barbequing, frying, grilling, roasting, and toasting. Though in barbeque samples the content of PAH was higher, there were significant quantities present in other high-temperature cooked foods, in concentrations at least 10 times higher than the detection limit of 0.01 μg/kg. Perello et al. (2009) estimated the concentrations of 16 individual PAH compounds, including the set of eight carcinogenic PAH, in 680 food samples in the area of Catalonia (Spain). The highest concentrations were found after pan-frying, with concentrations up to 35 μg/kg total PAH. This is attributed to the oil for frying, which had a high initial load of PAH, and to the grilling practice, which involved the use of a hot plate and was not on charcoal.

The 64th Joint FAO/WHO Expert Committee on Food Additives in 2005 reported that 33 polyaromatic hydrocarbons could be present in food. The Committee concluded that only 15 of them are clearly genotoxic *in vitro* and *in vivo*. These are benz(*a*)anthracene, benzo(*a*)pyrene, benzo(*b*)fluoranthene, benzo(*g,h,i*)perylene, benzo(*j*)fluoranthene, benzo(*k*)fluoranthene, chrysene, cyclopenta(*cd*)pyrene, dibenz(*a,h*)anthracene, dibenzo(*a,e*)pyrene, dibenzo(*a,h*)pyrene, dibenzo(*a,i*) pyrene, dibenzo(*a,l*)pyrene, indeno(1,2,3-*cd*)pyrene, and 5-methylchrysene. Some simple PAH compounds, like anthracene, fluoranthene, naphthalene, phenanthrene, and pyrene, among others, are also reported in significant concentrations in grilled, fried, and roasted food (White et al., 2008), but these are not considered particularly genotoxic. It is suggested to use benzo(*a*)pyrene (BaP) as a marker for the occurrence and the effect of the carcinogenic PAHs in food. Examinations of PAH profiles in food and the carcinogenicity of coal tar mixtures in mice (Culp et al., 1998) may lead to the assumption that the carcinogenic effect of total PAH in foods is 10 times that contributed by BaP alone. The Commission Regulation 1881/2006

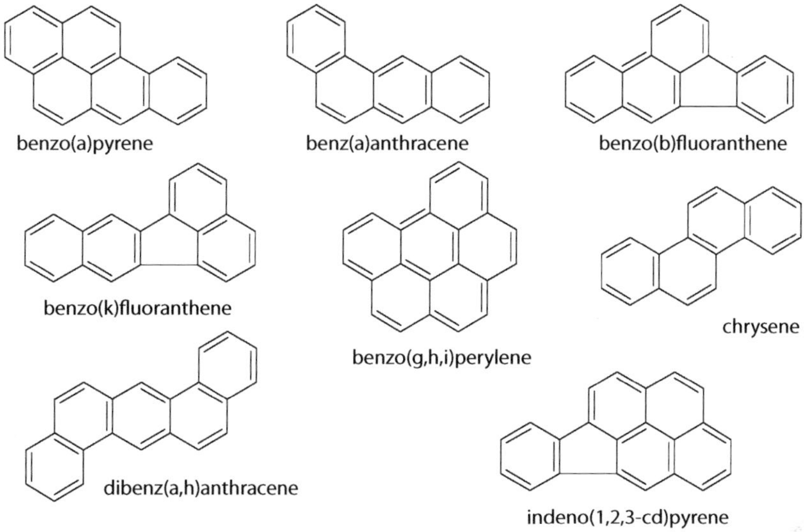

FIGURE 11.6 Eight polyaromatic hydrocarbons (PAH) used as indicators of carcinogenicity in food.

of EC (EC, 2006) has a limit of 2 μg/kg benzo(*a*)pyrene in oils and cooked meat at the point of consumption. In 2008 the EFSA Panel on Contaminants in the Food Chain also proposed to add benzo(*c*)fluorene to the primary list of PAH contaminants, since it was found in a few food samples. This EFSA panel concluded also that there are eight PAH compounds that could be used as possible indicators of carcinogenicity in food: benzo(*a*)pyrene, benz(*a*)anthracene, benzo(*b*)fluoranthene, benzo(*k*)fluoranthene, benzo(*g*,*h*,*i*)perylene, chrysene, dibenz(*a*,*h*)anthracene, and indeno(1,2,3-*cd*)pyrene (Figure 11.6). This set of eight compounds was found in almost 42% of 9714 food samples, while BaP alone was found in almost 50% of the food samples. The mean dietary exposure in European countries is estimated to be 3.1–4.3 ng/kg bodyweight per day for BaP alone and 23.6–35.6 ng/kg bodyweight per day for the set of eight PAH. Considering that the benchmark dose yield of 10% carcinogenicity (BDML10) is 0.07 and 0.49 mg/kg bodyweight per day for BaP and the eight PAH, respectively, the lowest margin of exposure (MOE) is 16,000 and 13,800, respectively (EFSA, 2008).

11.3.7 *Trans* Isomers

Vegetable shortening oils are partially hydrogenated to produce a more saturated product that is more stable upon heating (Warner and Knowlton, 1997). The double bonds in unsaturated fatty acid esters are reduced, and some of them are converted from the *cis* to *trans* configuration (Khor and Mohd Esa, 2008). Hydrogenated soybean oil with 0.1% linolenic acid is more stable upon heating than normal soybean oil with 2.3% linolenic acid (Tompkins and Perkins, 2000). Refining, bleaching, and deodorization can lead also to the formation of *trans* isomers. The range of total *trans* fatty acids in refined soybean oil, corn oil, sunflower oil, high-oleic sunflower oil, low-erucic rapeseed oil and high-erucic rapeseed oil, was between 0.15% and 6.03% (De Greyt et al., 1998). It was reported that linoleic and linolenic acid esters are primarily considered as substrates for the production of thermally induced *trans* isomers (Grandgirard et al., 1984). The common species of *trans* isomers are 18:1 9-*trans*, 18:2 9-12 *cis-trans* or *trans-cis* and 18:3 9-12-15 *cis-cis-trans* or *trans-cis-cis* (Sebedio and Ratnayake, 2008; Bansal et al., 2009).

The type of frying oil and time/temperature combinations during deep-fat frying are also considered a reason for the production of *trans* isomers. Even when lipids are in concentrations less than 1% in nonhydrogenated oils, this level could reach 50% in frying oils due to this small partially hydrogenated fraction (Aro et al., 1998). Romero et al. (2000) reported minimal formation of elaidic acid (18:1 9 *trans*) in extra virgin olive oil and high-oleic sunflower oil. The final concentrations of total trans lipids in palm oil, blended oil (palm, sesame and peanut oil), and sunflower oil after 40 frying session of potatoes were 8.79, 9.46, and 13.40 mg/g, respectively (Bansal et al., 2009). Elaidic acid appeared after five frying sessions in sunflower oil and after 15 sessions in palm oil and blended oil, becoming the dominant *trans* isomer until the last frying session (Bansal et al., 2009). According to Moreno et al. (1999), the amount of *trans* isomers in sunflower oil was 1.10% when heated at 200°C for 40 min, and 11.45% at the extreme temperature of 300°C for the same heating time. Aladedunye and Przybylski (2009) observed that the content of *trans* lipids in canola oil increased by 2.4%–3.3% after 49 frying sessions of french fries at 185°C and by 5.9% after the same number of frying sessions at 215°C (Figure 11.7).

Dietary intake of *trans* isomers in humans increases the levels of total cholesterol and LDL-cholesterol, compared with *cis* monounsaturated and polyunsaturated lipids. They also tend to decrease HDL-cholesterol levels compared with saturated lipids (Morin, 2005). A positive association was observed between the intake of

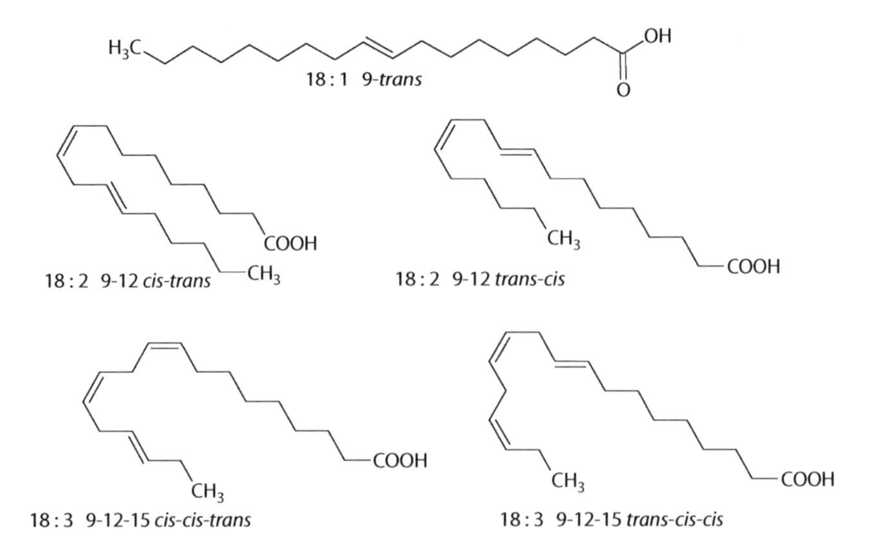

FIGURE 11.7 Basic *trans* isomers of fatty acids.

trans isomers and the risk of developing diabetes type 2, since elaidic acid produced higher blood insulin levels than oleic acid at the same blood sugar levels (Stender and Dyerberg, 2003). High intakes of trans lipids may increase the risk of colorectal neoplasia (Vinikoor et al., 2008), and may be also promoting other forms of cancer such as prostate, colon, and breast (Thompson et al., 2008; Smith et al., 2009). Positive relationships have been also observed with the prevalence of asthma, allergies, and eczema in teenagers (EFSA, 2004). Public health specialists suggest that consumers avoid the intake of food with trans lipids and advise the food industry to achieve technological reduction of the actual levels in processed foods (Mozaffarian et al., 2006; Gebauer et al., 2007). French fries and snacks contribute to the major dietary intake 20%–40%, followed by bakery products (1%–30%) and meat and dairy products (1%–6%) (EFSA, 2004). Food Standards Australia New Zealand (FSANZ, 2009) reported that, in a total number of 349 samples of processed foods (without primary sources of trans lipids), 33% contained nondetectable levels of *trans* isomers, 49% contained less than 2%, and 18% contained more than 2%. The mean daily intake of trans lipids in Europe is 1.2–6.7 g/day, corresponding to 0.5%–2.1% of total calorie intake. Elaidic acid contributes to the 54%–82% of the total intake of trans lipids. Intake is lower in the Mediterranean countries due to the use of olive oil in several culinary applications (EFSA, 2004). Most dietary references for trans lipids suggest that they should not *account* for more than 1% of the daily calorie intake (FAO/WHO, 2003).

11.3.8 Other Substances

A limited number of studies show that residues of pesticides or veterinary drugs in the food may be stable during the frying, with restricted increase or decrease in the initial concentration.

Nitrofuran antibiotics, though prohibited in many countries, have widespread use in the global food industry, and there is a focused attention on the toxicity and stability of metabolites from these drugs. Cooking at high temperatures causes tissue dehydration that will eventually increase the concentration of nitrofurans in meat. Rose et al. (1995, 1999) reported that cooking influences the level of risk posed by residues of veterinary drugs, with varying degrees of stability during heating. McCracken and Kennedy (1997) demonstrated that 3-amino-2-oxazolidinone residues were stable in the meat, liver, and kidney of furazolidone-treated pigs following cooking by microwave, grilling, and frying. Cooper and Kennedy (2007) applied various cooking methods (frying, grilling, roasting, and microwaving) to the muscle tissue and liver of nitrofuran-treated pigs. The residues remained after cooking, demonstrating that these metabolites are largely resistant to conventional cooking techniques and will continue to pose a health risk.

Perello et al. (2009) investigated the cooking-induced changes in the levels of polybrominated diphenyl ethers (PBDE) and hexachlorobenzene (HCB) in various cooked foods. There were some variations in the concentrations of these substances before and after cooking, depending not only on the cooking process, but mainly on the cooked food item. Cooking processes enhanced HCB levels in fried hake and fried lamb, while there were limited differences between raw and fried samples for other meat tissue foods. On the contrary, there was a reduction in fried potatoes or rice due to washing and soaking processes before frying. A year before, Perello et al. (2008) reported that the frying process is of a limited value as a means of reducing metal concentrations in foods.

11.4 CULINARY PROCESSES AND PREVENTIVE MEASURES

Though the actual frying process is the cooking of food in a heated fat, the overall culinary processes include several steps prior to frying (purchasing food supplies, food storage, food preparations) and some after frying (hot holding, serving). All the processes before and after frying are presented in the flowchart of Figure 11.8.

Based on the hazard description presented above, a selection of preventive measures could be proposed that could be applied in a professional food service establishment. Several of these measures can be also applied in domestic culinary applications.

11.4.1 Purchasing Oils or Fats for Frying (Step 1 in Figure 11.8)

Though all types of oils and fats can be used for frying, the procurement of products with saturated or monounsaturated fatty acid esters is suggested. Such oils are palm oil, olive oil, canola oil, and high-oleic sunflower oil. Vegetable shortening oils could be of moderate saturated or monounsaturated profile, depending on the mix of oils and the degree of hydrogenation (if applied). Soybean oil for frying is usually partially hydrogenated. Using fats with a low polyunsaturated profile will prevent the formation of decadienal and polar compounds. Placing the iodine value at less than 90 g I_2/100g in product description sheets would be a useful criterion. Selection of oils or fats free of partially hydrogenated triglycerides will prevent the increase of the dietary intake in trans lipids. The oil supplier should provide a report of the analysis for fatty acid methyl esters (FAME), together with the product descriptions. The initial quality of the oils or fats is important for the prevention of the formation of polar compounds and decadienal during frying. Oil supplies must be packed in airtight containers that are, preferably, dark or opaque and comply with the regulations for food contact material (fat contact in particular). The transporting vehicle must not let the oil containers be exposed to sunlight or to heat. Expiring dates should not be overdue or within a couple of months. The supplier must provide data for the peroxide value of the oils or fats for frying. Though there are official standards for the peroxide value for every country, those of Codex Alimentarius are suggested (expressed in meq active oxygen per kg): less than 10 for animal fats or vegetable oils, less that 15 for olive oil and sesame oil, and less than 20 for extra virgin olive oil (Codex Standards 033, 211, and 280). The peroxide value can be also determined at the point of delivery with semi-quantitative portable instruments. Some commercial oils for frying may contain antioxidant additives like ascorbyl palmitate or stearate (up to 500 ppm). In this case, the additives must be mentioned on the label of the oil container. These additives may reduce the formation of free fatty acids and their dimers, but this is only a partial solution for the formation of polar compounds. Dimethylsiloxane may be used as an antifoaming agent in oils for deep-fat frying

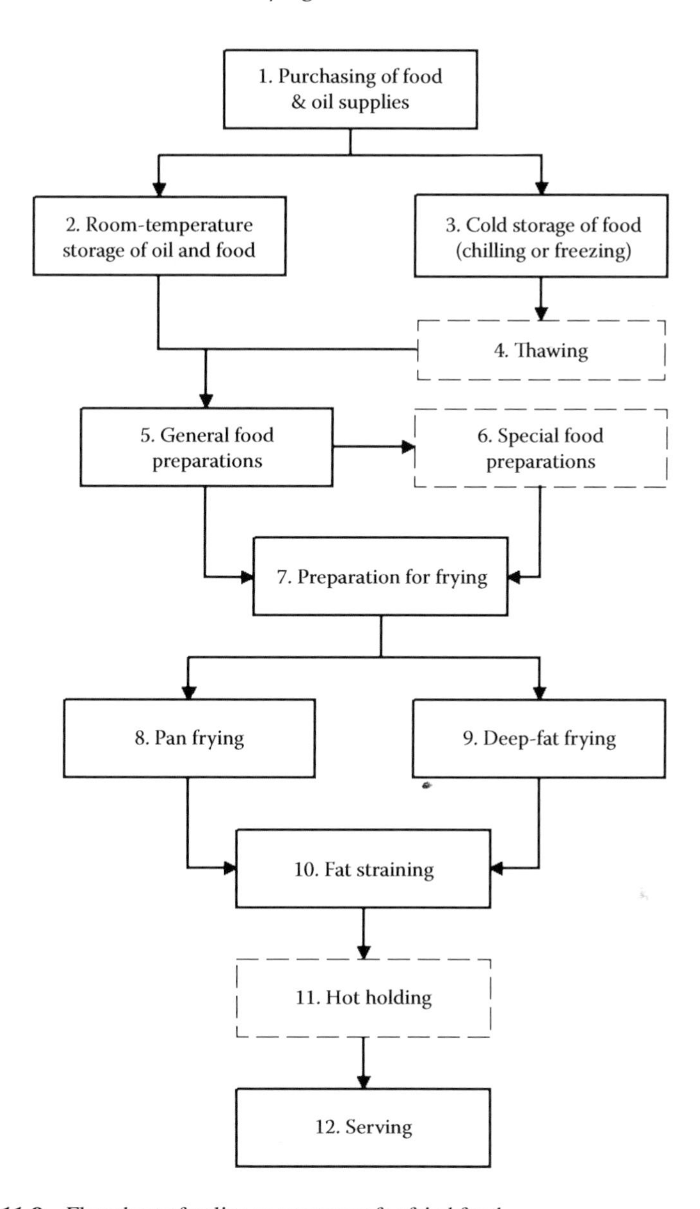

FIGURE 11.8 Flowchart of culinary processes for fried food.

(up to 10 ppm), and must also be mentioned on the label of the oil container. The effectiveness of this additive to reduce formation of acrylamide or polar compounds is limited. Ordering virgin olive oil may be quite an expensive practice for professional food service units. However, the natural antioxidants of olive oil may prevent the formation of heterocyclic amines in fried food, as well as the formation of polar compounds and decadienal in the frying oil. Though there is plenty of literature about the positive aspects of oils enriched with extracts of natural antioxidants, such products have limited commercial availability. Supplies of vegetable oils should be also free from pesticide residues. There are legal requirements for maximum limits of residues (MLR) within the European Union that are to be enforced. There are also legal requirements for the maximum levels of polyaromatic hydrocarbons in oils and fats. The supplier is obliged to carry reports for the levels of residues and contaminants in oils or fats, demonstrating that there is conformity with the current legislation. It is advisable to monitor the levels of 3-chloropropane-1,2-diol (3-MCPD) in refined oils as well, though the official limits in oils and the official mitigation strategies are not set yet.

11.4.2 Purchasing Foods for Frying (Step 1 in Figure 11.8)

All types of carbohydrate or protein foods are suitable for frying. Together with the main ingredients for meals of fried food, we may need to buy flour, bread crumbs, milk, or eggs to make blanketing or batters for the food to be fried. Materials rich in natural antioxidants such as wine, virgin olive oil, herbs, and spices may be useful for marinades. All food supplies must comply with the legislation in force for residues from pesticides and veterinary drugs. The food supplies must also comply with the legislation in force for PAH contaminants. There is particular concern about meat and dairy products from ruminant mammals that may contain trans lipids from heat-processed feeds. Soy sauce, beer yeast (dry or fresh), and gluten from corn or syrup may contain chloroprapanols. They should be monitored for 3-MCPD by the supplier. However, it is suggested that such products are not be used before or during frying.

11.4.3 Storage of Foods and Oils at Room Temperature (Step 2 in Figure 11.8)

Oils or fats for frying must be stored in cool (<20°C), dry, dark places. The containers must remain well closed and unnecessary transfusions should be avoided. In particular, transfusions to merge the content of two containers are considered bad practice. Monitoring of the peroxide value with portable devices on a monthly schedule is advised. Foods stored in room temperature must be in sealed containers or packages. The humidity must be less than 60% and the temperature less than 20°C. High humidity with moderate temperature may be a reason for the growth of molds and, furthermore, starch or protein hydrolysis. This will result in the formation of acrylamide or heterocyclic amines during frying. The ambient air in the storage area should be clean and not contaminated with gas emissions from the kitchen, heaters, vehicles, etc. Labeling elements of containers must remain visible and the stock rotation to be applied ("first in, first out").

11.4.4 Storage of Food in Chilling and Freezing Temperatures (Step 3 in Figure 11.8)

Normally, perishable foods (meat and dairy) are stored in temperatures below 5°C. There are several chilling temperatures suggested for the storage of vegetables.

Particular concern should be exercised in the storage of starchy vegetables like potatoes. Storage of fresh potatoes at temperatures above 6°C will avoid the formation of reducing sugars that will take part in acrylamide formation during frying. Potatoes with sprouts should be discarded because of the production of sugars. Precatered potatoes can be stored at normal chilling conditions. All products at freezing temperatures must be well packed or else they will be dehydrated. Dehydrated meat proteins may result in the formation of heterocyclic amines during frying. Labeling elements of containers must remain visible, and also stock rotation to be applied ("first in, first out"). Particularly for marinades or precatered products (cook-chill operations), the date and the quantity of the food when placed into the refrigerator must be recorded. Excess storage of these products in a refrigerator may result in carbohydrate or protein hydrolysis.

11.4.5 Thawing Frozen Products (Step 4 in Figure 11.8)

Usually, thawing of frozen products is performed under chilling conditions for 8 to 24 h. During thawing, the melting ice must be dripping out of the food container. If the food products remain long in the drip, there will be carbohydrate or protein hydrolysis by bacterial or enzyme activity. For the same reason, thawing in water tanks is not advisable. If the products are in small pieces and contain no ice crust, they can be directly deep fried (e.g., precatered potatoes and croquettes); however, this is very subjective concerning the evaluation of the ice crust. The ice crust is not desirable because it will release water in the frying medium and reduce the frying temperature.

11.4.6 General Food Preparations (Step 5 in Figure 11.8)

Before frying, several foods must be washed, peeled, sliced, or ground. Vegetable washing in food service units is often performed together with disinfection processes. In special water tanks, some tablets of hypochloride-based disinfectants are dissolved, and the vegetables remain for 3 to 5 min. Rinsing the vegetables after this process should be thorough so as to leave no hypochloride residues. All peeling, slicing, or grinding equipment must be made of materials with food contact properties that do not allow migration of minerals or other substances to the food (e.g., food-grade stainless steel, Teflon boards). All equipment should be washed and rinsed thoroughly so that no detergent residues will contaminate the food to be fried.

11.4.7 Special Food Preparations (Step 6 in Figure 11.8)

Application of marinades with sources of antioxidants like red wine, olive oil, herbs, and spices will result in reduced formation of heterocyclic amines in fried meat. Blanching of potatoes or soaking them in water will reduce the levels of asparagine or reducing sugars, respectively. Eventually, this will result in reduced formation of acrylamide in foods. Soaking of potatoes in citrate or phosphate buffers is good to prevent enzymic browning, to leach out sugars and amino acids, and to give volume to the potato with osmotic absorption. Blanketing the food with flour (wheat or corn), bread crumbs, or a batter mix (usually containing eggs or milk, flour, and occasionally beer or a soda drink) will protect the food from excessive heating during frying. The carbohydrate-based blanketing will be fried, and the food inside this will be boiled, resulting in, for example, the "croquette" with a crunchy outside (the fried crust) and a soft inside. This will reduce the formation of heterocyclic amines during meat, fish, or cheese frying but may increase the formation of acrylamide. Dough-based products for frying (e.g., donuts or dumplings) may need some time for leavening and during that time the dough raises in volume. This increases the frying surface over the amount of food to be fried, creating products with more of the desirable fried crust. The fried crust protects the food from heat but may be a source of acrylamide. Acidifying the dough with citrate or acetate before leavening may result in the reduction of acrylamide formation during frying, but may enhance the formation of chloropropanols.

11.4.8 Preparation for Frying (Step 7 in Figure 11.8)

All foods to be fried should be free of excess water (drip). Drips from meat or vegetable contain iron and other minerals that may function as prooxidants. Moreover, the water in the frying oil will reduce the frying temperature and will create bubbles that increase oxidation. So foods to be fried are supposed to be strained in colanders until there is no drip. Salting freshly cut potatoes will cause a starchy drip that will remove some reducing sugars. Blanketed foods should not be dripping batter or crumbs because these will be over-fried, creating undesirable effects. Used oils for frying are periodically filtered to remove over-fried food particles. There are filters made of cotton or paper that can remove even micro particles from oil. Over-fried food particles may contain acrylamide and accumulate oxidized triglycerides that promote further oxidation with radical reactions. However, this method does not assure reduction of total polar compounds, decadienal, or further formation of acrylamide. In other cases, the used frying oil is treated with absorbents, the so-called "fry powders," before filtering. These "fry powders" contain minerals (silicates, silica oxide, magnesium oxide, aluminum oxide, or diatomic earth) that can absorb the polar compounds and regenerate the oil. Some other commercial powders may contain antioxidants to be released into the oil. However, this practice can solve only part of the problem while formation of decadienal or acrylamide is advancing. They are considered malpractice by several consumer organizations, and therefore the fast-food service companies try to avoid them. Used frying oil that is too dark and releases off-flavors should be discarded. This is a subjective criterion and sometimes at this point there is already significant production of total polar compounds and other undesirable by-products. There are portable devices that can measure the polar compounds on site. There are strip tests with a "pass or fail" mode and conductivity meters calibrated on scale of total polar materials. These tests are based on the principle that when TPM is up to 20%, the oil can be replenished; and when it is 20%–24%, the oil can be discarded. Usually, replenishment is performed when the oil volume is reduced by 15% to 25%.

11.4.9 Pan Frying (Step 8 in Figure 11.8)

In pan frying, the heat capacity of the pan is more important than the heat capacity of the oil. If the frying is very shallow, heat conduction from the pan may overrule convection in oil (e.g., in Asian stir frying or sauté practice). If the frying pan has a thick steel bottom, it better protects the food from the heat source and creates more uniform frying. The temperature of the frying pan must be regulated and should not reach more than 160°C when frying in polyunsaturated oils and more than 175°C when frying in monounsaturated oils or saturated fats. The limit of 175°C is proposed for the reduction of acrylamide by the CIAA Acrylamide Toolbox. Continuous or periodic stirring while frying is required to prevent over-frying of the food sides in contact with the pan. Nonstick pans should not have scratches on the Teflon surface. Pan frying is supposed to be a batch operation so restrict the consecutive frying to about three sessions of maximum 10 min frying time, without replenishment. Of course, this depends on the type of food to be fried. If the fried food produces lots of particles that will be over-fried and makes the oil look dark with off-flavors, one frying session is enough. The remaining oil should be discarded and not be used again. The pan after frying must be washed, thoroughly rinsed, and let to dry up before next frying.

11.4.10 Deep-Fat Frying (Step 9 in Figure 11.8)

In deep-fat frying, the food is cooked by heat convection in the oil. Oils have a high heat capacity and can retain a temperature level with minor assistance from a heat source. Professional deep fryers have an oil capacity of 4–20 L for bench-top models and 15–80 L for floor models, while domestic deep fryers have an oil capacity of 1–3 L. Monounsaturated or saturated oils are suggested for deep-fat frying because they have more stability for long heating periods. Fryer design must match the type of product to be fried; for example, french fries are prepared in fryers with deep and narrow baskets, and donuts in wide and more shallow fryers specially designed for this product. The steel alloy of the oil tank and food basket must comply with the regulations for food contact materials. The ratio of food weight to oil volume varies from 1/3 g/mL in domestic fryers up to 1/10 g/mL in big professional fryers. Domestic fryers and small bench-top fryers of restaurants have a cover that reduces the contact with ambient oxygen, but may distill the volatile substances like decadienal. Common fryers have an electric power modulator, calibrated in approximate temperature units, and a countdown timer. Sophisticated fryers have an oil temperature sensor and a programmable modulator of the frying time with the electric power or the temperature. There is also automatic motion of the frying metal basket and sound signals for the end of the frying period or alert conditions. All bench-top or floor fryers have a tab at the bottom to empty the oil tank. Frying temperature should not be higher than 175°C, a temperature that will protect the food from acrylamide formation. Frying times are usually a few minutes and rarely surpass 10 min, depending on the food to be fried. Frying oils are expected to support more than 20 consecutive frying sessions; however, intermediate replenishment may be required. As mentioned before, the color and the odor of the frying oil are subjective criteria, while the on-site measurement of TPM can be more useful. In practice, there can be a calibration between the limit of the TPM value at the point of replenishment or rejection (20% and 24%, respectively) with the oil turnover rate. The oil turnover rate, expressed in hours, is defined as the oil capacity of the frying tank over the average oil consumption per hour. It is an indicator of the stress the oil is subjected to in a daily operation. As an example: If a 25 L fryer can produce 15 kg of french fries per hour that absorb 12% of the oil, it will be replenished with an additional 9 L for 5 h of operation. Since the fryer will be refilled with oil, the turnover rate is 5 h. If, at this turnover rate, the TPM is 20% or less, the fryer can still be replenished. If the TPM is 20%–24%, the frying oil must be discarded, and the oil tank thoroughly washed, rinsed, and dried.

11.4.11 Fat Straining (Step 10 in Figure 11.8)

After pan frying or deep-fat frying, the food may be placed in colanders or on tissue paper to strain out the excess oil on the surface of the food. This will reduce the intake of hazards that are formed in the frying oil such as polar compounds, decadienal, and *trans* isomers.

11.4.12 Hot Holding (Step 11 in Figure 11.8)

In fast food or in shelf-service outlets, the fried products may be placed in stand-by hot units until served. Usually they are kept warm (>65°C) on hot plateaus or under infrared light bulbs. Fried products are also exposed to ambient oxygen and sunlight. During this operation, lipid oxidation may still continue, and rancidity flavors appear, if products remain more than 30–60 min, depending on the type of food. A plain sensory test will determine whether the fried food has been standing for long in hot holding. Replenishment of food containers with new batches of fried foods is not advised. Fried food containers must comply with the food contact material regulations.

11.4.13 Serving (Step 12 in Figure 11.8)

In fast-food outlets, the fried products are served in disposable paper packs, laminated with polymers or aluminum foil. These packs, together with attached disposable forks or toothpicks, must comply with the food contact material regulations. If there is a notion that several of the above proposed measures did not result in minimizing the food hazards to acceptable levels, the fried food must be discarded and not served (the "If in doubt, throw it out" principle for food professionals).

11.5 DETERMINATION OF CRITICAL CONTROL POINTS

The Hazard Analysis and Critical Control (HACCP) preventive system for the food sector that was presented in the late 1960s became widespread in the 1970s and 1980s, and a legal obligation in the 1990s. The worldwide standard of HACCP is the Recommended International Code of Practice, General Principles of Food Hygiene of Codex Alimentarius: CAC/RCP 1-1969 rev. 4 2003 (CAC, 2009). The literature of HACCP application on frying is limited. Cordoba et al. (2000) presented some critical control points that could be used for fried fish fingers but upon microbiological criteria. Soriano et al. (2002) presented an HACCP system applied to deep-fat frying in sunflower oil at 19 university restaurants on the basis of associated hazards in frying. Vorria et al. (2004) presented a model HACCP system for safety assurance of fried foods, taking into account all types of food hazards (chemical, microbiological, physical). Ghidurus and Boskou (2005) presented a generic HACCP model for fried meatballs, including all types of food hazards as well. Sobukola et al. (2009) presented a "farm to fork" HACCP system for the production of a traditional African fried snack, taking into account only microbiological hazards.

Based upon the standard decision diagram by Codex Alimentarius to determine critical control points, a tailor-made decision tree is designed for culinary applications before, during, and after frying. This decision diagram takes into account all types of parameters, such as chemical precursors, temperatures, and culinary practices that may lead eventually to the formation or increase of the previously mentioned hazards. For example, parameters that may induce further development of a hazard include the presence of amines and reducing sugars for acrylamide or proteins for heterocyclic amines or hydrolyzed vegetable proteins for chloropropanol, etc. It can also be a technical parameter like the storage temperature of the potatoes for acrylamide, the frying temperature for TMP and PTG, or the application of a marinade for heterocyclic amines. Table 11.3 presents the decision-making flow for this decision diagram.

According to the decision-making flow in Table 11.3, the combinations of answers that may lead to a CCP are only two, containing the following operands:

Yes–No–Yes–skip–No and **Yes–No–No–Yes–No**

In Table 11.4 there is a table to determine the CCPs before, during, and after frying, in the cases for which the answer to question Q1 is yes. A number of for CCPs are determined for four processing steps: chilling storage of potatoes, preparation for frying, pan frying, and deep-fat frying. In Annex II the monitoring and validation

TABLE 11.3
Decision-Making Process for the Determination of CCPs before, during, and after Frying

Q1 Is there a parameter to induce further development of the hazard?	
If **Yes** go to the **next**	If **No stop** here
Q2 Are the prerequisite programs sufficient to eliminate this parameter?	
If **Yes** it is not a CCP **stop** here	If **No** go to the **next**
Q3 Is this process able to reduce this parameter to acceptable levels?	
If **Yes** go to the **Q5** question	If **No** go to the **next**
Q4 Is it possible for the hazard to develop unacceptable levels in this process?	
If **Yes** go to the **next**	If **No stop** here
Q5 Is there a next process that may reduce this hazard to acceptable levels?	
If **Yes** it is not a CCP **stop** here	If **No** it is a **CCP**

TABLE 11.4
Decision Process for the Determination of CCPs Before, During, and After Frying

Process	Hazards	Parameter	Q1	Q2	Q3	Q4	Q5	CCP
1a. Purchasing oil or fat for frying	CP	Refine oils or fats	Y	Y	—			
	DD, TPM	Polyunsaturated FA	Y	N	N	N	—	
	DD, TPM	Initial oxidation	Y	Y	—			
	PAH	Initial contamination	Y	Y	—			
	Trans	Hydrogenated oils or fats	*Y*	*Y*	—			
	Other	Pesticide residues	Y	Y	—			
1b. Purchasing food for frying	AA	Reducing sugars and amines	Y	N	N	N	—	
	CP	Foods with initial content	Y	Y	—			
	HA	Protein-rich food	Y	N	N	N	—	
	PAH	Initial contamination	Y	Y	—			
	Trans	Ruminant trans	*Y*	*Y*	—			
	other	Residues of pesticides and veterinary drugs	Y	Y	—			
2. Room storage	AA	Moisture in the food	Y	Y	—			
	HA	Moisture in food	Y	Y	—			
	TPM	Oils exposed to oxygen, heat, or light	Y	Y	—			
3. **Chilling or freezing storage**	AA	Potatoes stored at temperature <6°C	**Y**	**N**	**Y**	>	**N**	**CCP$_1$**
	AA	Sprouting of potatoes	**Y**	**N**	**Y**	>	**N**	**CCP$_1$**
	HA	Meat tissue dehydration in deep freezer	Y	Y	—			
	AA & HA	Hydrolysis of foods by overdue storage	Y	Y	—			

TABLE 11.4 (Continued)
Decision Process for the Determination of CCPs Before, During, and After Frying

Process	Hazards	Parameter	Q1	Q2	Q3	Q4	Q5	CCP
4. Thawing	AA & HA	Hydrolysis of foods immersed in melting ice	Y	Y	—			
	TPM	Ice crust for frozen foods that will be directly fried without thawing	Y	Y	—			
5. General preparations	CP	Chlorine residues	Y	Y	—			
6. Special preparations	AA	No soaking, no blanching, no blanketing,	Y	Y	—			
	HA	No marination	Y	Y	—			
	CP	Leavening of acid dough	Y	Y	—			
7. Preparation for frying	TPM	Drip of foods to be fried	Y	Y	—			
	DD, PAH, TPM, *Trans*	Replenishment of oil with a high turnover	**Y**	**N**	**N**	**Y**	**N**	**CCP_2**
	TPM	Ineffective use of “fry-powders” for deep-fat frying	**Y**	**N**	**Y**	>	**N**	**CCP_2**
	TPM	Ineffective filtration of used oils for deep-fat frying	**Y**	**N**	**Y**	>	**N**	**CCP_2**
8. Pan frying	AA	Frying temperature >175°C	**Y**	**N**	**Y**	>	**N**	**CCP_3**
	AA, PAH, TPM	Accumulation of over-fried particles	Y	Y	—			
	TPM	Frying temperature >160°C for polyunsaturated oils	**Y**	**N**	**Y**	>	**N**	**CCP_3**
	AA, DD, PAH, TPM, *Trans*	Consecutive frying sessions	**Y**	**N**	**Y**	>	**N**	**CCP_3**

(*Continued*)

TABLE 11.4 (Continued)
Decision Process for the Determination of CCPs Before, During, and After Frying

Process	Hazards	Parameter	Q1	Q2	Q3	Q4	Q5	CCP
9. Deep-fat frying	AA	Frying temperature >175°C	**Y**	**N**	**Y**	>	**N**	**CCP_4**
	AA, PAH, TPM	Accumulation of over-fried particles	Y	Y	—			
	TPM	Frying temperature >160°C for polyunsaturated oils	**Y**	**N**	**Y**	>	**N**	**CCP_4**
	AA, DD, PAH, TPM, *Trans*	Consecutive frying with high turnover rate	**Y**	**N**	**Y**	>	**N**	**CCP_4**
10. Fat straining	AA, CP, DD, PAH, TPM, *Trans*	If not applied, the surface oil will increase the dietary intake	Y	Y	—			
11. Hot holding	TPM	Prolonged hot holding	Y	Y	—			
12. Serving	All the above	Doubtful previous operations	Y	Y	—			

Note: Qn refers to the questions in Table 11.3; Y, yes; N, no; >, skip; —, stop; CCP_n, Critical Control Point with serial number; AA, acrylamide; CP, chloropropanols; DD, decadienal; HA, heterocyclic amines; PAH, polyaromatic hydrocarbons; TPM, total polar materials; *Trans*, *trans* isomers of lipids; the numbers of processes refer to Figure 11.8.

methods for these CCPs are proposed. The critical limits for acrylamide are proposed by the CIAA Acrylamide Toolbox (2009). The critical limit of 160°C for polyunsaturated oils is empiric, and therefore arbitrary, but can be reviewed by the validation methods proposed. It is presented more to place emphasis on the sensitivity of polyunsaturated oils in thermoxidation. The critical limits for oil turnover rate are not presented with figures since they must be calculated by taking into account the oil capacity of the deep fryer, the type of fried food, the oil uptake of the fried food, the frying sessions and, of course, the TPM value.

The remaining process steps are plain control points, referring to good catering practices. According to the terminology of the EN ISO 22000:2005 food safety management standard, the processes that are not CCPs are considered prerequisite programs (PrPs). Prerequisite programs are standard programs associated with

TABLE 11.5
Monitoring and Validation of CCPs for Frying

CCP	Process	Hazards	Critical Limit	Monitoring	Corrections	Records	Validation
CCP_1	3. Chilling storage of potatoes	AA	Storage temperature >6°C, no sprouts	2–3 times per day temperature monitoring Visible observation	Do not fry or bake affected lots Discard affected lots	Temperature monitoring sheets Production records Destruction records	Lab analysis of reducing sugars every 3 months
CCP_2	7. Preparation for frying	DD, PAH, TPM, *Trans*	Oil turnover rate	At every replenishment	Replenish after filtration or use of "fry powders"	Oil consumption records Filtration records	On-site analysis of TPM once or twice a week
CCP_3	8. Pan frying	AA, DD, PAH, TPM, *Trans*	Frying temperature <175°C and <160°C for polyunsaturated oils 3–4 consecutive frying sessions	Examine frying pan temperature 3 times per day Count of frying operations in the same oil	Discard affected batches	Temperature monitoring sheets Production records	On-site analysis of TPM in oil of the frying pan once or twice per day Lab analysis of fried food for TPM, PTG, AA, PAH, & *trans* 2 or 3 times per year
CCP_4	9. Deep-fat frying	AA, DD, PAH, TPM, *trans*	Frying temperature <175°C and <160°C for polyunsaturated oils Oil turnover rate	Constant or periodic monitoring of fryer temperature At every replenishment	Discard affected batches Clean and refill the fryer	Temperature monitoring sheets Production records Oil consumption records	On-site analysis of TPM in oil of the fryer once or twice per day Lab analysis of fried food for TPM, PTG, AA, PAH, & *Trans* 2 or 3 times per year

Note: CCP_n, Critical Control Point with serial number; AA, acrylamide; CP, chloropropanols; DD, decadienal; HA, heterocyclic amines; PAH, polyaromatic hydrocarbons; TPM, total polar materials; *Trans*, *trans* isomers of lipids; the numbers of processes refer to Figure 11.8.

food production like cleaning operations, basic culinary processes, storage temperature control, traceability labeling and FIFO, chemical criteria for supplies, control of raw materials at purchasing, etc. Considering that these PrPs are supposed to be effective, the answer to question Q2 was yes. For these PrPs, the preventive measures described previously must be applied. However, standard monitoring practices for these measures are required as well. There was not a single case where the answer in question Q5 was yes. That means that there is positive trend to increase frying hazards throughout the culinary processes. Once the parameter for a frying hazard appears, it is not possible to reduce it in subsequent processes.

Three CCPs for frying in university restaurants were determined by Soriano et al. (2002): receiving, storage, and frying. The monitoring method for receiving is not feasible since suppliers do not provide chemical reports at every delivery. During storage, control of foreign materials is usually a prerequisite program and is not frying specific. For the CCP at frying, Soriano et al. (2002) have set eight monitoring procedures, of which some could be considered redundant since they are covered by prerequisites, like cleaning of the fryer or use of stainless steel utensils. Monitoring of frying oil with organoleptic control is subjective, except if very specific guidelines are provided (with pictures and vivid descriptions of odors). Soriano et al. (2002) have also set critical limits for the level of frying temperature (<180°C), the daily turnover rate (15%–25%), and the heating time (no exact value set). As a result of this application, the TPM values had an average decrease of 25% among 13 restaurants.

Vorria et al. (2004) identified 27 CCPs for 12 of the 15 processing steps for potato chips and french fries. Of course, these CCPs concern all types of chemical, physical, and microbiological hazards. The frying-specific CCPs (and critical limits, respectively) are at three processing steps: (1) receiving oil for frying (antifoaming agents <0.04 ppm), (2) storage of oils (peroxide value <3 meq/kg), and (3) deep-fat frying (temperature 165°C–185°C, optimum temperature 177°C, oil turnover rate 5–10 h). The critical limits set by Vorria et al. (2004) for the first two CCPs cannot be monitored for every lot of received oils. The validation limits proposed for the CCP of deep-fat frying are acrylamide concentration in fried food at <0.5 ppm, and *trans* fatty acid consumption at 2.7–12.8 g/day (the food consumption rate is not mentioned).

Ghidurus and Boskou (2005) determined CCPs for 6 of the 12 processing steps for fried meatballs, involving all types of chemical, physical, and microbiological hazards. The CCP at oil receiving concerns the oil rancidity and at deep-fat frying concerns the formation of polar compounds, polymerized triglycerides, and decadienal. The first is monitored with visual inspection of the transportation vehicle, oil containers, and organoleptic assessment of the oil for frying. The second CCP is monitored with temperature (critical limit <180°C) and organoleptic assessment (excessive foam, darkening and smoking in the oil tank). For the validation of the CCP at the frying process, the levels 24% TPM, 12% PTG, and 20mg% decadienal in the frying oil are proposed.

11.6 CONCLUSIONS

There are several types of chemical hazards produced during the frying of food. Though microbiological hazards may be eliminated and physical hazards can be controlled by good catering practices, the chemical hazards seem to have a throughput potential. Acrylamide, heterocyclic amines, and polyaromatic hydrocarbons are hazards appearing in cooked food probably since humans discovered fire. Our knowledge about the levels of these hazards and the implications for health has increased in the last decades. But also the consumption of foods processed in high heat has increased, leading to higher exposure levels. Chloropropanols, *trans* lipids, and residues of agrochemicals are technological hazards, reflecting the status of the modern food industry, appearing in raw materials and increasing during frying. Polar compounds and volatile aldehydes are hazards associated with the heat degradation of the frying oil. Therefore, they are very good parameters for the safety and quality management of frying operations.

The frying processes, the interrelationships between oil uptake and oil quality and the impact on nutrition are a very complex issue. No overall agreement exists concerning the direct relationship between frying and health risks. Is fried food really hazardous? It should not be, because it is a popular food. Current studies propose mitigation strategies to minimize fried food intake but not to eliminate it from our diet. On the other hand, well-designed HACCP systems for frying may reduce the dietary intake of frying hazards. Fried foods are nutritious foods that can provide dairy and meat proteins, fat-soluble vitamins, polyunsaturated and monounsaturated lipids, phytosterols, antioxidants, and many others. Analysis of risk over benefit is required to compare hazard intake over nutrient intake for improvement of the overall quality of fried foods.

REFERENCES

Aladedunye, F.A. and Przybylski R. 2009. Degradation and nutritional quality changes of oil during frying, *J. Am. Oil Chem. Soc.* 86, 149–156.

Andrikopoulos, N.K., Boskou, G., Dedoussis, G.V.Z., Chiou, A., Tzamtzis, V.A., and Papathanasiou, A. 2003. Quality assessment of frying oils and fats from 63 restaurants in Athens, Greece. *Food Serv. Techn.* 3, 49–59.

Andrikopoulos, N.K., Chiou, A., Mylona, A., Boskou, G., and Dedoussis, G.V.Z. 2004. Monitoring of 2,4-decadienal in oils and fats used for frying in restaurants in Athens, Greece. *Eur. J. Lipid Sci. Technol.* 106, 671–679.

Andrikopoulos, N.K., Dedousis, G.V.Z., Falirea, A., Kalogeropoulos, N., and Hatzinicola, H. 2002a. Deterioration of natural antioxidant species of vegetable edible oils during the domestic deep-frying and pan-frying of potatoes. *Int. J. Food Sci. Nutr.* 53, 351–363.

Andrikopoulos, N.K., Dedoussis, G.V.Z., Tzamtzis, V.A., Chiou, A., and Boskou, G. 2002b. Evaluation of some medium polarity materials isolated as frying by-products from edible vegetable oils by RP-HPLC. *Eur. J. Lipid Sci. Technol.* 104, 110–115.

Aro, A., Van Amelsvoort, J., Becker, W., Van Erp-Baart, M. A., Kafatos, A., Leth, T. et al. 1998. Trans fatty acids in dietary fats and oils from 14 European countries: The TRANSFAIR study. *J. Food Comp. Anal.* 11, 137–149.

Bansal, G., Zhou, W., Tan, T.W., Neo, F.L., and Lo, H.L. 2009. Analysis of trans fatty acids in deep frying oils by three different approaches. *Food Chem.* 116, 535–541.

Bansal, G, Zhou, W., Barlow, P.J., Lo, H.L., and Neo, F.L. 2010. Performance of palm olein in repeated deep frying and controlled heating processes. *Food Chem.* 121, 338–347.

Bastida, S. and Sanchez-Muniz, F.J. 2001. Thermal oxidation of olive oil, sunflower oil and a mix of both oils during forty discontinuous domestic fryings of different foods. *Food Sci. Technol.* 7, 15–21.

Becalski, A., Lau, B.P., Lewis, D., and Seaman, S.W. 2003. Acrylamides in foods: Occurrence, sources, and modelling. *J. Agric. Food Chem.* 51, 802–08.

Bhattacharya, A.B., Sajilata, M.G., Tiwari, S.R., and Singhal, R.S. 2008. Regeneration of thermally polymerized frying oils with adsorbents. *Food Chem.* 110, 562–570.

Bheemreddy, R.M., Chinnan, M.S., Pannu, K.S., and Reynolds, A.E. 2002. Active treatment of frying oil for enhanced fry-life. *J. Food Sci.* 67, 1478–1484.

Billek, G. 2000. Health aspects of thermoxidized oils and fats. *Eur. J. Lipid Sci. Technol.* 102, 587–593.

Blekas, G. and Boskou, D. 1986. Effect of esterified sterols on the stability of heated oils. In: *The Shelf Life of Food and Beverages*. Charalambous G., ed., 641–645, Amsterdam, Elsevier.

Bordas, M., Moyano, E., Puignou, L., and Galceran, M.T. 2004. Effect of temperature, time and precursors. *J. Chromatogr.* B 802, 11–17.

Boskou, G., Salta, F. N., Chiou, A., Troullidou, E., and Andrikopoulos, N.K. 2006. Content of *trans,trans*-2,4-decadienal in deep-fried and pan-fried potatoes. *Eur. J. Lipid Sci. Technol.* 108, 109–115.

Busquets, R., Bordas, M., Toribio, F., Puignou, L., and Galceran, M.T. 2004. Occurrence of heterocyclic amines in several home-cooked meat dishes of the Spanish diet. *J. Chromatogr.* B 802, 79–86.

Buttery, R.G. 1989. Importance of lipid-derived volatiles to vegetable and fruit flavor. In: *Flavor Chemistry of Lipid Foods.* Min D.B., Smouse T.H., eds., 156–165, Champaign, IL, AOCS Press.

Cabre, A., Girona, J., Vallve, J.C., Heras, M., and Masana, L. 2003. Cytotoxic effect of the lipid peroxidation product 2,4-decadienal in vascular smooth muscle cells. *Atherosclerosis* 169, 245–250.

Cabre, A., Girona, J., Vallve, J.C., and Masana, L. 2004. Aldehydes mediate factor induction: A possible mechanism linking lipid peroxidation to thrombotic events. *J. Cell. Physiol.* 198, 230–236.

CAC, Codex Alimentarius Commission. 2009. *Food Hygiene. Basic Texts*. 4th edition. Rome, FAO/WHO, 2009.

Cerbulis, J., Parks, O.W., Liu, R.H., Piotrowski, E.G., and Farrell, H.M. Jr. 1984. Occurrence of diesters of 3-chloro-1,2-propanediol in the neutral lipid fraction of goats' milk. *J. Agric. Food Chem.* 32, 474–476.

Che Man, Y.B. and Jaswir, I. 2000. Effect of rosemary and sage extracts on frying performance of refined, bleached and deodorized (RBD) palm olein during deep-fat frying. *Food Chem.* 69, 301–307.

Chen, Y.C. and Chen, B.H. 2003. Determination of polycyclic aromatic hydrocarbons in fumes from fried chicken legs. *J. Agric. Food Chem.* 51, 4162–4167.

Chiou, A., Kalogeropoulos, N., Salta, F.N., Efstathiou, P., and Andrikopoulos, N.K. 2009. Pan-frying of french fries in three different edible oils enriched with olive leaf extract: Oxidative stability and fate of microconstituents. *LWT—Food Sci. Technol.* 42, 1090–1097.

Choe, E. and Min, D.B. 2007. Chemistry of deep-fat frying oils. *J. Food Sci.* 72, R77–R86.

Chung, J., Lee, J., and Choe, E. 2004. Oxidative stability of soybean and sesame oil mixture during frying of flour dough. *J. Food Sci.* 69, 574–578.

CIAA 2009. The CIAA Acrylamide "Toolbox" rev. 12, Brussels Confederation of the Food and Drink Industries of the EU.

Collier, P.D., Cromie, D.D.O., and Davies, A.P. 1991. Mechanism of formation of chloropropanols present in protein hydrolysates. *J. Am. Oil Chem. Soc.* 68, 785–790.

Cooper, K.M. and Kennedy, D.G. 2007. Stability studies of the metabolites of nitrofuran antibiotics during storage and cooking. *Food Addit. Contam.* 24, 935–942.

Cordoba, M.G., Jordano, R., and Cordoba, J.J. 2000. Microbial hazards analysis in commercial processing of prepared and frozen hake fish fingers. *Food Sci. Technol. Int.* 6, 307–314.

Costa, L.G., Deng, H., Gregotti, C., Manzo, L., Faustman, E.M., Bergmark, E. et al. 1992. Comparative studies on the neuro- and reproductive toxicity of acrylamide and its epoxide metabolite glycidamide in the rat. *Neurotoxicology* 13, 219–224.

Crews, C., Brereton, P., and Davies, A., 2001. Effect of domestic cooking on the levels of 3-monochloropropandiol in foods. *Food Addit. Contam.* 18, 271–280.

Croon, L.B., Rogstad, A., Leth, T., and Kiutamo, T. 1986. A comparative study of analytical methods for quality evaluation of frying fat. *Fette Seifen Anstrich.* 88, 87–91.

Cuesta, C., Sanchez-Muniz, F.J., Garrido-Polonio, C., Lopez-Varela, S., and Arroyo, R. 1993. Thermooxidative and hydrolytic changes in sunflower oil used in frying with a fast turnover of fresh oil. *J. Am. Oil Chem. Soc.* 70, 1069–1073.

Culp, S.J., Gaylor, D.W., Sheldon, W.G., Goldstein, L.S., and Beland, F.A. 1998. A comparison of the tumours induced by coal tar and benzo[a]pyrene in a 2-year bioassay. *Carcinogenesis* 19, 117–124.

Dearfield, K.L., Abernathy, C.O., Ottley, M.S., Brantner, J.H., and Hayes, P.F. 1988. Acrylamide: Its metabolism, developmental and reproductive effects, genotoxicity, and carcinogenicity. *Mutat. Res.* 195, 45–77.

De Greyt, W., Kint, A., Kellens, M., and Huyghebaert, A. 1998. Determination of low trans levels in refined oils by Fourier transform infrared spectroscopy. *J. Am. Oil Chem. Soc.* 75, 115–118.

DGF, Deutsche Gesellschaft für Fettforschung. 2000. Recommendations for frying oils. 3rd International Symposium on Deep-Fat Frying, optimal operations. March 20–21. Hagen/Westphalia: Germany. *Eur. J. Lipid Sci. Technol.* 102, 594.

Dobarganes, M.C. and Márquez-Ruiz, G. 2003. Oxidized fats in foods. *Curr. Opin. Clin. Nutr. Metab. Care* 6, 157–163.

Dobarganes, C., Márquez-Ruiz, G., and Velasco, J. 2000. Interactions between fat and food during deep-frying. *Eur. J. Lipid Sci. Technol.* 102, 521–528.

Doledzal, M., Dvorakova, L., Zelinkova, Z., and Velisek, J. 2008. Analysis of potato product lipids for 3-MCPD esters. *Proceedings, 6th Eurofed Lipid Congress*, p. 325, Athens.

Doleschall, F., Recseg, K., Kemeny, Z., and Kovari, K. 2003. Comparison of differently coated SPME fibres applied for monitoring volatile substances in vegetable oils. *Eur. J. Lipid Sci. Technol.* 105, 333–338.

Drummond, L. 2005. Deep-Frying in New Zealand—A Review and Technology Update. The National Heart Foundation of New Zealand.

Dybing, E., Farmer, P.B., Andersen, M., Fennell, T.R., Lalljie, S.P.D., Müller, D.J.G., Olin, S. et al. 2005. Human exposure and internal dose assessments of acrylamide in food. *Food Chem. Toxicol.* 43, 365–410.

EC 1997. Opinion on 3-monochloro-propane-1,2-diol (3-MCPD), expressed on 16 December 1994. In: *Food Science and Techniques: Reports of the Scientific Committee for Food (36th series),* European Commission, Brussels, pp. 31–33.

EC 2001. Commission Regulation No. 466/2001. Setting maximum levels for certain contaminants in foodstuffs. Official Journal L77/1, 16.3.2001.

EC 2006. Commission Regulation (EC) No 1881/2006 of 19 December 2006, setting maximum levels for certain contaminants in foodstuffs. Official Journal L 364, 20.12.2006.

EFSA 2004. European Food Safety Authority. Opinion of the EFSA Scientific Panel on dietetic products, nutrition and allergies [NDA] related to the presence of trans fatty acids in foods and the effect on human health of the consumption of trans fatty acids. *EFSA*. 81, 1–49.

EFSA 2008. Scientific opinion of the panel on contaminants in the food chain polycyclic aromatic hydrocarbons in food. *EFSA*. 724, 1–114.

Esterbauer, H. 1993. Cytotoxicity and genotoxicity of lipid oxidation products. *Am. J. Clin. Nutr.* 57, 779–786.

Esterbauer, H., Zollner, H., and Schaur, R. 1990. Aldehydes formed by lipid peroxidation: Mechanism of formation, occurrence and determination. In: *Membrane Lipid Oxidation*, Pelfrey, C., ed., 239–268, Boca Raton, FL, CRC Press.

FAO/WHO 2002a. Evaluation of certain food additives and contaminants. 57th Report of the Joint FAO/WHO Expert Committee on Food Additives. Technical Report Series No 909. Geneva, WHO.

FAO/WHO 2002b. Health Implications of Acrylamide in Food. Report of the Joint FAO/WHO Consultation, Geneva, WHO.

FAO/WHO 2003. Diet, nutrition and the prevention of chronic diseases. Report of the Joint FAO/WHO Expert Consultation. WHO Technical Report Series No. 916, 55–60, Geneva, WHO.

FAO/WHO 2005. Joint Fao/Who Expert Committee on Food Additives. Report of the 64th meeting, Rome, FAO.

Firestone, D. 2004. Regulatory requirements in the frying industry. In: *Frying Technology and Practices*, Gupta, M.K., Warner, K., White, P.J., eds., 200–216, Champaign, IL, AOCS Press.

Friedman, M. and Levin, C.E. 2008. Review of methods for the reduction of dietary content and toxicity of acrylamide. *J. Agric. Food Chem.* 56, 6113–6140.

Fritsch, W.C. 1981. Measurements of frying fat deterioration. *J. Am. Oil Chem. Soc.* 58, 272–281.

FSANZ, Food Standard Australia New Zealand, 2009. Risk Assessment Report. Trans Acids in the New Zealand and Australia Food Supply. July 2009.

Gebauer, S.K., Psota, T.L., and Kris-Etherton, P.M. 2007. The diversity of health effects of individual trans fatty acid isomers. *Lipids* 42, 787–799.

Gertz, C. 2000. Chemical and physical parameters as quality indicators of used frying fats. *Eur. J. Lipid Sci. Technol.* 102, 566–572.

Gertz, C., Klostermann, S., and Kochhar, S.P. 2000. Testing and comparing oxidative stability of vegetable oils and fats at frying temperature. *Eur. J. Lipid Sci. Technol.* 102, 543–551.

Ghidurus, M. and Boskou, G. 2005. Hazards associated with frying. Implementation on a generic HACCP model for ground meat products. *Proceedings, 26th ISF World Congress Modern Aspects of Fats and Oils*, Prague, The International Society for Fat Research (ISF), p. 19.

Girona, J., Vallvé, J.C., Ribalta, J., Heras, M., Olivé, S., and Masana, L. 2001. 2,4-Decadienal downregulates TNF-α gene expression in THP-1 human macrophages. *Atherosclerosis* 158, 95–101.

Gordon, M.H. and Kourimska, L. 1995. The effects of antioxidants on changes in oils during heating and deep frying. *J. Sci. Food Agric.* 68, 347–353.

Grandgirard, A., Sebedio, J.L., and Fluery, J. 1984. Geometrical isomerization of linolenic acid during heat treatment of vegetable oils. *J. Am. Oil Chem. Soc.* 61, 1563–1568.

Granvogl, M., Koehler, P., Latzer, L., and Schieberle, P. 2008. Development of a stable isotope dilution assay for the quantitation of glycidamide and its application to foods and model systems. *J. Agric. Food Chem.* 56, 6087–6092.

Hamlet, C.G. and Sadd, P.A. 2004. Chloropropanols and their esters in cereal products. *Czech J. Food Sci.* 22, 259–262.

Hamlet, C.G. and Sadd, P.A. 2005. Effects of yeast stress and pH on 3-mono-chloro-propanediol (3-MCPD)-producing reactions in model dough systems, *Food Addit. Contam.* 22, 616–623.

Hamlet, C.G., Sadd, P.A., Crews, C., Velisek J., and Baxter D.E. 2002. Occurrence of 3-chloro-propane-1,2-diol (3-MCPD) and related compounds in foods: A review. *Food Addit. Contam.* 19, 619–631.

Hamlet, C.G., Sadd, P.A., and Gray, D.A. 2004. Generation of monochloropropanediols (MCPDs) in model dough systems. 1. Leavened doughs, *J. Agric. Food Chem.* 52, 2059–2066.

Holownia, K.I., Chinnin, M.S., Erickson, M.C., and Mallikarjunan, P. 2000. Quality evaluation of edible film-coated chicken strips and frying oils. *J. Food Sci.* 65, 1087–1090.

International Agency for Research in Cancer 1994. Acrylamide. *IARC Monogr. Eval. Carcinog. Humans* 60, 389–433.

Janoszka, B., Warzecha, L., Błaszczyk, U., and Bodzek, D. 2004. Organic compounds formed in thermally treated high-protein food. Part I. Polycyclic aromatic hydrocarbons. *Acta Chromatogr.* 14, 115–128.

Johansson, M., Fredholm, L., Bjerne, L., and Jägerstad, M. 1995. Influence of frying fat on the formation of heterocyclic amines in fried beefburgers and pan residues. *Food Chem. Toxicol.* 33, 993–1004.

Kalantzakis, G., Blekas, G., Pegklidou, K., and Boskou, D. 2006. Stability and radical-scavenging activity of heated olive oil and other vegetable oils. *Eur. J. Lipid Sci. Technol.* 108, 329–335.

Kaliora, A.C., Andrikopoulos, N.K., Dedoussis, G.V.Z., Chiou, A., and Mylona, A. 2003. Medium polarity lipids from fried oils promote LDL oxidation, in vitro. *Ital. J. Food Sci.* 4, 511–520.

Kalogeropoulos, N., Salta, F.N., Chiou, A., and Andrikopoulos, N.K. 2007. Formation and distribution of oxidized fatty acids during deep- and pan-frying of potatoes. *Eur. J. Lipid Sci. Technol.* 109, 1111–1123.

Keating, G.A. and Bogen, K.T. 2004. Estimates of heterocyclic amine intake in the US population. *J. Chromatogr.* B 802, 127–133.

Khor, G.L. and Mohd Esa, N. 2008. Trans fatty acids intake: Epidemiology and health implications. In: *Trans Fatty Acids*, Dijkstra, A.J., Hamilton, R.J., Hamm, W., eds., 25–45, Oxford, Blackwell Publishing.

Kim, M. and Choe, E. 2004. Effects of burdock (*Arctium lappa* L.) extracts on autoxidation and thermal oxidation of lard. *Food Sci. Biotechnol.* 13, 460–466.

Kusaka, H., Nagano, S., and Ohta, S. 1985. On functions of silicone oil in frying oil. Influence of silicone oil on convection of frying oil. *Yukagaku* 34, 187–190.

Lamboni, C. and Perkins, E.G. 1996. Effects of dietary heated fats on rat liver enzyme activity. *Lipids* 31, 955–962.

Lan, C.M., Kao, T.H., and Chen, B.H. 2004. Effects of heating time and antioxidants on the formation of heterocyclic amines in marinated foods. *J. Chromatogr.* B 802, 27–37.

Lantz, I., Ternite, R., Wilkens, J., Hoenicke, K., Guenther, H., and van der Stegen, G.H. 2006. Studies on acrylamide levels in roasting, storage and brewing of coffee. *Mol. Nutr. Food Res.* 50, 1039–1046.

Larsen, J.C. 2009. 3-MCPD esters in food products. Summary Report of a Workshop held in February 2009 in Brussels, Belgium. ILSI Europe Report Series.

Li, C.T., Lin, Y.C., Lee, W.J., and Tsai, P.J. 2003. Emission of polycyclic aromatic hydrocarbons and their carcinogenic potencies from cooking sources to the urban atmosphere. *Environ. Health Persp.* 111, 483–487.

Lopachin, R.M. and Lehning, E.J. 1994. Acrylamide-induced distal axon degeneration: A proposed mechanism of action. *Neurotoxicology* 15, 247–260.

Loureiro, A.P.M., Campos, I.P.D.A., Gomes, O.F., Di Mascio, P., and Medeiros, M.H.G. 2004. Structural characterization of diastereoisomeric ethano adducts derived from the reaction of 2′-deoxyguanosine with *trans,trans*-2,4-decadienal. *Chem. Res. Toxicol.* 17, 641–649.

Loureiro, A.P.M., Di Mascio, P., Gomes, O.F., and Medeiros, M.H.G. 2000. *Trans,trans*-2,4-decadienal-induced 1,N2-etheno-2-deoxyguanosine adduct formation. *Chem. Res. Toxicol.* 13, 601–609.

Mancini-Filho, J., Smith, L.M., Creveling, R.K., and Al-Shaikh, H.F. 1986. Effects of selected chemical treatments on quality of fats used for deep frying. *J. Am. Oil Chem. Soc.* 63, 1452–1456.

Marmesat, S., Velasco, J., and Dobarganes, M.C. 2008. Quantitative determination of epoxy acids, keto acids and hydroxy acids formed in fats and oils at frying temperatures. *J. Chromatography* A 1211, 129–134.

Márquez-Ruiz, G., Guerel, G., and Dobarganes, M.C. 1998. Applications of chromatographic techniques to evaluate enzymatic hydrolysis of oxidized and polymeric triglycerides by pancreatic lipase in vitro. *J. Am. Oil Chem. Soc.* 75, 119–126.

Maskan, M. and Bagci, H.I. 2003. The recovery of used sunflower seed oil utilized in repeated deep-fat frying process. *Eur. Food Res. Technol.* 218, 26–31.

Mazza, G. and Qi, H. 1992. Effect of after-cooking darkening inhibitors on stability of frying oil and quality of french fries. *J. Am. Oil Chem. Soc.* 69, 847–853.

McCracken, R.J. and Kennedy, D.G. 1997. The bioavailability of residues of the furazolidone metabolite 3-amino-2-oxazolidinone in porcine tissues and the effect of cooking upon residue concentrations. *Food Addit. Contam.* 14, 507–513.

Mestdagh, F., Maertens, J., Cucu, T., Delporte, K., Van Peteghem, C., and De Meulenaer, B. 2008. Impact of additives to lower the formation of acrylamide in a potato model system through pH reduction and other mechanisms. *Food Chem.* 107, 26–31.

Moreno, M., Olivares, D.M., Lopez, F.J.A., Adelantado, J.V.G., and Reig, F.B. 1999. Determination of unsaturation grade and trans isomers generated during thermal oxidation of edible oils and fats by FTIR. *J. Molec. Struct.* 482–483, 551–556.

Morin, O. 2005. Trans fatty acids: New developments. *Oleag., Crops Gras, Lipides* 12, 414–421.

Mottram, D.S., Wedzicha, B.L., and Dodson, A.T. 2002. Food chemistry: acrylamide is formed in the Maillard reaction. *Nature* 419, 448–449.

Mozaffarian, D., Katan, M.B., Ascherio, A., Stampfer, M.J., and Willett, W.C. 2006. Medical progress—trans fatty acids and cardiovascular disease. *New Engl. J. Med.* 354, 1601–1613.

Murkovic, M. 2004. Chemistry, formation and occurrence of genotoxic heterocyclic aromatic amines in fried products. *Eur. J. Lipid Sci. Technol.* 106, 777–785.

Murkovic, M., Friedrich, M., and Pfannhauser, W. 1997. Heterocyclic aromatic amines in fried poultry meat. *Z. Lebensm. Unters. Forsch.* A 205, 347–350.

Murkovic, M., Steinberger, D., and Pfannhauser, W. 1998. Antioxidant spices reduce the formation of heterocyclic aromatic amines in fried meat. *Z. Lebensm. Unters. Forsch.* A 207, 477–480.

Nappez, C., Battu, S., and Beneytout, J.L. 1996. *Trans,trans*-2,4-decadienal: Cytotoxicity and effect on glutathione level in human erythroleukemia (HEL) cells. *Cancer Lett.* 99, 115–119.

Negroni, M., D'Agustina, A., and Arnoldi, A. 2001. Effects of olive oil, canola, and sunflower oils on the formation of volatiles from the Maillard reaction of lysine with xylose and glucose. *J. Agric. Food Chem.* 49, 439–445.

Nielsen, P., Vahl, M., and Gry, J. 1984. HPLC profiles of mutagens in lean ground pork fried at different temperatures. *Z. Lebensm. Unters. Forsch.* A 187, 451–456.

O'Brien, J., Renwick, A.G., Constable, A., Dybing, E., Muller, D.J.G., Schlatter, J. et al. 2006. Approaches to the risk assessment of genotoxic carcinogens in food: A critical appraisal. *Food Chem. Toxicol.* 44, 1613–1635.

Oguri, A., Suda, M., Totsuka, Y., Sugimura, T., and Wakabayashi, K. 1998. Inhibitory effect of antioxidants on formation of heterocyclic amines. *Mutat. Res.* 402, 237–245.

Overvik, E., Kleman, M., Berg, I., and Gustafsson, J.A. 1989. Influence of creatine, amino acids and water on the formation of the mutagenic heterocyclic amines found in cooked meat. *Carcinogenesis* 10, 2293–2301.

Pearson, A.M., Chen, C., Gray, J.I., and Aust, S.D. 1992. Mechanism(s) involved in meat mutagen formation and inhibition. *Free Rad. Biol. Med.* 13, 161–167.

Perello, G., Marti-Cid, R., Castell, V., Llobet, J.M., and Domingo, J.L. 2009. Concentrations of polybrominated diphenyl ethers, hexachlorobenzene and polycyclic aromatic hydrocarbons in various foodstuffs before and after cooking. *Food Chem. Toxicol.* 47, 709–715.

Perello, G., Marti-Cid, R., Llobet, J.M., and Domingo, J.L. 2008. Effects of various cooking processes on the concentrations of arsenic, cadmium, mercury and lead in foods. *J. Agric. Food Chem.* 56, 11262–11269.

Persson, E., Graziani, G., Ferracane, R., Fogliano, V., and Skog, K. 2003. Influence of antioxidants in virgin olive oil on the formation of heterocyclic amines in fried beefburgers. *Food Chem. Toxicol.* 41, 1587–1597.

Pokorny, J. 1989. Flavor chemistry of deep-fat frying in oil. In: *Flavor Chemistry of Lipid Foods*, Min D.B., Smouse T.H., eds., 113–115, Champaign, IL, AOCS Press.

Poumeyrol, G. 1986. Etude de l'altération des huiles de friture utilisées au restauration collective par dosage des composés polaires. *Rev. Franc. Corps Gras* 33, 263–268.

Prevot, A., Desbordes, S., Morin O., and Mordret, F. 1988. Volatiles and sensory effects from frying oils. In: *Frying of Food: Principles, Changes, New Approaches*, Varela G., Bender A.E., Morton I.D., eds., 155–165, Chichester, U.K., Ellis Horwood Ltd.

Purcaro, G., Navas, J.A., Guardiola, F., Conte, L.S., and Moret, S. 2006. Polycyclic aromatic hydrocarbons in frying oils and snacks. *J. Food Protect.* 69, 199–204.

Reynolds, T. 2002. Acrylamide and cancer: tunnel leak in Sweden prompted studies. *J. Nat'l. Cancer Inst.* 94, 876–78.

Riemersma, R.A., Perkins, D., Brown, A.J., and Brown, J. 1994. Linoleic acid and coronary artery disease. *Am. J. Clin. Nutr.* 59, 949–950.

Romani, S., Bacchiocca, M., Rocculi, P., and Dalla Rosa, M. 2008. Effect of frying time on acrylamide content and quality aspects of french fries. *Eur. Food Res. Technol.* 226, 555–560.

Romero, A., Bastida, S., and Sanchez-Muniz, F.J. 2006. Cyclic fatty acid monomer formation in domestic frying of frozen foods in sunflower oil and high oleic acid sunflower oil without oil replenishment. *Food Chem. Toxicol.* 44, 1674–1681.

Romero, A., Cuesta, C., and Sanchez-Muniz, F.J. 2000. Trans fatty acid production in deep-fat frying of frozen foods with different oils and frying modalities. *Nutr. Res.* 20, 599–608.

Rose, M.D., Bygrave, J., and Sharman, M. 1999. Effect of cooking on veterinary drug residues in food. Part 9. *Nitroimidazoles. Analyst* 124, 289–294.

Rose, M.D., Shearer, G., and Farrington, W.H.H. 1995. The effect of cooking on veterinary drug residues in food. 1. Clenbuterol. *Food Addit. Contam.* 12, 67–76.

Saguy, I.S. and Dana, D. 2003. Integrated approach to deep-fat frying: engineering, nutrition, health and consumer aspects. *J. Food Eng.* 56, 143–152.

Sakata, M., Takahashi, Y., and Sonehara, M. 1985. Quality of fried foods with palm oil. *J. Am. Oil Chem. Soc.* 62, 449–454.

Salta, F.N., Mylona, A., Chiou, A., Boskou, G., and Andrikopoulos, N.K. 2007. Oxidative stability of edible vegetable oils enriched in polyphenols with olive leaf extract. *Food Sci. Tech. Int.* 13, 413–421.

SCF (2001), European Commission, Health and Consumer Protection Directorate-General. Opinion of the Scientific Committee on food on 3-monochloro-propane-1,2-diol (3-MCPD) updating the SCF opinion of 1994. Adopted on 30 May 2001.

Schlatter, J., Baars, A.J., DiNovi, M., Lawrie, S., and Lorentzen, R. 2001. 3-Chloro-1,2-propanediol. In: *WHO Food Additives Series 48*. Safety evaluation of certain food additives and contaminants, Prepared by the 57th meeting of the Joint FAO/WHO, 401–432, Rome, FAO.

Schwarz, K. 2000. Quick tests for used frying fats and oils. *Eur. J. Lipid Sci. Technol.* 102, 573.

Seal, C.J., de Mul, A., Eisenbrand, G., Haverkort, A.J., Franke, K., Lalljie, S.P.D. et al. 2008. risk-benefit considerations of mitigation measures on acrylamide content of foods—a case study on potatoes, cereals and coffee. Report of the ILSI Europe Process Related Compounds Task Force, Brussels.

Sebedio, J.L. and Ratnayake, W.M. 2008. Analysis of *trans* mono- and polyunsaturated fatty acids. In: *Trans Fatty Acids*, Dijkstra, A.J., Hamilton, R.J., and Hamm, W., eds., 102–128, Oxford, Blackwell Publishing.

See, S.W. and Balasubramanian, R. 2008. Chemical characteristics of fine particles emitted from different gas cooking methods. *Atmosph. Environ.* 42, 8852–8862.

Sharp, D. 2003. Acrylamide in food. *The Lancet* 361, 361–362.

Shaw, I. and Thomson, B. 2003. Acrylamide food risk. *The Lancet* 361, 434–435.

Shin, H.S. and Ustunol, Z. 2004. Influence of honey-containing marinades on heterocyclic aromatic amines formation and overall mutagenicity in fried beef steak and chicken breast. *Food. Chem. Toxicol.* 69, 147–153.

Skog, K., Augustsson, K., Steineck G., Stenberg M., and Jagerstad M. 1997. Polar and non-polar heterocyclic amines in cooked fish and meat products and their corresponding pan residues. *Food Chem. Toxicol.* 35, 555–565.

Smith, B.K., Robinson, L.E., Nam, R., and Ma, D.W. 2009. Trans-fatty acids and cancer: A mini-review. *Br. J. Nutr.* 102, 1254–1266.

Sobukola, O.P., Awonorin, O.S., Idowu, A.M., and Bamiro, O.F. 2009. Microbial profile and critical control points during processing of "robo" snack from melon seed (*Citrullus lunatus* thumb) in Abeokuta, Nigeria. *African J. Biotechnol.* 8, 2385–2388.

Soriano, J.M., Molto, J.C., and Manes, J. 2002. Hazard analysis and critical control points in deep-fat frying. *Eur. J. Lipid Sci. Technol.* 104, 174–177.

Spiteller P., Kern W., and Reiner G. 2001. Aldehydic lipid peroxidation products derived from linoleic acid. *Biochim. Biophys. Acta.* 1531, 188–208.

Stadler, R.H. and Goldmann, T. 2008. Chapter 20: Acrylamide, chloropropanols and chloropropanol esters, furan. In: *Comprehensive Analytical Chemistry series, volume 51, Food Contaminants and Residue Analysis*, Picó Y., ed., 705–732, Amsterdam, Elsevier B.V.

Stadler, R.H., Blank, I., Varga, N., Robert, F., Hau, J., Guy, Ph. et al. 2002. Food chemistry: Acrylamide from Maillard reaction products. *Nature* 419, 449–450.

Stadler, R.H., Theurillat, V., Studer, A., Scanlan, F., and Seefelder, W. 2007. The formation of 3-MCPD in food and potential measures of control. In: *DGF Symposium: Thermal Processing of Food: Potential Health Benefits and Risks*, Eisenbrand, G., ed., 141, Weinheim, Wiley-VCH.

Stender, S. and Dyerberg, J. 2003. *The Influence of Trans Fatty Acids on Health*, 4th ed. Danish Nutrition Council, DNC Publication No 34.

Sugimura, T. 1995. History, present and future, of heterocyclic amines, cooked food mutagens. In: *Heterocyclic Amines in Cooked Foods: Possible Human Carcinogens*, Adamson R. et al., eds., 214–231, Princeton, NJ, Princeton Scientific.

Sugimura, T. 1997. Overview of carcinogenic heterocyclic amines. *Mutat. Res.* 376, 211–219.

Summa, C., Wenzl, T., Brohee, M., De La Calle, B., and Anklam, E. 2006. Investigation of the correlation of the acrylamide content and the antioxidant activity of model cookies. *J. Agric. Food Chem.* 54, 853–859.

Svejkovska, B., Novotny, O., Divinova, V., Reblova, Z., Dolezal, M., and Velisek, J. 2004. Esters of 3-chloropropane-1,2-diol in foodstuffs. *Czech J. Food Sci.* 22, 190–196.

Takeoka, G., Perrino, C. Jr., and Buttery, R. 1996. Volatile constituents of used frying oils. *J. Agric. Food Chem.* 44, 654–660.

Takeoka, G.R., Full, G.H., and Dao, L.T. 1997. Effect of heating on the characteristics and chemical composition of selected frying oil and fat. *J. Agric. Food Chem.* 45, 3244–3249.

Tareke, E., Rydbeg, P., Karlsson, P., Eriksson, S., and Törnqvist, M. 2000. Acrylamide: A cooking carcinogen? *Chem. Res. Toxicol.* 13, 517–22.

Tareke, E., Rydberg, P., Karlsson, P., Eriksson, S., and Törnqvist, M. 2002. Analysis of acrylamide, a carcinogen formed in heated foodstuffs. *J. Agric. Food Chem.* 50, 4998–5006.

Thompson, A.K., Shaw, D.I., Minihane, A.M., and Williams, C.M. 2008. Trans-fatty acids and cancer: The evidence reviewed. *Nutr. Res. Rev.* 21, 174–188.

Tilson, H.A. 1981. The neurotoxicity of acrylamide: An overview. *Neurobehav. Toxicol. Teratol.* 3, 445–461.

Tompkins, C. and Perkins, E.G. 2000. Frying performance of low-linolenic acid soybean oil. *J. Am. Oil Chem. Soc.* 77, 223–229.

Valenzuela, A., Sanhueza, J., and Nieto, S. 2004. Cholesterol oxidized products in foods: Potential health hazards and the role of antioxidants in prevention. *Grasas y Aceites* 55, 312–320.

Velasco, J., Berdeaux, O., Márquez-Ruiz, G., and Dobarganes, M.C. 2002. Sensitive and accurate quantitation of monoepoxy fatty acids in thermoxidized oils by gas–liquid chromatography. *J. Chromatogr.* A 982, 145–152.

Velasco, J., Marmesat, S., Bordeaux, O., Márquez-Ruiz, G., and Dobarganes, C.M. 2004. Formation and evolution of monoepoxy fatty acids in thermoxidized olive and sunflower oils and quantitation in used frying oils from restaurants and fried-food outlets. *J. Agric. Food Chem.* 52, 4438–4443.

Velasco, J., Marmesat, S., and Dobarganes, M.C. 2008. Chemistry of frying. In: *Advances in Deep-Fat Frying of Foods*, Sumnu S.G., Sahin S., eds., 33–56, Boca Raton, FL, CRC Press.

Velısek, J., Calta, P., Crews, C., Hasnip, S., and Doledzal, M. 2003. 3-Chloropropane-1,2-diol in models simulating processed foods: Precursors and agents causing its decomposition. *Czech J. Food Sci.* 21, 153–161.

Velısek, J., Davıdek, T., Davıdek, J., and Hamburg, A. 1991. 3-Chloro-1, 2-propanediol derived amino alcohol in protein hydrolysates. *J. Food Sci.* 56, 136–138.

Velısek, J., Davıdek, J., Hajslova, J., Kubelka, V., Janıcek, G., and Mankova, B. 1978. Chlorohydrins in protein hydrolysates. *Zeitschr. Lebensm.-Untersuch.-Forschung* 167, 241–244.

Velısek, J., Davıdek, J., Kubelka, V., Bartosova, J., Tuekova, A., Hajslova, J. et al. 1979. Formation of volatile chlorohydrins from glycerol (triacetin, tributyrin) and hydrochloric acid. *Lebensm.-Wissensch.-Technol.—Food Sci. Technol.* 12, 234–236.

Velısek, J., Davıdek, J., Kubelka, V., Janıcek, G., Svobodova, Z., and Simicova, Z. 1980. New chlorine-containing organic compounds in protein hydrolysates. *J. Agric. Food Chem.* 28, 1142–1144.

Vinikoor, L.C., Schroeder, J.C., Millikan, R.C., Satia, J.A., Martin, C.F., Ibrahim, J. et al. 2008. Consumption of trans-fatty acid and its association with colorectal adenomas. *Am. J. Epidemiol.* 168, 289–297.

Vitaglione, P. and Fogliano, V. 2004. Use of antioxidants to minimize the human health risk associated to mutagenic/carcinogenic heterocyclic amines in food. *J. Chromatogr.* B 802, 189–199.

Vorria, E., Giannou, V., and Tzia, C. 2004. Hazard analysis and critical control point of frying—Safety assurance of fried foods. *Eur. J. Lipid Sci. Technol.* 106, 759–765.

Warner, K and Knowlton, S. 1997. Frying quality and oxidative stability of high-oleic corn oils. *J. Am. Oil Chem. Soc.* 74, 1317–1322.

Weisschar, R. 2008. 3-MCPD esters in edible fats and oils—a new and worldwide problem. *Eur. J. Lipid Sci. Technol.* 110, 671–672.

Widmark, E. 1939. Presence of cancer-producing substances in roasted food. *Nature* 143, 984–984.

Wegmüller, F. 1998. Die Qualität von Fritierölen dielektrisch erfassen. *Mitt. Gebiete Lebensm. Hyg.* 89, 301–307.

White, S., Fernandes, A., and Rose, M. 2008. Investigation of the formation of PAHs in foods prepared in the home and from catering outlets to determine the effects of frying, grilling, barbecuing, toasting and roasting. Food Standard Agency (FSA)–Central Science Laboratory (CSL), FD 06/13.1.

Wu, C.M. and Chen, S.Y. 1992. Volatile compounds in oils after deep frying or stir frying and subsequent storage. *J. Am. Oil Chem. Soc.* 69, 858–65.

Wu, S.C. and Yen, G.C. 2004. Effects of cooking oil fumes on the genotoxicity and oxidative stress in human lung carcinoma (A-549) cells. *Toxic. In Vitro* 18, 571–580.

Zelinkova, Z., Svejkovska, B., Velisek, J., and Doledzal, M. 2006. Fatty acid esters of 3-chloropropane-1,2-diol in edible oils. *Food Addit. Contam.* 23, 1290–1298.

Zhu, X., Wang, K., Zhu, J., and Koga, M. 2001. Analysis of cooking oil fumes by ultraviolet spectrometry and gas chromatography mass spectrometry. *J. Agric. Food Chem.* 49, 4790–4794.

Index

D

E

Q

R